Agroecology and Small Farm Development

Editors

Miguel A. Altieri
Division of Biological Control
University of California, Berkeley
Albany, California

Susanna B. Hecht
Graduate School of Planning
University of California, Los Angeles
Los Angeles, California

CRC

CRC Press
Boca Raton Ann Arbor Boston

Library of Congress Cataloging-in-Publication Data

Agroecology and small farm development/editors, Miguel A. Altieri,
Susanna B. Hecht.
p. cm.
Includes bibliographical references.
ISBN 0-8493-4885-4
1. Farms, Small—Developing countries. 2. Agricultural ecology-
-Developing countries. 3. Human ecology—Developing countries.
4. Agriculture—Social aspects—Developing countries. I. Altieri,
Miguel A. II. Hecht, Susanna B.
HD1476.D44A36 1989
306.3'49'091724—dc20 89-37686
CIP

Direct all inquiries to CRC Press, Inc., 2000 Corporate Blvd., N.W., Boca Raton, Florida 33431.

International Standard Book Number 0-8493-4885-4

Library of Congress Card Number 89-37686
Printed in the United States

PREFACE

Two main lessons can be gleaned from decades of international rural development research: first, that small farm development is extremely complex and involves a great deal more than just technical issues, and second, that there is no single discipline or methodological approach that can adequately address the range of biological, cultural, and politico-economic processes that surround small farm development. Agroecology, with its emphasis on reproduction of the household and regeneration of the agricultural resource base, provides an agile framework for analyzing and understanding the diverse factors affecting small farms. It also provides methodologies aimed at promoting the development of technologies closely tailored to the needs and circumstances of specific peasant communities.

Throughout the developing world, small farmers face numerous ecological and socio-economic constraints, They are impoverished by their limited assets and access to land. They are also impoverished by the low productivity of their land given the above constraints, and the low quality of their natural resource base. In these situations, agroecologists can provide guidelines for enhancing and sustaining the productivity of resource-poor agriculture, while conserving and improving the resource base. However, these technical contributions developed by agroecologists, though important, cannot by themselves address the root causes of rural poverty. A much broader development agenda is required that addresses the issues of resource governance, political power, and resource distribution in rural areas.

In this book we chose to explore the ways in which agroecologists can enhance agricultural productivity to benefit rather than marginalize small producers. We are well aware that the biological and cultural heterogeneity that characterizes small farms in Africa, Asia, and Latin America permit neither simple analyses nor simple solutions. This is why we have incorporated a diversity of environmental contexts, from mountain zones to the lowland tropics, and included chapters that range from conservation biology to development economics, as well as agroecological studies. While each chapter contains a unique theme, all chapters are tied together by the issue of rural development. The importance of ecological and social process and outcome and how both affect the sustainability and productivity of small farms has been stressed. Thus, while genetic erosion of traditional rice cultivars in Asia has differing origins than the loss of native potato germplasm in the Andes, the effects on small farm agricultural sustainability are similar. African and Amazonia agroforestry systems have distinct structural features, but they share common agroecological principles.

Agroecology differs from other approaches such as Farming Systems Research, whose main concern is technology transfer. Agroecology focuses on defining the principles on which to base an agroecosystem design, leaving the specific technological forms to be determined by the local cultural and economic milieu. The central idea is to encourage farmer discrimination, selection and adoption of the technologies of use to them, while validating those local practices that have emerged over centuries of agricultural activities. We hope that the diversity of perspectives and approaches to rural development presented in these chapters permits a reevaluation of strategies and tactics for small farm development.

Miguel A. Altieri
University of California
Berkeley

Susanna B. Hecht
University of California
Los Angeles

January, 1990

THE EDITORS

Miguel A. Altieri received his Ph.D. in Entomology/Pest Management from the University of Florida in 1979 for work on biological control of corn pests through weed manipulation. As a postdoctoral fellow, he researched the interactions between weeds and beneficial and pest insects in soybeans as well as on the effects and use of plant volatiles on *Trichogramma* behavior to control *Heliothis zea* populations at the University of Georgia. From 1981 to the present as associate professor and entomologist at the Division of Biological Control, University of California, Berkeley, he is researching methods to enhance biological pest control in annual agricultural systems and orchards.

This research was expanded to include the design and management of self-sustaining cropping systems in terms of pest control, nutrient cycling, and yield sustainability. The principles derived from this work have been intensively applied by research and action groups involved in improving peasant agriculture in Latin America.

In addition, he teaches courses in agricultural ecology, rural development, and integrated pest management in developing countries, agroforestry systems, insect ecology and biological control, and weed science. He also conducts international short courses in the Unites States and Latin America on biological control and agroecology.

Dr. Susanna Hecht is an Associate Professor in the Graduate School of Planning at UCLA. She has worked extensively in the Amazon Basin, analyzing the dynamics and soil impacts of the livestock industry and the ways this land use triggered deforestation. These efforts led to an interest in alternative forms of resource management, including those of indigenous people and forest dwellers such as rubber trappers. She is the author, along with Alexander Cockburn, of *Fate of the Forest, Developers, Destroyers and Defenders of the Amazon.* She has edited several books on tropical development, including *Amazonia Land Use and Agricultural Research* (CIAT), *Development or Destruction: The Livestock Sector in Latin America* (Westview, forthcoming), as well as dozens of academic articles.

CONTRIBUTORS

Janis B. Alcorn
Agency for International Development
ANE/TR/ENR
Washington, D.C.

Miguel A. Altieri
Division of Biological Control
University of California, Berkeley
Albany, California

Jacqueline A. Ashby
Coordinator IPRA Project
CIAT
Cali, Colombia

Doyle Baker
IRA/NCRE Project
(Messa) Yaounde, Cameroun

Stephen B. Brush
Department of Applied Behavioral Sciences
University of California, Davis
Davis, California

Robert J. H. Chambers
Institute of Development Studies
University of Sussex
Brighton, England

Gordon R. Conway
Representative
The Ford Foundation
New Delhi, India

Kenneth A. Dahlberg
Professor, Department of Political Science
Western Michigan University
Kalamazoo, Michigan

Alain de Janvry
Professor and Chairman
Department of Agricultural and Resource Economics
University of California at Berkeley
Berkeley, California

Cornelia Butler Flora
Professor
Department of Sociology
Virginia Polytechnic Institute and State University
Blacksburg, Virginia

Louise Fortmann
Professor
Forestry and Resource Management
University of California at Berkeley
Berkeley, California

Charles A. Francis
Professor, Extension Crops Specialist
Department of Agronomy
University of Nebraska
Lincoln, Nebraska

Stephen R. Gliessman
Director, Agroecology Program
Associate Professor, Environmental Studies
University of California, Santa Cruz
Santa Cruz, California

Nicolo Gligo
Coordinator
Joint ECLAC/UNEP Development
and Environment Unit
Economic Commission for Latin America
and the Caribbean United Nations
Santiago, Chile

Susanna B. Hecht
Graduate School of Planning
University of California, Los Angeles
Los Angeles, California

Steven Helfand
Graduate Student
Department of Agricultural and Resource Economics
University of California at Berkeley
Berkeley, California

Janice Jiggins
Rural Development Consultant
6673 MD
Andelst, Netherlands

Roger Kirkby
Cropping Systems Agronomist/Regional Coordinator
Centro Internacional de Agricultura Tropical (CIAT)
Debre Zeit, Ethiopia

Gerald G. Marten
Research Associate
Environment and Policy Institute
East-West Center
Honolulu, Hawaii

Jennifer A. McCracken
Research Associate
Sustainable Agriculture Programme
International Institute for Environment and Development
London, England

Laura C. Merrick
Research Assistant Professor
Department of Plant, Soil, and Environmental Sciences
University of Maine
Orono, Maine

Gary P. Nabhan
Assistant Director for Research
Desert Botanical Garden
Phoenix, Arizona

P.K.R. Nair
Professor of Agroforestry
Department of Forestry, IFAS
University of Florida
Gainesville, Florida

David W. Norman
Team Leader ATIP
Department of Agricultural Research
Ministry of Agriculture
Gaborone, Botswana

Paul Richards
Reader in Anthropology
University College London
London, England

Mark Slater
Research Assistant
Desert Botanical Garden
Phoenix, Arizona

Victor M. Toledo
Centro de Ecologia
Universidad Nacional Autonoma de México
México, D.F., México

Larry Yarger
Agricultural Development Specialist
Food for the Hungry, International
Alamos, Sonora, Mexico

TABLE OF CONTENTS

The Physical and Social Context of Small Farm Agriculture

1 CROP GENETIC DIVERSITY AND ITS CONSERVATION IN TRADITIONAL AGROECOSYSTEMS

Laura C. Merrick

I. INTRODUCTION

The stability and sustainability of traditional agricultural production are based on crop diversity.[1-4] The general strategy followed by farmers is to hedge against risk by planting several crop species and several varieties within species in both space and time. This chapter focuses on the genetic diversity within individual crops grown primarily for food production in traditional agroecosystems. Recently, the high degree of variability present within crop genetic resources in traditional farming systems has become increasingly threatened principally due to the adoption of genetically uniform modern cultivars.[3,5-7] *In situ* crop genetic conservation should be linked to rural development projects that seek to improve traditional farming systems. In this way a strategy that simultaneously encourages preservation of crop genetic resources while allowing low-income farmers to produce subsistence and cash crops without dependence on external inputs and seed supplies can be achieved.[4,8]

II. GENETIC DIVERSITY WITHIN CROPS IN TRADITIONAL FARMING SYSTEMS

A. Varietal Diversity

Most studies of traditional agroecosystems indicate that these systems are frequently characterized by a high degree of within-crop diversity despite differences in crop breeding systems in regard to the relative ease with which distinct varieties can be maintained.[2] With vegetatively propagated crops such as potatoes (*Solanum* spp.) or cassava (*Manihot esculenta*), each individual variant can be maintained as genetically distinct over time. Individual farmers in the Andean region of Peru and Bolivia cultivate as many as 30 to 40 distinct clones of potatoes.[9,10] An average of 12 cassava cultivars per garden was found in a survey of over 200 gardens of the Aguaruna Jívaro in the Peruvian Amazon basin — although many more cassava cultivars were likely to be present in each garden since each sample plot consisted of only about 5% of the total area planted to cassava.[11] Varietal diversity is also common in traditionally managed fields of primarily self-pollinated crops such as rice (*Oryza* spp.) or common beans (*Phaseolus vulgaris*).[12-14] For example, Cambodian landrace populations of rice typically consist of five to seven varieties.[12] Bean landraces in Malawi commonly consist of mixtures of at least 12 — some with over 50 — distinct seed types (Figure 1), while those in Peru have been found to consist of about 15 unique types and in Mexico, about 4 to 6.[13,14] Even with cross-pollinated crops, within-crop diversity is often high, with such factors as spatial and temporal isolation and deliberate human selection maintaining the genetic integrity of different varieties.[15,16] An average of four strains of maize (*Zea mays*) is typical in traditional agroecosystems in both Mexico and Guatemala.[2,15] Tremendous variation is often evident in landrace populations of squash (*Cucurbita* spp.) in Latin America[16,17] and pearl millet (*Pennisetum americanum*) in Africa,[18] although the variation of those cross-pollinated crops is generally not classified into distinct varieties. In some instances, several different species comprise a given crop and may be cultivated simultaneously by individual farmers. Two species of cultivated rice — *Oryza sativa* and *O. glaberrima* — are commonly grown together in fields in West Africa.[12,19,20] In northwestern Mexico, typically two species of squash are grown by individual farmers: *Cucurbita argyrosperma* (= *C. mixta*) and *C. moschata* at lower elevations; *C. ficifolia* and *C. pepo* at higher elevations; and *C. argyrosperma* and *C. pepo* at intermediate elevations.[18] Varieties of different potato species representing a range of chromosome ploidy levels from diploids to pentaploids are cultivated in various combinations by Andean farmers: *Solanum tuberosum* var. *andigena*, *S. tuberosum* var. *tuberosum*, *S. stenotomum*, *S. phureja*, *S. X ajanhuiri*, *S. X curtilobum*, and *S. X juzepczukii*.[9,10] Within-crop diversity is just one of a number of common strategies used to maintain

FIGURE 1. The most striking indication of variability in Malawian common bean *(Phaseolus vulgaris)* landraces is the diversity of seed colors, patterns, shapes, and sizes. The seed types pictured here — collected in a market in Karonga — were included in a study of regional variation of the crop in Northern Malawi.[13,49,79] (Photo by M. Wayne Adams, Michigan State University.)

stable yields in traditional agroecosystems[8,21] and there are exceptions to the general trend presented above. For example, certain intensively cropped traditional gardens surveyed in Guatemala, Puerto Rico, and Indonesia are based on a high degree of interspecific diversity, rather than on intraspecific diversity of any given crop.[22] However, in contrast, the level of diversity within a single crop — cassava — typically far exceeds the amount of interspecific diversity otherwise present in Aguarana Jívaro gardens.[11]

B. Advantages to Farmers

Different varieties within landrace populations are commonly distinguished by farmers by such characteristics as: color of different plant parts, yield, culinary quality, resistance to diseases or pests, storage properties, altitude for optimum growth, phenology, and growth habit.[2,9,11,13,16,18,23,24] Unlike selection and breeding criteria applied to modern crop varieties, yield may or may not be an overriding concern in the selection of traditional varieties by farmers.[11,13] When several different varieties of a crop are produced, a broad range of goals can be achieved simultaneously. For example, farmers in the highlands of Guatemala balance culinary quality with storage quality by cultivating more than one type of maize: flint textured corn is more resistant to insect damage while in storage than are softer dent or flour corns, but the latter are preferred for the preparation of tortillas, a staple food product; as a result, some of each may be grown.[15] In Jamaica, three varieties of the cush-cush yam, *Dioscorea trifida*, are cultivated, each differing in flesh color, in maturation time (from 6 to 9 months) and in degree of tuber "hardness". The purple variety is the quickest in maturation and is valued for culinary quality in that its soft texture requires less cooking time. However, it deteriorates much more rapidly in storage than do harder yellow and white varieties, and thus again, all three are grown.[2] Certain crop varieties may be grown for commercial purposes without being consumed locally; other varieties of the crop may be grown exclusively for subsistence.[9,18]

One reason for maintenance of diversity within landrace populations is that quite often a crop is utilized locally for multiple purposes. Different varieties of the crop may be selected for separate, specific uses; sometimes several products are harvested at different stages of plant development. In Africa, sorghum (*Sorgum bicolor*) provides multiple products: stems are used as fodder, fuel, or building material; sap is extracted to provide a sweet or a red pigment for dye; grains are eaten raw, boiled, or ground into a flour; the plants flavor beer, provide medicine, or are used to make malt. Within the same village, uses of sorghum are often mutually exclusive and different varieties are maintained by selection and isolation of seed stocks.[25] In northwestern Mexico, immature fruits, mature fruits, and seeds are the principal parts of the squash plants that are consumed, but minor or localized uses also include consumption of flowers, tendrils, leaves, or vine tips; intact rinds utilized as containers; or mature fruits used as fodder for domestic livestock. Different squash fruit types may be preferred for particular purposes such that farmers may be selecting for different plant types simultaneously within their landrace populations, e.g., long-necked types when fruits are harvested when immature to be consumed as a vegetable; thick-rinded types for better keeping quality during storage or for use as containers.[16,18]

An advantage that the possession of diverse genotypes of a single crop provides is the opportunity to more effectively exploit different microhabitats that occur in the spatial heterogeneity prevalent in traditional farmers' fields.[2,8,18,21] In the intensely steep elevational gradients of the Andes, individual families commonly cultivate potatoes in three or four different fields which may span perhaps as much as 2000 ft in elevation. Due to differing intensities of threats such as frost or late blight in the different fields, potato varieties characterized by specific resistances or tolerances are often preferentially planted in certain fields.[21] Similarly, among the traditional K'ekchi Mayans in the highlands of Guatemala, the ecological zone of a field — which is in large part determined by elevation — is the primary consideration for selection of particular maize varieties for planting.[15] The Caxihua of the Amazon basin match physiological characteristics of cassava cultivars with specific soil conditions, although in contrast, varietal differences in ecological performance characteristics were not found to be important considerations for selection among the many cassava cultivars grown by the Aguaruna Jívaro in another area of eastern Peru.[11,26] Varieties of the same crop may either be grown in mixed plantings or separated in different fields. In northwestern Mexico, the squash species *Cucurbita argyrosperma* and *C. moschata*, both of which are usually intercropped with maize, are sometimes spatially separated in fields.[18] *C. moschata* tends to require a better fertility and moisture regime and matures as much as a month later than *C. argyrosperma*. As a consequence, the latter species may be planted in areas with poorer quality soil or on steeper slopes, while the former is planted in more fertile, level sites.[16,18] In western areas of the Nuba Mountain region of Sudan, red and white sesame *(Sesamum indicum)* varieties which differ in maturation time and in height and branching habit are planted separately in adjacent fields and harvested sequentially. In contrast, in the southern Nuba region, several distinct sesame varieties are planted as a mixture within single fields.[27] The differential response to diseases and pests that may be represented in the crop population when different crop genotypes are grown in mixture may enhance the stability of crop yields.[3,12,13,23,28-30] Thus in the mountains of central Mexico, farmers typically plant two or three different cultivars of potatoes per hill to slow the spread of fungal pathogens.[23]

Phenological and physiological differences that distinguish different crop varieties may confer selective advantage under environmental conditions that are particular to certain seasons. When several genotypes of a crop are available, there may be more than one production cycle of the crop per year.[2,8,18,31-32] Mayan farmers of the Yucatan Peninsula of Mexico grow a rich assortment of the squash *Cucurbita moschata*. Certain cultivars of *C. moschata* are known as long cycle varieties and mature over 7 to 8 months, whereas others are short-cycle types, selected for maturity in 3 to 4 months.[31] In West Africa, farmers transplant one sorghum race with a 90-day maturity into fields of receding flood waters at the beginning of the dry season to mature on residual moisture and then plant another race with a 180-day maturity during the rainy season.[32] Even when the crop is produced during a single season, having a range of phenological responses among cultivars is frequently advantageous.[2,3,9,13,14,21,28,33] Whereas bush-type bean cultivars that germinate and ripen uniformly are desirable for modern, mechanized production systems, the mixtures of distinct seed types prevalent in Mexican and Malawian bean landraces have been measured to differ in rates of emergence by 5 to 7 and 10 to 21 days, respectively. Likewise, the beans in the landraces are predominantly vine-type cultivars which tend to flower asynchronously and ripen over long periods of time.[13,14] The range in phenological responses represented within the landrace populations are often highly adaptive under conditions of rainfed agriculture where environmental conditions during the growing season are unpredictable.[2,13,14,33] The maturation times for the three or four maize strains grown by certain traditional Mexican farmers may differ by as much as 2 or 3 months.[2,33] The highest yielding or better tasting strain may be the longest maturing and if it fails to produce well in a bad year, a crop may still be harvested from the earlier maturing strains.[33] Or with crop failure early in the season of a preferred longer-maturing strain, the less desirable yet quicker-maturing strains can still be planted.[2,15]

C. Patterns of Distribution of Crop Genetic Diversity

Four major factors are often correlated with geographic regions exhibiting high levels of crop genetic diversity: great antiquity and continuity of agriculture; ecological and physical diversity of the environment; cultural diversity; and extent of hybridization and introgression of crop plants with wild or weedy relatives.[3] For the vast majority of cultivated plants, their regions of greatest genetic diversity are concentrated in the Third World, often in mountainous regions of the tropics.[22,24,28] When the Russian geneticist N.I. Vavilov first recognized and described geographical centers of crop diversity, the latter were thought to be regions of domestication for the crops in question.[34] The tubers known as oca *(Oxalis tuberosa)*, ulluco *(Ulluca tuberosus)*, and anu *(Tropaeolum tuberosum)*; a legume known as tarwi *(Lupinus mutabilis)*; and the grains known as quinua *(Chenopodium quinoa)* and canihua *(C. palidicaule)* are examples of crops for which the center of diversity and domestication coincide: all of the latter crops are native to the Andes and very little cultivation of them has ever occurred outside that region.[35-38] However, it is now recognized that for many crops, genetically diverse landrace populations may occur in regions far from the hypothesized area of crop origin.[3,12,22,24,28] As an example, the degree of heterogeneity documented within common bean landrace populations in Malawi often exceeds that found within landraces of beans in Peru and Mexico, although the common bean originated in Latin America and was introduced to Malawi only about 400 years ago.[13,14] Human-mediated selection, natural selection, or hybridization among diverse crop genotypes or between crops and their wild relatives are the major factors which are thought to have promoted genetic diversity in crops subsequent to migration from the region of domestication.[3,12,13,16,23]

Since many traditional agroecosystems are located in centers of crop diversity, it is not uncommon for wild or weedy relatives of crop plants to be growing in and around traditional agroecosystems. Cycles of natural hybridization and introgression periodically occur between the crops and their weedy counterparts.[9,10,16,18,19,23,24,39-46] The significance of the role that wild or weedy relatives can play in terms of enhancement of the genetic diversity of co-occurring crops is exemplified by the tomato *(Lycopersicon esculentum)*, a crop which is characterized by extreme genetic uniformity throughout most of the world.[44,47] Despite the fact that the tomato is thought to have been domesticated in Mexico, landraces from western South America are notably genetically diverse. Spontaneous hybridization between the crop plants and one of their wild relatives native to the region — *L. pimpinellifolium* — is hypothesized to be the primary source of variablity for the South American tomatos landraces.[44,47] Genes derived from spontaneous hybridization with wild relatives are thought to have contributed to the increased local adaptation of cultivated plants after the crop's introduction to new environments such as when potatoes were moved into colder climates at the higher altitudes of the Andes[10,45] or the mountains of central Mexico;[23] sorghum into more humid areas of Africa;[25] or squash into arid regions of northwestern Mexico.[41]

The existence of certain cosmopolitan varieties of native cultivars that are present over widespread geographic areas indicate that trade, whether by exchange or sale, is an important source of diversity by way of introducing new crop genotypes into landrace populations.[8,9,11,15,18,23,42,48,49] Yet, unlike modern production systems where seed or tuber stocks are almost always purchased for each growing season, there are a number of diversity enhancing factors which may occur in traditional farming systems because seeds or tubers are harvested, saved, and planted again by farmers. Even crops that are normally propagated clonally may periodically undergo sexual reproduction, a process which can serve as a means for introducing new genes into local landrace populations.[10,11,23,43,45,48] For the typically asexually propagated crops, cassava,[11] white potato,[10,23,45] and sweet potato *(Ipomoea batata)*,[48] the occurrence of sexually produced seedlings which are on occasion incorporated into crop populations is thought to be an important factor promoting genetic diversity in landrace populations within traditionally managed fields. Although common beans are predominantly inbred, a low rate of outcrossing occurring among diverse bean cultivars in Malawian landrace populations has been shown to produce a tremendous array of phenotypes which are evident in segregating progeny, thus providing new variants for farmers to select among.[49] In some localities, traditional farmers encourage intervarietal or interspecific crosses among crop plants or between crops and weeds such as in Mexico among maize varieties or between maize and teosinte[46] or among squash species;[31] the hybrid progeny are considered desirable due to hybrid vigor or preferred combinations of traits. Weedy plants that closely resemble their cultivated counterparts — whether they are escaped, naturalized cultivars or hybrids between cultivated and wild plants — at times may represent a resource which can be utilized and incorporated directly into crop populations when seed stocks of local cultivars are lost due to crop failure through unusually harsh ecological conditions, as has been documented for weedy potatoes in Bolivia and weedy squash in Mexico.[10,50]

III. GENETIC EROSION

A. Loss of Crop Genetic Diversity

Despite the continued cultivation of traditional varieties in some localized regions of the world, the adoption of high-yielding, genetically uniform modern cultivars has resulted in a widespread abandonment of the genetically variable, indigenous varieties by traditional farmers.[3,5,6,51] The adoption of modern agricultural practice has also hastened the disappearance of wild or weedy crop relatives.[52] Since the genetic variability of traditional varieties and wild and weedy crop relatives provide much of the resource base for modern plant breeding, the increasing rate abandonment of native varieties by traditional farmers — termed "genetic erosion" — on a worldwide scale has been cause for considerable concern.[3,6,12,28,53] Moreover, the planting of vast areas with genetically uniform cultivars, a practice which characterizes modern agricultural systems, makes agricultural productivity extremely vulnerable to yield-limiting factors such as epidemics of diseases or pests.[12,29,30,53,54]

In addition to direct replacement of traditional varieties by modern ones, a number of other factors have also contributed to loss of crop genetic resources within areas where traditional agriculture has long been practiced: acculturation; abandonment of indigenous farming practices; replacement of mixed cropping systems with large scale monocultures; and displacement of peasants by political or economic changes.[6,7,9,11,18,28,42,55] Other factors have tended to decrease genetic variation found within crops on the scale of villages and fields: decrease in the number of growers; decrease in crop size per field and in planting frequency; loss of seed saving and seed collection skills such that variants are no longer recognized and maintained; and changes in the crop's vulnerability to pests and weeds or loss of cross-compatible wild or weedy crop relatives through clean cultivation or use of herbicides; collapse of several distinct landraces into a single one, etc.[11,15,36,42]

B. Extent of Genetic Erosion

The extent of genetic erosion of locally adapted varieties varies from crop to crop and from region to region.[5,7,51] An example of a crop that as yet is not substantially threatened throughout much of its center of diversity by virtue of direct replacement with modern varieties is squash: for the most part, farmers in Latin America continue to grow landraces both for subsistence and commercial distribution.[18,56] In contrast, another vegetable crop — tomato — is rapidly being replaced: there has been widespread adoption of modern tomato cultivars and hybrids on a worldwide basis, principally due to market demand.[47] Even 15 years ago, traditional varieties of bread-wheat (*Triticum aestivum*) native to the Near East had virtually disappeared from that region.[51] In Southeast Asia, there has been extensive replacement of rice landraces native to lowland areas, whereas there has been comparatively less replacement of indigenous rice varieties adapted to the uplands. However, the overall threat to that crop's resource base caused by the resultant genetic erosion is made all the more serious by the fact that, in general, rice landraces native to the lowlands are much more genetically diverse than those in the uplands.[7] Culinary preference, cultural significance, agronomic performance, market access, and cost of seed and associated inputs or new technology are all factors that influence the rate and amount of abandonment of traditional cultivars.[7,18,37,47,55,57] In some instances, modern varieties may be incorporated into local agroecosystems while farmers continue to cultivate native varieties as well.[7,9,12] In some areas of the Andes, farmers grow modern varieties of potatoes for market, but due to the latter's poor culinary and storage qualities and limited regenerative capability relative to native varieties, farmers continue to produce landraces for home consumption.[7,9] In Thailand, traditional rice farmers plant modern semi-dwarf varieties during the dry season and sow a wide range of traditional varieties during the monsoon season. They have thus established a system that maximizes the productivity of irrigated modern varieties during dry months and the stability of the traditional varieties in the wet season when pest outbreaks are common.[12]

More marginalized areas far from urban and market centers have tended to have much lower rates of adoption of improved varieties.[7] In some regions, continued cultivation of landraces may be based on their dependability under environmental conditions that are considered marginal for agriculture in general.[42] In desert zones of Africa, it is thought that there is relatively little threat of replacement of local races of pearl millet because virtually no other crop has such extreme drought tolerance.[20] The ability of pearl millet to reliably and rapidly produce a harvestable crop has meant that it continues to be grown in Malawi, despite local culinary preference for maize.[58] Similarly, quinua, with its exceptionally high protein content and general nutritional value, has persisted as a local crop primarily at high elevations in the Andes because it can be produced under harsh environmental conditions where few other crops prosper.[37] Frost-resistant triploid and pentaploid species of potatoes grown at high elevations in Peru for the production of "chuño" are thought to be less affected by genetic erosion than cultivated diploid, triploid, and tetraploid *Solanum* species more common at lower elevations.[57] Yet even these hardy landraces may eventually disappear from their native areas. As new crop varieties are bred with adaptation to increasingly more marginal environmental conditions, landraces once grown in such areas disappear.[12,53] For example, Japanese barnyard millet *(Echinochloa utilis)* a protein-rich grain that is tolerant of cool weather, has remained as a minor crop in certain regions of Japan where rice is frequently subject to cool temperature damage or total loss. However, the recent introduction of improved rice varieties with better cold tolerance and associated improved methods for rice production has lead to an increased rate of abandonment of the millet.[59] Often adoption of modern crop varieties means that the nutritional status of indigenous communities declines substantially: older varieties often provide much higher nutritional quality than improved varieties which have been bred instead with high-yield characteristics or marketability rather than nutritional quality as the most important features.[7,18,23,35,36,42,60,61]

C. Response of Farmers

As the economic crisis deepens in most developing countries, and rural populations become increasingly improverished, farmers in many areas of the world are renewing use of the traditional varieties and low-input management practices needed for subsistence agriculture.[62] Farmers may discover that improved varieties are less dependable than the landraces they have replaced, even when the improved varieties are introduced along with associated new technologies and inputs required to enhance their yield potential.[12,63,64] Despite the opportunities for higher yields offered by modern varieties, their adoption can represent a considerable economic risk to farmers due to the potential for genetic vulnerability caused by their relative genetic uniformity.[3,7,12,28,30,53,54,61] For example, in Asian countries where there has been widespread adoption of modern short-stature varieties of rice (e.g., in India, the Philippines, Indonesia, Sri Lanka, Taiwan, Vietnam, and South Korea), farmers have experienced epidemics of rice blast (a pathogen) or of insects — leafhoppers — which serve as vectors for devastating viruses, causing tremendous drops in yield and often necessitating substantial imports of rice. In contrast, farmers have not been plagued by pest outbreaks in Bangladesh, Thailand, and Burma because more than two thirds of the acreage in each of the latter countries is still planted with traditional varieties.[12] In Greece, farmers revived cultivation of a locally adapted native wheat variety after an introduced, modern wheat variety failed. Unlike its poorly adapted modern counterpart, the landrace of wheat produced reliable yields under the soil conditions of the area.[63] Similarly, in eastern Mexico, after hybrid maize varieties introduced as part of a major development project failed to yield due to lack of adaptation to soil conditions and to the occurrence of serious insect outbreaks to which the hybrids where susceptible, the development project provided farmers with seeds of a local landrace in order to ensure continued maize production.[64] However, increasingly, landraces are no longer available to farmers for readoption unless conscious efforts have been made to preserve them.[3,4,6,8,28,42,55]

IV. CONSERVATION OF CROP GENETIC RESOURCES

A. *Ex Situ* Storage in Gene Banks

The main focus of efforts to prevent the loss of crop genetic diversity has been the strategy of *ex situ* conservation.[5,53,65,66] Landraces and wild relatives of crops are collected from their native habitats and the seed or vegetative material is placed in storage for future use in plant breeding programs to develop crops more resistant to diseases, insect pests, poor soils, and harsh weather.[66,67] Although *ex situ* conservation methods have contributed to the germplasm preservation and genetic improvement of certain crops,[53] they do not provide a panacea for conserving crop genetic resources.[68] Despite the tremendous number of plants now stored in gene banks through national and international efforts to collect and store them, a number of problems with reliance on *ex situ* conservation strategies have been acknowledged.[53,66,67,69,70] The vast majority of gene banks deal only with those major crops which have national and international agricultural research programs supporting them. Minor crops, i.e., those without value in world trade due to the very localized scale of their use, are generally neglected in gene bank associated conservation efforts.[69] An example is tef *(Eragrostis tef)*, a type of millet which is produced almost solely in traditional agroecosystems in Ethiopia where it occupies more acreage there than any other crop.[20] Minor crops can contribute substantially on a local or regional scale to the diet and welfare of people who grow them and have potential to be introduced and developed as new crops in other regions.[35,37,55,71,72] Even when crops are placed in gene banks, only a fraction of their genetic diversity is represented.[53,69] Wild relatives are typically very poorly represented in gene bank collections,[53] despite recognition of their often considerable importance for crop improvement.[73-75] Loss of genetic resources within gene banks can be substantial due to such factors as inadequate control of reproduction, selective deterioration of particular genotypes during storage, and lack of adaptation to the locations used for grow-out. Lack of evaluation or even basic characterization of material in gene banks further limits their usefulness.[53,66,70]

Gene banks involve a centralization of the crop resource base and, as a consequence, political problems have surfaced.[53,76] As Wilkes[66] stated: "The centers of genetic variability are moving from natural systems and primitive agriculture to gene banks and breeders' working collections with the liabilities that a concentration of resource (power) implies". Considerable international controversy has occurred centering around the question of free access to genetic resources.[53,76] In industrial countries, breeders develop new commercial varieties often using valuable genes derived from landraces or wild species originally collected in the Third World. The new commercial varieties are then sold back to the Third World at considerable profit.[76,77] The basic orientation in regard to landraces and wild crop relatives conserved in most *ex situ* collections is that the role of the latter is to serve primarily as potential breeding material for the genetic improvement of modern cultivars.[70] However, as described in the sections above, landraces and associated wild or weedy relatives can directly provide farmers with distinct advantages and thus deserve to be considered as valuable resources for continued use in traditional agroecosystems.[4,8]

B. Towards a Strategy of *In Situ* Conservation

In situ conservation has been suggested as an alternative or complement to *ex situ* conservation since it allows for continued, dynamic adaptation of plants to the environment.[4,6,8,21,42,52,78] *In situ* conservation of crop plants in traditional agroecosystems allows for crops to continue to evolve in their original cultural ecological context, the human-modified systems in which they have evolved.[42,55]

Recommendations for *in situ* conservation of crop germplasm have emphasized the development of a system of village-level landrace custodians (a farmer-curator system) whose purpose would be to continue to grow a limited sample of endangered landraces native to the region.[6,42,78,79] The sites should be located in centers of crop genetic diversity where native agriculture is still practiced and where sympatric close wild or weedy crop relatives may periodically hybridize with the crop plants; the areas would be maintained either by enlisted local farmers or by government research efforts.[6,8,78,79] However, the setting aside of land to be used strictly for conservation of crops and their close relatives may be considered to be a luxury in countries where farmland is already at a premium.[4,52,53] Other supporters of *in situ* conservation have proposed that farmers should be subsidized for their conservation efforts, but have stressed that agricultural "improvements" be deliberately excluded.[75,78,79] Such efforts are more likely to succeed if members of a culture identify their own reasons for maintaining their crop heritage; conservation measures will be most effective when native farmers are cognizant of and involved in their planning and implementation.[42]

Some researchers agree that *in situ* preservation is the preferred strategy for conserving wild plant resources, but consider that *in situ* preservation of landraces would require a return to or the preservation of microcosms of primitive agricultural systems — an unacceptable and impractical proposition.[6,52,53,70,80] However, Altieri and Merrick[4] argue that for conservation of crop genetic diversity to succeed among small farmers, the process must be linked to rural development efforts that give equal importance to local resource conservation and to food self-sufficiency and market participation. Although most traditional agroecosystems are under some process of modernization or drastic modification, conservation efforts can still be linked to rural development projects in Third World countries.[4,8,42,81] Rural development programs such as the Projecto de Agrobiología de Cochabamba in Bolivia and Centro de Educación y Tecnología in Chile are attempting to help peasants recover some of their production autonomy by reducing reliance on purchased inputs through a variety of measures, one component of which is basing production on locally adapted crop varieties.[4]

Citizen grassroots and non-governmental organizations are beginning to mobilize to conserve local genetic resources.[4,42,62,82,83] In the arid region of northwestern Mexico and southwestern United States, the non-profit organization Native Seeds/SEARCH is promoting both *in situ* and *ex situ*

FIGURE 2. These ears of White Mountain Apache corn, used for flour and harvested from a single dry-farmed field in New Mexico, provide an example of a crop being maintained through joint efforts of traditional farmers and Native Seeds/SEARCH, a non-profit organization based in Arizona which promotes both *in situ* and *ex situ* conservation of crops and wild plants native to southwestern United States and northwestern Mexico. (photo by Matt Liebman, University of Maine.)

conservation of species indigenous to the region (Figure 2). Farmers are encouraged to grow traditional crops and are provided with seed of varieties that have been lost.[42,82] Even in the context of storage of collections of crop genetic diversity in gene banks, farmers should be made aware that reciprocal exchanges of seeds with gene banks are possible and that they have unconditional access to seed samples held in the gene banks, which they can obtain if they lose their seeds.[42] The national gene bank in Ethiopia works cooperatively with local farmers in efforts to conserve, evaluate, and utilize native crops; with assistance from nongovernmental development agencies and other national programs, plans are underway for extending cooperation in the future.[83]

REFERENCES

1. **Chang, J. H.,** Tropical agriculture: crop diversity and crop yields, *Econ. Geogr.*, 53, 241, 1977.
2. **Clawson, D. L.,** Harvest security and intra-specific diversity in traditional tropical agriculture, *Econ. Bot.*, 39, 56, 1985.
3. **Harlan, J. R.,** Our vanishing genetic resources, *Science*, 188, 618, 1975.
4. **Altieri, M. A. and Merrick, L. C.,** *In situ* conservation of crop genetic resources through maintenance of traditional farming systems, *Econ. Bot.*, 41, 86, 1987.
5. **Frankel, O. H. and Bennett, E., Eds.,** *Genetic Resources in Plants — Their Exploration and Conservation*, International Biological Programme Handbook #11, Blackwell, Oxford, 1970, 554.
6. **Wilkes, H. G. and Wilkes, S. K.,** The green revolution, *Environment*, 14, 32, 1972.
7. **Brush, S. B.,** Genetic diversity and conservation in traditional farming systems, *J. Ethnobiol.*, 6, 151, 1986.
8. **Oldfield, M. L. and Alcorn, J. B.,** Conservation of traditional agroecosystems, *BioScience*, 37, 199, 1987.
9. **Brush, S. B., Carney, H. J., and Huamán, Z.,** Dynamics of Andean potato agriculture, *Econ. Bot.*, 35, 70, 1981.
10. **Johns, T. and Keen, S. L.,** Ongoing evolution of the potato on the Altiplano of western Bolivia, *Econ. Bot.*, 40, 409, 1986.
11. **Boster, J. S.,** Classification, cultivation, and selection of Aguaruna cultivars of *Manihot esculenta* (Euphorbiaceae), in *Ethnobotany in the Neotropics*, Prance, G. T. and Kallunki, J. A., Eds., *Adv. Econ. Bot.*, 1, 34, 1984.
12. **Chang, T. T.,** Conservation of rice genetic resources: luxury or necessity, *Science*, 224, 251, 1984.
13. **Martin, G. B. and Adams, M. W.,** Landraces of *Phaseolus vulgaris* (Fabaceae) in northern Malawi. I. Regional variation, *Econ. Bot.*, 41, 190, 1987.
14. **Kaplan, L.,** What is the origin of the common bean, *Econ. Bot.*, 35, 240, 1981.
15. **Johannessen, C. L.,** Domestication process of maize continues in Guatemala, *Econ. Bot.*, 36, 84, 1982.
16. **Merrick, L. C.,** Systematics, evolution, and ethnobotany of a domesticated squash, its wild relatives, and allied species in the genus *Cucurbita*, unpubl. Ph.D. dissertation, Cornell University, Ithaca, 1990.
17. **Brunken, J., de Wet, J. M. J., and Harlan, J. R.,** The morphology and domestication of pearl millet, *Econ. Bot.*, 31, 163, 1977.
18. **Merrick, L. C.,** Wild and cultivated cucurbits from the Sierra Madre Occidental of Northwest Mexico and the Rio Balsas Valley of Southwest Mexico, Report to the International Board for Plant Genetic Resources, Rome, 1987, 55.
19. **Chu, Y. E. and Oka, H. I.,** Introgression across isolating barriers in wild and cultivated *Oryza* species, *Evolution*, 24, 344, 1970.
20. **Harlan, J. R.,** Genetic resources of some major field crops in Africa, in *Survey of Crop Genetic Resources in their Centers of Diversity: First Report*, Frankel, O. H., Ed., FAO/IBP, Rome, 1973, 45.
21. **Brush, S. B.,** Farming the edge of the Andes, *Nat. Hist.*, 86, 32, 1977.
22. **Hawkes, J. G.,** *The Diversity of Crop Plants*, Harvard University Press, Cambridge, MA, 1983, 184.
23. **Ugent, D.,** The potato in Mexico: geography and primitive culture, *Econ. Bot.*, 22, 108, 1968.
24. **Harlan, J. R.,** *Crops and Man*, American Society of Agronomy, Madison, WI, 1975, 295.
25. **De Wet, J. M. J., Harlan, J. R., and Price, E. G.,** Variability in *Sorghum bicolor*, in *Origins of African Plant Domestication*, Harlan, J. R., de Wet, J. M. J., and Stemler, A., Eds., 1976, 465.
26. **Kensinger, K. M.,** Studying the Cashinahua, in *The Cashinahua of Eastern Peru: Studies in Anthropology*, Vol. 1, Dwyer, J., Ed., Haffenreffer Museum of Anthropology, Brown University, Providence, RI, 1975, 9.
27. **Bedigan, D. and Harlan, J. R.,** Nuba agriculture and ethnobotany, with particular reference to sesame and sorghum, *Econ. Bot.*, 37, 384, 1983.
28. **Wilkes, H. G.,** The world's crop plant germ plasm — an endangered resource, *Bull. At. Sci.*, 33, 8, 1977.
29. **Adams, M. W., Ellingbae, A. H., and Rossineau, E. C.,** Biological uniformity and disease epidemics, *BioScience*, 21, 1067, 1971.
30. **Plucknett, D. L. and Smith, N. J. H.,** Sustaining agricultural yields, *BioScience*, 36, 40, 1986.
31. **Zizumbo-V., D.,** Aspectos etnobotánicos de las calabazas silvestres y cultivadas (*Cucurbita* spp.) de la Peninsula de Yucatán, *Bol. Escuela Ciencias Antrop. Univ. Yucatán*, 13, 15, 1986.

32. **Harlan, J. R., de Wet, J. M. J., and Price, E. G.,** Comparative evolution of cereals, *Evolution*, 27, 311, 1973.
33. **Hernandez-X., E.,** *Exploracíon Etnobotánica y su Metodología*, Chapingo, Mexico, 1970, 67.
34. **Vavilov, N. I.,** The origin, variation, immunity and breeding of cultivated plants, *Chron. Bot.*, 13, 1, 1951.
35. **King, S. R.,** Four endemic Andean tuber crops: promising food resources for agricultural diversification, *Mountain Res. Dev.*, 7, 43, 1987.
36. **Gade, D. W.,** Vanishing crops of traditional agriculture: the case of tarwi *(Lupinus mutabilis)* in the Andes, *Proc. Assoc. Am. Geogr.*, 1, 47, 1969.
37. **Cusack, D. F.,** Quinua: grain of the Incas, *Ecologist*, 14, 21, 1984.
38. **León, J.,** Primitive genetic resources in Latin America, in *Survey of Crop Genetic Resources in their Centers of Diversity: First Report*, Frankel, O. H., Ed., FAO/IBP, Rome, 1973, 71.
39. **de Wet, J. M. J. and Harlan, J. R.,** Weeds and domesticates: evolution in the man-made habitat, *Econ. Bot.*, 29, 99, 1975.
40. **Harlan, J. R.,** The possible role of weed races in the evolution of cultivated plants, *Euphytica*, 14, 173, 1965.
41. **Merrick, L. C. and Nabhan, G. P.,** Natural hybridization of wild *Cucurbita sororia* group and domesticated *C. mixta* in southern Sonora, Mexico, *Cucurbit Genet. Coop. Rep.*, 7, 73, 1984.
42. **Nabhan, G. P.,** Native crop diversity in Aridoamerica: conservation of regional gene pools, *Econ. Bot.*, 39, 387, 1985.
43. **Rabinowitz, D., Linder, C. R., Begazo-O., D., Murguia-S., H., Ortega-D., R., and Schmiediche, P.,** Ecological mechanisms of genetic interchange between cultivars and their wild relatives: traditional potato farming in the Peruvian Andes, *Ecol. Monogr.*, in press.
44. **Rick, C. M.,** The role of natural hybridization in the derivation of cultivated tomatoes of western South America, *Econ. Bot.*, 12, 346, 1958.
45. **Ugent, D.,** The potato, *Science*, 170, 1161, 1970.
46. **Wilkes, H. G.,** Hybridization of maize and teosinte, in Mexico and Guatemala and the improvement of maize, *Econ. Bot.*, 31, 254, 1977.
47. **Esquinas-A., J. T.,** *Genetic Resources of Tomatoes and Wild Relatives*, International Board for Plant Genetic Resources Secretariat, Rome, 1981, 65.
48. **Yen, D. E.,** Natural selection and human selection in the Pacific sweet potato, in *Evolution and Environment*, Drake, E. T., Ed., Yale University Press, New Haven, CT, 1968, 387.
49. **Martin, G. B. and Adams, M. W.,** Landraces of *Phaseolus vulgaris* (Fabaceae) in northern Malawi. II. Generation and maintenance of variability, *Econ. Bot.*, 41, 204, 1987.
50. **Nabhan, G. P., Merrick, L. C., and Rea, A. M.,** The gourd family among Tepiman cultures: cultural biogeography and comparative folk taxonomies, *Biotica*, submitted.
51. **Frankel, O. H., Ed.,** *Survey of Crop Genetic Resources in their Centers of Diversity: First Report*, FAO/IBP, Rome, 1973, 169.
52. **Prescott-Allen, R. and Prescott-Allen, C.,** The case for *in situ* conservation of crop genetic resources, *Nat. Resour.*, 23, 15, 1982.
53. **Plucknett, D. L., Smith, N. J. H., Williams, J. T., and Anishetty, N. M.,** *Gene Banks and the World's Food*, Princeton University Press, Princeton, NJ, 1987, 247.
54. **N.A.S.,** *Genetic Vulnerability of Major Crops*, National Academy of Sciences, Washington, D.C., 1972, 307.
55. **Nabhan, G. P.,** Cultivation and culture: cultivated crops of native cultures in southwestern North America — their current status, *Ecologist*, 9, 259, 1979.
56. **Whitaker, T. W. and Knight, R. J.,** Collecting cultivated and wild cucurbits in Mexico, *Econ. Bot.*, 34, 132, 1980.
57. **Ochoa, C.,** Potato collecting expeditions in Chile, Bolivia and Peru, and the genetic erosion of indigenous cultivars, in *Crop Genetic Resources for Today and Tomorrow*, Frankel, O. H. and Hawkes, J. G., Eds., Cambridge University Press, Cambridge, 1975, 167.
58. **Appa Rao, S., Mengesha, M. H., Sibale, P. K., and Rajagopal Reddy, C.,** Collection and evaluation of pearl millet (*Pennisetum*) germplasm from Malawi, *Econ. Bot.*, 40, 27, 1986.
59. **Yabuno, T.,** Japanese barnyard millet (*Echinochloa utilis*, Poaceae) in Japan, *Econ. Bot.*, 41, 484, 1987.
60. **Wilkes, H. G.,** Native crops and wild food plants, *Ecologist*, 312, 1977.
61. **Doyle, J.,** *Altered Harvest: Agriculture, Genetics, and the Fate of the World's Food Supply*, Viking Press, 1985, 502.
62. **Altieri, M. A. and Anderson, M. K.,** An ecological basis for the development of alternative agricultural systems for small farmers in the Third World, *Am. J. Altern. Agric.*, 1, 30, 1986.
63. **WGBH,** *Seeds of Tomorrow*, transcript of NOVA program #1212, WGBH Educational Foundation, Boston, MA, 1985, 22.
64. **Ewell, P. T. and Poleman, T. T.,** *Uxpanapa: Resettlement and Agricultural Development in the Mexican Tropics*, Intern. Agric. Econ. Study 82-7, Cornell University, Ithaca, NY, 1982, 218.
65. **Frankel, O. H. and Hawkes, J. G., Eds.,** *Crop Genetic Resources for Today and Tomorrow*, Cambridge University Press, Cambridge, 1975, 492.
66. **Wilkes, H. G.,** Current status of crop germplasm, *CRC Crit. Rev. Plant Sci.*, 1, 133, 1983.
67. **Plucknett, D. L., Smith, N. J. H., Williams, J. T., and Anishetty, N. M.,** Crop germplasm conservation and developing countries, *Science*, 220, 163, 1983.
68. **Oldfield, M. L.,** *The Value of Conserving Genetic Resources*, Dept. Interior, U.S. Government Printing Office, Washington, D. C., 1984, 274.
69. **Lyman, J. M.,** Progress and planning for germplasm conservation of major food crops, *Plant Genet. Resour. Newsl.*, 60, 3, 1984.
70. **Frankel, O. H. and Soulé, M. E.,** *Conservation and Evolution*, Cambridge University Press, Cambridge, 1981, 327.
71. **Vietmeyer, N. D.,** Lesser-known plants of potential use in agriculture and forestry, *Science*, 232, 1379, 1986.
72. **N.A.S.,** *Underexploited Tropical Plants with Promising Economic Value*, National Academy of Sciences, Washington, D.C., 1975, 188.
73. **Prescott-Allen, R. and Prescott-Allen, C.,** *Genes From the Wild: Using Genetic Resources For Food and Raw Materials*, International Institute for Environment and Development, London, 1983, 101.
74. **Harlan, J. R.,** Genetic resources in wild relatives of crop plants, *Crop Sci.*, 16, 329, 1976.
75. **Myers, N.,** *A Wealth of Wild Species: Storehouse for Human Welfare*, Westview Press, Boulder, CO, 1983, 215.
76. **Kloppenburg, J., Jr. and Kleinman, D. L.,** The plant germplasm controversy: analyzing empirically the distribution of the world's plant genetic resources, *BioScience*, 37, 190, 1987.
77. **Wolf, E. C.,** Conserving biological diversity, *The State of the World, A Worldwatch Institute Report on Progress Toward a Sustainable Society*, W. W. Norton, New York, 1985, 124.
78. **Iltis, H. H.,** Freezing the genetic landscape — the preservation of diversity in cultivated plants as an urgent social responsibility of the plant geneticist and plant taxonomist, *Maize Genet. Coop. Newsl.*, 48, 199, 1974.
79. **Barnes-McConnell, P.,** Keepers of the trust, *Diversity*, 12, 22, 1987.
80. **Ingram, G. B. and Williams, J. T.,** *In situ* conservation of wild relatives of crops, in *Crop Genetic Resources: Conservation and Evaluation*, Holden, J. H. W. and Williams, J. T., Eds., George Allen and Unwin, London, 1984, 163.
81. **Alcorn, J. B.,** Development policy, forests and peasant farms: reflections on Huastec-managed forests' contributions to commercial production and resource conservation, *Econ. Bot.*, 38, 389, 1984.

82. U. S. Congress, OTA, *Grassroots Conservation of Biological Diversity in the United States — Background Paper #1*, Office of Technology Assessment, Washington, D.C., 1986, 67.
83. *R.A.F.I.*, African genetic resources workshop, summary report, Rural Advancement Fund International, 1987, 10.

2 THE ECOLOGY AND MANAGEMENT OF TRADITIONAL FARMING SYSTEMS

Stephen R. Gliessman

I. INTRODUCTION

Throughout most of the developing world today, local or traditional knowledge continues to form the basic foundation of agriculture and small farm management. Such knowledge reflects experience gained from past generations, yet continues to develop in the present as the ecological and cultural environment of the populations involved goes through a continual process of adaptation and change. Studies of traditional agriculture are beginning to show us the great value such systems have for contributing to the development of ecologically sound management practices that are understandable and acceptable to rural peoples.[1-4] These systems often make use of locally available resources rather than rely on costly inputs imported from distant sources. They allow for the simultaneous satisfaction of local needs together with a significant contribution to demands on a larger scale. Most importantly, production takes place in ways that focus more on the long-term sustainability of the system, rather than an over-emphasis on the maximization of yields. The ability of these systems to keep the land productive on a permanent basis reduces the need for the development of new lands. An agroecological approach to understanding how such systems function can provide information which can significantly contribute to establishing sustainable small farm systems.

By considering a particular farm management system in the context of a larger ecosystem, the impacts or results of any particular management practice can be understood. The study of agricultural ecosystems (agroecosystems) places such impacts in the context of nutrient cycles, population dynamics, and energetics. A thorough knowledge of the interplay between physical factors of the environment, their interaction with or modification by biological factors, and the incorporation of the management of particular resources for human use, can permit the establishment of an understanding of both the agroecological potentials and limitations of the farming system under study. The link to ecosystem studies can be expanded to include the idea that many traditional or indigeneous agroecosystems even "mimic" the structure and function of the naturally occurring ecosystems of an area.[5] The examination of a variety of traditional agroecosystems in Middle America can provide a valuable means of beginning to test these ideas.

II. ECOLOGY AND MANAGEMENT OF TRADITIONAL AGROECOSYSTEMS

A. Multiple-Crop Agroecosystems

The use of crop associations or multiple cropping can take place in many cropping systems to great ecological and agronomic advantage.[6,7] More than one crop occupies the same piece of land either simultaneously or in some type of rotational sequence during the season. Production can be increased, more efficient use of resources takes place, and the land can be occupied productively more continuously. The importance of multiple cropping is being recognized,[8] and the need for intensive agroecological studies of such mixed cropping has become more evident.[9]

A multiple cropping system that has been studied in some detail is the traditional maize, bean, and squash polyculture. There is evidence that intercropping maize and beans *(Phaseolus vulgaris)* has been practiced in Central America since pre-Hispanic times, and continues to form an important part of the patterns of food production in this region today.[10] In a series of studies done in Tabasco, Mexico, it was found that maize yields could be stimulated as much as 50% beyond monoculture yields when planted with beans and squash (see Table 1).[6] There was some yield reduction for the two associated crop species, but the summed yields for the crops planted together were higher than for an equivalent amount of land planted to the crops in monoculture (overyielding). Studies that provide an understanding of the ecological mechanisms of the yield increase are important for establishing a strong basis for recommending more widespread use of the cropping system. On the one hand, it

TABLE 1
Yields of Polyculture Maize/Bean/Squash as Compared to Monocultures Planted at Four Different Densities, Cardenas, Tabasco, Mexico[6]

	Monoculture densities				Polyculture
Densities of maize[a]	33,000	40,000	66,000	100,000	50,000
Yield (kg/ha)[b]	990	1,150	1,230	1,170	1,720
Densities of beans	56,800	64,000	100,000	133,200	40,000
Yield (kg/ha)	425	740	610	695	110
Densities of squash	1,200	1,875	7,500	30,000	3,330
Yield (kg/ha)	15	250	430	225	80

[a] Densities expressed as number of plants/ha.
[b] Yields for maize and beans expressed as dried grain, squash as fresh fruits.

TABLE 2
Mean Fresh Weight (g/0.2 m sq) of the Aboveground Weed Biomass in Maize, Bean, and Squash Monocultures as Compared to a Polyculture of All Three

Crop system	Mean fresh weight of weeds: Total	Mean fresh weight of weeds: Monocots	% light transmission
Maize monoculture	61.8	16.8	36.06 ± 03.8
Squash monoculture	61.4	20.8	51.40 ± 31.2
Bean monoculture	79.8	14.9	74.50 ± 22.7
Polyculture	55.0	19.0	44.70 ± 25.5

Note: Percent light transmission was determined by measurements at the soil surface in each crop system with a LI-COR quantum radiometer and PAR sensors. Biomass figures are the means of three samples taken in each plot, and light figures are the means of 10 readings taken at 50 cm intervals in each plot.[18]

appears that beans in polyculture with maize nodulate more and potentially are more active in biologically fixing nitrogen which could be made directly available to the maize.[11] Net gains of nitrogen have been observed when the crops are associated, despite the removal of this element with the harvest.[12] This contributes to both the long-term reduction in dependence on external purchased inputs of fertilizer, as well as an overall more stable basis for managing resources within the system.

At the same time, study of the management practices employed in this polyculture has demonstrated the ecological basis upon which the practices can function. For example, despite the lower squash yields in the mixed planting, farmers insist that the crop system benefits from the squash presence through the control of weeds.[13] Thick, broad, horizontal leaves cast a dense shade that blocks sunlight, while leachates in rains washing the leaves contain allelopathic compounds that potentially inhibit weed growth. Herbivorous insects are at a disadvantage in the intercrop system,[14] and the presence of beneficial insects is promoted.[15] Leaving weeds in the intercropped system can be advantageous as well.[16] *Chenopodium ambrosioides*, for example, has the potential for inhibiting plant pathogenic nematodes through the release of toxic root exudates.[17] *Lagascea mollis* can allelopathically control the invasion of weeds detrimental to the crops if it is allowed to form a dense cover after the critical establishment stage in crop development (the first 3 to 4 weeks), thus avoiding inhibition of the crop as well.[13]

Understanding the dynamics of resource use and distribution in a mixed crop system is very important as well. In a planting in Costa Rica, weed biomass was determined and compared in mature monocultures of maize, beans, and squash, and in the polyculture of all three.[18] Plots were set up to test the hypothesis that the vegetative canopy of a polyculture system has a more complex structural diversity, hence an enhanced ability to capture incoming solar radiation. As a result, it was hypothesized that the increased shading possible in the polyculture should lead to a reduction in weed biomass. The plots were planted in May, followed by a uniform weeding 6 weeks later, with the final sampling of aboveground live biomass of the weeds taken 5 weeks after that.

As can be seen in Table 2, there definitely are trends in weed weight distribution according to the crop type, although variability in sample weights led to a lack of statistical significance between treatments. At the time of the sampling, the squash monoculture cover had only partially covered the plots, bean cover was fairly limited, and the densest cover was observed in the maize monoculture and the polyculture. Total weed biomass under the maize and squash systems were the same. Biomass from the bean plots was 23% greater and weed biomass from the polyculture system was approximately 10% less than the maize or squash systems. Monocots accounted for a lower proportion of the bean plot weed weight (about 10% of the total biomass) than the maize (27% monocots), squash or polyculture (34 and 35% monocots). However, actual mean weights of the monocot biomass varied little between treatments. The percent light transmission (calculated as the percent of transmitted light reaching the soil surface in each crop system as compared to that reaching an unplanted soil surface adjacent to the plots) again showed similar trends, but high variability in this sampling again ruled out statistical significance. The highest light transmission was recorded in the bean monoculture, correlating to the lower cover produced by the bean crop. The lowest transmission occurred in the maize monoculture. The trend for lower weed biomass in the polyculture, then, is only partially related to light reception by the crop canopy. Many other factors are coming into play in such an interactive system, and detailed and long-term research is necessary to determine their relative importance.[9]

B. Diverse Home Gardens

One of the agroecosystems that seems to be extremely

TABLE 3
Species Types and Characteristics of Home Garden Agroecosystems at Upland (Tepeyanco, Tlaxcala) and Lowland (Cupilco, Tabasco) Sites in Mexico

Characteristics	Cupilco	Tepeyanco
Average garden size	0.70 ha	0.34 ha
No. of useful species per garden	55	33
Diversity (bits)	3.84	2.43
Leaf area index	4.5	3.2
% cover	96.7	85.3
% light transmission	21.5	30.5
Perennial species (%)	52.3	24.5
Tree species (%)	30.7	12.3
Ornamental plants (%)	7.0	9.0
Medicinal plants (%)	2.0	2.8

Note: Data from four gardens in Tepeyanco and three in Cupilco.[19]

TABLE 4
The Number of Plant Species and Individuals of Each Species in a Home Garden Agroecosystem on the Outskirts of Cañas, Guanacaste Province, Costa Rica, Listed According to Common Uses

Plant uses	# spp.	# indiv.	% spp.	% indiv.
Ornamental	36	517	48	21.6
Food	26	164	36	68.2
Medicinal	6	1	8	1.6
Firewood	5	17	7	1.7
Animal feed	1	51	1	6.7
Total	71	758		

Note: Total species and individual numbers are less than the sum of the columns due to the multiple function of some species.[23]

well adapted ecologically to the needs of small farms is the home garden or kitchen garden system.[19,20] It is structurally very diverse, with an overstory of trees and an understory of a mixture of herbs, shrubs, small trees, and vines. Such diversity permits year-round harvesting of food products, as well as a wide range of other products used on small farms in developing countries such as firewood, medicinal plants, spices, and ornamentals. In an ecological analysis of home gardens in both lowland and upland sites in Mexico, it was found that in quite a small area (0.3 to 0.7 ha) high diversity permitted a high degree of similarity between the managed agroecosystem and local natural systems. Relatively high species diversity for a cropping system was also achieved (Table 3).

In a home garden on the outskirts of Cañas, Guanacaste Province, Costa Rica, in an area of 1240 sq m, a total of 71 plant species were found. As can be seen in Table 4, the garden served as a source of food, firewood, medicine, and color and enjoyment for the household. Some of the plant species served more than one function. The Shannon-Weaver species diversity index for the garden was 3.55, a relatively high value for an agricultural system. To a certain extent, plants were also distributed in the garden depending on the uses. Trees were concentrated towards the back of the plot, providing shade for the work area at the back of the house, as well as providing a stabilizing border along a riverbank that parallels the back of the property. Annual food crops were concentrated towards the front of the garden in full sunlight. The large number of ornamental species were clustered in beds or containers around the walls of the house or along the pathway leading from the front of the property to the house. Animal pens behind the house in the shade of the trees contained two pigs, a goat, and a guinea pig. An undetermined number of chickens freely roamed throughout the plot, as did several small dogs and two cats. Mango was the principal tree species, with maize, squash, beans, papaya, bananas, and yuca (cassava) playing the most important roles in food production. The man of the household had full-time employment in the nearby town, so the garden played more of a supplemental role in the family economics.

Ecological as well as sociological studies are needed in order to further understand the structure and diversity of the garden. Home gardens are extremely variable in size and design. They respond to local variations in soil type, drainage patterns, cultural preferences, economic standing of the family, family size and age patterns, etc., reflecting a multiplicity of both ecological and cultural components. At the same time they are flexible, dynamic, and changing depending on the needs of the family.[21] In a home garden located in the Atlantic lowlands of Costa Rica, near the town of Puerto Viejo, Sarapiqui, mapping revealed considerable diversity and complexity within an area of approximately 3250 sq. m.[22] There were 26 species of trees, 16 perennial ornamentals, 8 annual/biennial crops, and 6 herbaceous species in the garden at the time of the study. The plants were distributed into what could be characterized as five functional areas as follows:

1. A low-diversity, regularly patterned planting of crops of potential cash value, including tuber crops, pineapple, and young coconuts
2. A high diversity, irregularly patterned planting of trees, shrubs, herbs and vines of many uses designed to satisfy domestic needs
3. A low diversity, widely spaced planting of trees, most often with low grass or bare soil below, often used for social or recreational purposes
4. A very high diversity, intercropped planting of ornamental herbs and shrubs planted very close to the house and cared for by the women in the household.
5. A moderate diversity, alternately planted fencerow surrounding the property primarily composed of fruit and firewood tree species.

The garden reflects an interaction between the need for

domestic food or use items, the desire or need for cash income, personal preference and enjoyment, and the constraints of time and space. A move into cash cropping, relatively new to this particular garden, is changing its structure dramatically. This trend will continue as trees mature, markets change, and the socio-economic status of the family changes. Home gardens seem to incorporate this flexibility and dynamism, at the same time maintaining a sound ecological basis to their design and management.

C. A Traditional Wetland Agroecosystem

Factors of the environment that are limiting to agricultural production, be they physical or biological, require special adaptations for agriculture to be successfully carried out. Agricultural development projects in much of the developing world have normally approached such limitations intent upon eliminating or altering them to fit the needs of the cropping systems being developed. This usually involves the use of large levels of external inputs of energy or materials. There are many and well-known examples of massive irrigation, drainage, or desalination projects that attempted to alter existing ecological conditions, but achieved only limited success when evaluated in terms of crop productivity and economic viability, at the same time having little applicability for meeting the needs of small farmers. In many areas of Mesoamerica, where heavy rainfall and low-lying topography combine to generate conditions of excess soil moisture, local farmers have found ways to accommodate this factor into the design and management of quite sustainable traditional agroecosystems.

A very interesting and productive use of wetland areas has been observed in Tabasco for the production of maize and other crops.[24] Since this practice is most often observed in communities comprised primarily of remnants of indigenous populations, it might indicate that the practice is not one that was recently introduced, but rather may have been handed down from generation to generation. In this system, maize is planted on higher ground around flood-prone areas during the wet season months of June—December, but as water levels drop during the dry season months of February and March, farmers follow the receding water line with another maize planting, known locally as the March planting (marceño). Similar systems have been observed as far north in Mexico as Veracruz.[25]

During much of the year the low areas are inundated to a depth of a few centimeters to as much as a meter. Marsh vegetation densely covers the area during the wet season, and is felled quickly with machetes as the water level recedes. A very dense mat of organic matter is produced by this process (10 to 20 cm), into which seed is planted with a pointed stick. About a week following planting, fire is used to burn part of this organic mat, as well as kill back any weed seedlings or sprouts of the marsh plants. The timing of burning is very important in order to ensure that the fire burns only the dry leaves on top of the mat and does not reach the soil. The maize seed, planted 10 to 15 cm below the surface of the soil, is not harmed by the fire. Local short-cycle varieties of maize (2 to 3 months from planting to harvest) are most frequently used. The practice of using seed from the previous harvest for the subsequent planting favors the use of local varieties. The use of a $2^1/_2$ month variety known as "mejen", a Maya word meaning precocious or early maturing, again shows the link to the past that many of the present day systems may have.

The maize grows very quickly in this system, and when fire is not used excessively and flooding is allowed to occur every year, weeding is usually not necessary. At approximately $2^1/_2$ months the mature maize stalks are "doubled-over" just below the maize ear, facilitating final drying of the grain for another 2 to 4 weeks before harvest. Yields between 4 to 5 tons/ha of dry grain are common, with some yields reaching 10 tons/ha. This is many times the average yield for mechanized production on lands that have been cleared and drained in the same region, but at a fraction of the input labor and cost.[26]

Following the harvest, all crop and non-crop residues end up on the soil surface. This contributes to a key element in the productivity of the system — the soil. Soil profiles demonstrate the presence of a thick organic soil to a depth of 30 to 40 cm below the surface. During inundation, organic matter produced by the marsh plants or left by the previous cropping cycle, is incorporated and conserved below water. Nutrient minerals that enter the system with surface drainage are captured by the highly productive aquatic sector of the ecosystem, leading to the formation of a soil that analysis has shown to have organic matter levels over 30%, total N as high as 3%, and high levels of other important plant nutrients.[27] The key element, then, in the management of this system is to take advantage of the benefits gained from a period of inundation. Apart from the benefits gained in terms of soil fertility, the flooded period is probably important for pest, disease, and weed control by interrupting the life cycles of such organisms. When the system is drained artificially in an attempt to extend the cropping season, the organic layer in the soil can be reduced to 5 cm in less than 2 years,[28] yields drop dramatically, and weeds invade the system. The importance for understanding the ecological basis for such a system is great, because as such systems are drained for modern agricultural projects, not only does production drop, but the indigenous knowledge of how to manage such systems is lost too. At the same time, a dependence on purchased inputs increases greatly. The agroecosystem loses both its applicability to small farm needs as well as the locally sustainable base it once had.

III. FUTURE DIRECTIONS

An agroecological focus on small farm development goes much beyond crop yields, delving deeply into the complex set of factors that make up the agroecosystem. Local, indigenous agroecosystems that have evolved under the diverse and often limiting conditions facing small farmers, are adapted to this set of factors. They have evolved through time as reduced external-input systems, with a greater reliance on renewable resources and an ecologically based management strategy. A research focus in agriculture that can take advantage of this knowledge and experience can permit us to explore the multiple bases upon which sustainability rests. It represents the blending of knowledge gained by ecologists studying the dynamics and stability of natural ecosystems with the knowledge of farmers and agronomists on how to manage the complexities of food-producing agroecosystems. From this can come the sustainability in the production base so critical for giving small farmers the stability and viability they need to both provide their own needs and contribute to meeting the needs of the greater society.

ACKNOWLEDGMENTS

This chapter is the result of ideas, support, and challenges from an innumerable set of friends and acquaintances. I want you all to know how appreciated your input has been over the years. I especially acknowledge the support of the UCSC Agroecology Program, the Alfred Heller Chair of Agroecology, and the Kellogg National Fellowship Program.

REFERENCES

1. **Klee, G.,** *World Systems of Traditional Resource Management,* Halstead, 1980, chap. 1.
2. **Gliessman, S. R.,** Resource Management in Traditional Tropical Agroecosystems in Southeast Mexico, in *Agricultural Sustainability in a Changing World Order,* Douglass, G., Ed., Westview Press, Boulder, CO, 1984, 191.
3. **Altieri, M. A.,** Towards a grassroots approach to rural development in the Third World, *Agric. Hum. Values,* 1, 45, 1984.
4. **Wilken, G. C.,** *Good Farmers: Traditional Agricultural Resource Management in Mexico and Central America,* University of California Press, Berkeley, 1988, chap. 1.
5. **Gliessman, S. R., R. Garcia E., and M. Amador A.,** The ecological basis for the application of traditional agricultural technology in the management of tropical agroecosystems, *Agro-Ecosystems,* 7, 173, 1981.
6. **Amador, M. F.,** Comportamiento de tres especies (Maiz, Frijol, Calabaza) en policultivos en la Chontalpa, Tabasco, México. Tesis Profesional, CSAT, Cárdenas, Tabasco, México, 1980.
7. **Gliessman, S. R.,** Multiple cropping systems: a basis for developing an alternative agriculture, in *Innovative Biological Technologies for Lesser Developed Countries — Workshop Proceedings,* OTA, Washington, D.C., 1985, 69.
8. **Francis, C. A.,** *Multiple Cropping Systems,* Macmillan, New York, 1986, chap. 1.
9. **Gliessman, S. R.,** Plant interactions in multiple cropping systems, in *Multiple Cropping Systems,* Francis, C. A., Ed., Macmillan, New York, 1986, 82.
10. **Pinchinat, A. M., Soria, J., and Bazan, R.,** Multiple cropping in Tropical America, in *Multiple Cropping,* Papendick, R. I., Sanchez, P. A., and Triplett, G. B., Eds., American Society of Agronomy Spec. Publ. 27, 1976, 51.
11. **Boucher, D. H.,** La nodulación del frijól en policultivo: el efecto de la distancia entre las plantas de frijól y maiz. *Agricultura Tropical (CSAT),* 1, 276, 1979.
12. **Gliessman, S. R.,** Nitrogen distribution in several traditional agroecosystems in the humid tropical lowlands of south-eastern Mexico, *Plant Soil,* 67, 105, 1982.
13. **Gliessman, S. R.,** Allelopathic interactions in crop-weed mixtures: applications for weed management, *J. Chem. Ecol.,* 9, 991, 1983.
14. **Risch, S.,** The population dynamics of several herbivorous beetles in a tropical agroecosystem: the effect of intercropping corn, beans and squash in Costa Rica, *J. Appl. Ecol.,* 17, 593, 1980.
15. **Letourneau, D.,** Population Dynamics of Insect Pests and Natural Control in Traditional Agroecosystems in Tropical Mexico, Ph.D. dissertation, University of California, Berkeley, 1983.
16. **Chacon, J. C. and Gliessman, S. R.,** Use of the "non-weed" concept in traditional tropical agroecosystems of south-eastern Mexico, *Agro-Ecosystems,* 8, 1, 1982.
17. **Garcia, R.,** *Chenopodium ambrosioides* L., planta con uso potenciál en el combate de nemátodos fitoparásitos, *Agricultura Tropical (CSAT),* 2, 92, 1980.
18. **Bergmark, C., Bergmann, B. A., Dixon, S., Vazquez, C., and Woodbridge, W.,** Weed recovery in monoculture corn, bean, squash, and polyculture systems, in *Course Book: OTS 85-4,* Jaffe, R. M. and Gliessman, S. R., Eds., 1986, 185.
19. **Allison, J.,** An Ecological Analysis of Home Gardens (Huertos Familiares) in Two Mexican Villages, M. A. thesis, Biology, University of California, Santa Cruz, 1983.
20. **Gliessman, S. R.,** The home garden agroecosystem: a model for developing more sustainable tropical agricultural systems, in *Global Perspectives in Agroecology and Sustainable Agriculture,* Allen, P. and Van Dusen, D., Eds., University of California, Santa Cruz, 1988, 445.
21. **Gonzalez-Jacome, A.,** Home gardens in central Mexico, in *Prehistoric Intensive Agriculture in the Tropics,* Farrington, I. S., Ed., BAR International Series 232, Oxford, England, 1985.
22. **Flietner, D.,** Don Ignacio's home garden, in *Coursebook: OTS 85-4,* Jaffe, R. M. and Gliessman, S. R., Eds., San Jose, Costa Rica, 57, 1986.
23. **Carter, J.,** Map of a home garden in Cañas, Costa Rica, in *Coursebook: OTS 85-4,* Jaffe, R. M. and Gliessman, S. R., Eds., San Jose, Costa Rica, 122, 1986.
24. **Orozco-Segovia, A. and Gliessman, S. R.,** The marceño in flood-prone regions of Tabasco, Mexico, in *Pasado y Presente en los Sistemas Agricolas en Mexico,* Gonzalez-Jacome, A. and Gliessman, S. R., Eds., SEP, Mexico, in press.
25. **Siemens, A.,** Reducing the risk: some indications regarding pre-hispanic wetland agricultural intensification from contemporary use of a wetland/terra firma boundary zone in central Vera Cruz, Mexico, in *Agroecology: Researching the Ecological Basis for Sustainable Agriculture,* Gliessman, S. R., Ed., Springer-Verlag Ecological Studies Series, New York, 1978.
26. **Amador, M. F. and Gliessman, S. R.,** The response of three species (corn, beans, and squash) in polyculture in the Chontalpa, Tabasco, Mexico, 1981.
27. **Gliessman, S. R.,** unpublished data, 1980.
28. **Mejia, A. and Gliessman, S. R.,** unpublished data, 1981.

3 NEW CROPS FOR SMALL FARMERS IN MARGINAL LANDS? WILD CHILES AS A CASE STUDY

Gary P. Nabhan, Mark Slater, and Larry Yarger

I. INTRODUCTION

Over the last decade there has been considerable scientific interest in domesticating wild plants with known economic products as alternatives to conventional crops less well adapted to marginal lands.[1-3] The rationale for this scientific effort has been based on the truism that there are many underexploited wild plant resources tolerant or resistant to heat, drought, cold, salinity, pathogens, and pests.[4] Recently, however, it has been amply demonstrated that when these plants are grown in intensive monocultural production, vulnerability to disease and other stresses increase radically. Along with the impetus for finding new nutritional and economic resources for marginal lands, there is a desire to improve the well being of small farmers, particularly those of indigenous cultures living in environments too difficult for conventional agriculture, where few other sources of cash or subsistence "income" exist.[6,7] As social scientists have demonstrated, not all new crops can be grown without the availability of considerable capital for machinery, processing, or crop protection. Certain new crops promoted today are specifically designed to fit highly mechanized, capital-intensive farming systems subsidized and controlled by transnational corporations.[6] Other proposed new crops may fit into more than one kind of farming system and scale. These proposed new crops may provide social and economic benefits to marginal peoples but effective production will depend upon how well matched the crops are to those people's known farming systems.[7]

If one chooses to explore the potential of new crops for small-scale agricultural development, what criteria should be used to fit an undomesticated plant into a new agronomic environment and an existing cultural/social system? The criteria chosen depend upon how well the goals of a rural development project are defined. Would the crop provide both cash income and direct nutritional benefits to small farmers, or only one of these? Would it be a crop that would have advantages only in a small-scale setting, so that competition would be less likely from well-endowed farmers with extensive land holdings? Will growing the crop help to halt the depletion of a wild plant resource by reducing harvest pressures in its natural habitat? Can the crop be grown in an area where water and soil fertility are limited, or should these physical resources be further modified to allow yields to increase substantially? Would the project provide a new product to a local or to an international market? Would the crop production be simply one component of an overall rural development scheme?

Any combination of these goals has been cited as valid reasons for arid land crop development projects. From our point of view, however, the following criteria collectively offer the best guidelines for ensuring that small farmers in marginal lands will benefit from a new crop development plan:

1. The focus of a project should be on improving the *objective* and *subjective* well being of a community, (the latter as its members perceive themselves) and not simply on finding a cheap production site for a new resource.[7,8]
2. Especially in marginal regions, the adaptation of a new crop should utilize existing community resources for production, processing, and marketing. The project should emphasize low external inputs and capital for production, particularly where water, soil fertility, and technological infrastructure have been chronically limited.[6,9]
3. The new crop's propagation, maintenance, or harvest may be labor-intensive compared to commodity crops; but the economic return per unit weight or per unit time should provide farm workers with better than minimum wage or other locally accepted, non-monetary rewards.[6,8,10]

4. The cultivation of the new crop should replace the wild harvest if the latter is depleting the resource in its natural habitat, and should attempt to provide a new source of income to wild harvesters. At the same time, the cultivation scheme should attempt to sustain both economic productivity and its ecological support services in the long run.[8,11]
5. The plant chosen should offer multiple products or be adaptable to intercrop, multicrop, or diversified farming systems, which may buffer small farmers from the unpredictability of any single market, reduce investment risks, and enhance dietary diversity.[6,12]
6. The crop(s) should offer farmers a source of cash income without eliminating other subsistence crops; it should also be suitable for home consumption and/or cottage-industry scale processing.[6,13]
7. Whenever possible, the small farmers should already have some familiarity, use, and appreciation for the wild plant being domesticated, or for some analog of it (e.g., a domesticated congener or source of similar products).[13] The cropping scheme should also fit into traditional or easily understood farming systems.
8. If possible, the farmers should be able to propagate the crop from locally available genetic resources, rather than being dependent upon exotic, patented or hybrid seeds.[14]

In the following case study, we will examine how well one new crop fits these criteria for small farm development in the arid subtropics of Mexico and adjacent Arizona. The case study does not present a linear history of the development of this crop up until the present moment, nor does it pretend to be a sure-fire "model" of an adapted cropping system that will inevitably be socially, economically, and ecologically successful.[15] In fact, it is still too early to determine whether this new perennial crop has the capacity to survive for more than 5 years under field cultivation, or whether there is a guaranteed market for the farmer. Nonetheless, it offers us an example to test whether the above criteria are helpful in sustaining small farm development in marginal lands. It draws upon our experiences studying the ecology and harvest of the plant in wild habitats in Sonora and Arizona; bringing it into experimental cultivation at the Gentry Agroecology Project in Arizona and at Escuela La Argentina in Sonora; and observing its incipient cultivation in rural communities in Sonora.

II. THE CHILTEPIN: A NEW CROP

A. Botanical Background

The chiltepin or chiltecpin is the northern wild race of *Capsicum annuum* L., the most common and wide-ranging chile pepper in the Americas.[16] The species *Capsicum annuum* also includes most domesticated chile peppers in international commerce today, which were domesticated from another wild race which inhabits east central Mexico.[17] The chiltepin is the folk name for the pungent round or oblong erect fruit of a wild woody perennial shrub which is sometimes bushy, sometimes scandent. It is the only non-domesticated chile that extends north of the Tropic of Cancer into arid subtropical and even desertic zones. Its northern limits occur in the Sonoran and Chihuahuan Deserts, and in adjacent oak woodlands, 50 km beyond the U.S./Mexico border in Arizona, New Mexico, and Texas. In this region, the chiltepin has been referred to by scientists as *C. baccatum* L., *C. frutescens* L., *C. annuum* L. var *minimum* (Miller) Heiser among other names. Although there remains some debate whether the currently accepted nomenclature for this taxon is *Capsicum annuum* L. var. *glabriusculum* (Dunal in D.C.) Pickersgill and Heiser, the epithet *aviculare* (Dierbach) D'Arcy and Eshbaugh is nomenclaturally acceptable and has been more widely adopted. Commonly growing near fences, ditches, and roadsides throughout much of Latin America, the chiltepin is essentially a colonizer with certain weedy characteristics. However, in desert regions, the chiltepin grows in the understory of trees in mountainous areas near watercourses and in canyons. These habitats provide higher humidity and richer soils than the surrounding desert.[18,19]

B. Ethnobotanical Context

Ironically, the chiltepin is most popular in the desert regions of Northwest Mexico and the Southwestern U.S. where wild chiles are less common than they are further south. Historically, Pima and Apache Indians relished the chiltepin even though they lived 150 km north of the plant's range.[19] The Papago and Pima (O'odham) Indians have traditionally made annual pilgrimages into the desert mountains of Southern Arizona and adjacent Mexico to gather them. Tarahumara Indians of the Sierra Madre in Chihuahua have constructed stone walls around chiltepines to protect them from goats and other grazers.[19] The chiltepin remains very popular among these people today.

The chiltepines' popularity is not restricted to indigenous peoples. It is common to find a jar of pickled or dried chiltepines on almost any dinner table in northern Mexico.[20] In addition to being used as a condiment, the chiltepin is used medicinally to treat sore throats and acid indigestion.[19]

Entrenched in the Sonoran's cuisine and folklore, it is doubtful that the chiltepin will decrease in popularity and be replaced by other domesticated peppers. Moreover, people of Mexican descent and recent immigrants from Mexico to the U.S. also prize the chiltepin. Not as readily available in the U.S. as in northern Mexico, there is a sentimental value given to the chiltepin by their state-side Mexican-American consumers.[20] The chiltepin seems to help them reinforce their Sonoran identity.

In July of most years, in the desert regions, small white flowers announce the coming of the coveted fruit. When ripe, the round, red fruit is either quickly devoured by birds or harvested by the people of the region. The fruits add flavor to monotonous diets and are rich sources of vitamins A, C, B_2, niacin, and potassium.[16,21] Compared to many domesticated cultivars, the chiltepin contains a higher percentage of capsaicin, the chemical that gives chiles their piquancy.[22] Chiltepines are also used as a preservative for *carne seca* or *carne machaca,* which are provided complete protection for spoiling due to chile antioxidants such as ascorbic acid, tocopheral, flavinoids, and capsaicin.[21]

Currently, the chiltepin is almost completely wild harvested. Each fall when the chiltepines ripen, families, farmhands, ranchers, and subsistence farmers venture into the sierras of Sonora, Chihuahua, and Sinaloa to collect the fruit.[19,23] The harvesters are not migrants, but local residents of ejidos and ranches who seek extra cash in a time of year when income is low.[23,24] As many as 100 individuals participate in the harvest in the Ures municipio (township) alone; at least another ten municipios are located in habitats where wild chiles are just as prolific, suggesting that perhaps 1000 Sonorans may be involved in the wild harvest for the export market. Buyers in Mexico from the U.S. purchase the chiles from the harvesters and load them into trucks to take back to urban centers where they are sold fresh, or as pickles, dried, or processed into hot sauce.

As much as 12 tons of chiltepines may be harvested from a single Sonoran municipio in a good year, but the total harvest may vary from perhaps 8 to as high as 50 tons. Because it is wild harvested in an unpredictable climate, the availability of the chiltepin from one year to the next is dependent on many uncontrollable factors. Weather, grazing, and harvesting methods are the major variables that control the chiltepines availability.[18,23,24] Drought and freeze years are especially hard on the northwesternmost wild chile populations. Not only is there more stress on the plant, but scarcity increases the demand for the chile in the market. Consequently, harvesters compete for the few available chiles often uprooting plants or severely damaging the plants.[23,24] As early as World War II, Dr. Howard Scott Gentry commented on the pressure put on chiltepins by wild harvesting:

> Recently, the natives of southern Sonora have started collecting the fruit for export to the United States . . . if commercial exploitation of the plant is successful, it is probable that the great majority of plants will die under persistent breaking, and what has been a common plant will become rare. Or perhaps it will be cultivated.[25]

C. Incipient Cultivation

Although scientists such as Gentry[25] and Felger and Nabhan[26] earlier proposed cultivation of this wild crop, Sonoran farmers independently came to the same idea, and initiated field plantings in the early 1980s.[24] Several chiltepin plantations of 1 ha or more have been planted in Sonora's Costa de Hermosillo, in the Rio Moctezuma, and Rio San Miguel watersheds, with mixed success. Whether propagated from seed or from cuttings, the plants grown by these farmers came from a restricted but unselected genetic base. Initial field plots with chiltepines planted at high densities have suffered from mortality or fruit damage due to a moth larvae (Lepidoptera), charcoal rot *(Macrophomina phaseoli)*, or three other diseases, as well as from too much direct sunlight and temperature extremes. Ironically, some of these same stresses were not found to affect wild chiltepin plants growing within 2 km of the field plots.

Because of these agronomic problems not evident in chile populations in wild habitats, growers have requested technical assistance from government agronomists and pathologists at INIFAP, researchers at the Universidad de Sonora and the University of Arizona, private consultants, or from us. Thus, much of our own research has been in response to requests and needs articulated by farmers themselves. Once farmers themselves have identified the production of a new crop as desirable, scientific support is more valued as appropriate.

In response to agronomic problems mentioned above, chiltepin growers have embraced one of two distinctive approaches to problem solving. One approach, advanced by a Sonoran agribusinessman, Alfredo Noriega, is to select and manage chiltepines much like any commercial chile crop under intensive irrigated monoculture. He and his agronomists have undertaken recurrent selection for disease resistance, high first-year yield and responsiveness to agronomic inputs (water, fertilizer, and pesticides). He has developed a chiltepin that has few characteristics of the original materials, weighing three times as much as the wild type did 8 years ago, and yielding five times the harvest weight per plant (up to 1.5 kg from 0.25 kg in 1980). He currently is growing 62 ha of these chiltepines in open fields at a density of 10,000 plants per hectare.

Due to disease build up at these densities, and damage to the plants during harvesting, Noriega has chosen to manage the chiltepin as an annual crop. Plowing the chiles under after the last harvest of the season, he transplants seedlings the next spring. His successes in selecting higher yielding plants with larger fruit have created other problems: (1) the fruits no longer look or taste like native chiletepines, so there is reduced acceptance in the Sonora and Arizona marketplaces; and (2) the coordination of seasonal labor for harvesting 62 ha by hand is a massive task, costing 60% of the product's wholesale value, and leaving Noriega with the unfulfilled desire to mechanize the entire harvest.

The second approach has been taken by smaller-scale operations in rural areas where the existing labor force is underutilized, and where other sources of seasonal income are welcomed. Plots ranging from $^1/_4$ to 5 ha have been planted as part of multiple cropping endeavors. The multiple crops diversify the sources of income and food available to the poor rural dweller. There has been no dramatic selection away from the chiltepin's wild type, for these people are both familiar with and favor its distinctive "bouquet", as well as its perennial nature. Whether funded privately, or through development agencies, these producers expect to remain at the cottage-industry level of operation for both local consumption and export. The rest of this discussion will focus on their involvement with chiltepines, with the understanding that the farmer (agribusiness) approach does offer them some competition in the U.S. market.

Turning to the criteria that they cite as reasons for investing in chiltepin production, criteria 2, 3, 6, 7, and 8 have been explicitly mentioned, rather than 1, 4, or 5 per se. Yet by their own investment in chile production, they have implicitly decided that wild chile cultivation may improve their own well being, if not that of other community members (fieldworkers on wages, product canners, and bottlers) as well. Thus, unlike many other crops, wild chile development has been initiated at the grassroots level for locally defined reasons. Technical assistance has served to support these locally initiated projects, and/or interest in criteria 1, 4, and 5 has so far been compatible with farmers seeking sustainable production options.

D. Nurse Plant Ecology

Due to problems which developed from monocultural cultivation without large amounts of chemical inputs, it is pertinent to evaluate the chiltepin's requirements for survival in its natural habitat. Our observations of chiltepines in the wild repeatedly point to one overriding requirement: the chiltepin needs the protection of a nurse tree. Chiltepin bushes are almost exclusively found growing beneath trees and shrubs in the northern portions of its range. The tree creates a special microenvironment providing shade, increased humidity, and increased soil fertility and protects the chiltepin from grazing and trampling. Vegetation analyses indicate that the chiltepin grows beneath red-fruited nurse trees (*Celtis, Lycium, Solanum,* etc.) more often than is expected.[27] We hypothesize that wild chiles mimic the fruits of these trees and by doing this, "they are eaten by frugivorous birds that are attracted primarily by the overstory plant, which adventitiously disperse them to other feeding sites."[27] It is unclear whether the red-fruited shrubs provide benefits for the chiltepin other than shade and safe-sites for establishment.

In theory, an agroforestry system that utilizes the nurse tree/chiltepin relationship observed in the wild would require fewer inputs and have fewer problems with disease and pests than a monocultural planting.[28] A perfect mimic of the natural system would need little external input. However, periodic weeding, pruning, watering, and some level of pest management might be necessary to establish a cultivated system. In addition to the microhabitat, the nurse tree may also provide the farmer with many other products depending on the species of trees grown including firewood, fruit, organic mulches, medicine, and food. At the same time, the nurse tree could serve other functions such as being part of a shelterbelt, an erosion control barrier, or a hedge.

At the Gentry Agroecology Project at the Desert Botanical Garden in Phoenix, we are currently using the nurse tree/chiltepin association observed in the wild as a model for cultivation. In a series of replicated blocks, we are growing chiltepines beneath a nurse tree (*Celtis reticulata* Torrey), in artificial shade, and in the full sun. We are measuring the chiltepines' growth monthly along with observing its flowering and fruiting phenology. The results of the experiment will help us to recommend ways to plant the chiltepin in a suitable agroforestry design.

To date, we have observed higher mortality and lower growth rates in *both* the unshaded and nurse tree-shaded chiles than those in artificial shade. Direct solar radiation was the cause of death in the unshaded and nurse tree-shaded treatments. The young nurse trees were too small to provide adequate shade for the chiles.

Our attempts to grow chiltepines with nurse trees suggest that either the chiltepines should be planted beneath existing nurse trees or with quick growing nurse trees. Both of these options are currently being tested in Sonora. Near Agua Prieta, Antonio Heras-Duran is placing 600 chiltepines under existing nurse plants along a desert grassland wash. At Escuela La Argentina near Alamos, one of us (Larry Yarger) has planted chiltepines in terraces shaded by alley crops of *Leucaena,* a genus of subtropical legumes with species already naturalized or native there. Locally known as *guaje*, *Leucaena* provides an improved microclimate, a certain amount of nitrogen fixation, and several products including forage, fresh vegetables, and fuelwood. Utilizing a randomized complete block design, we are conducting a comparison between monocropped chiltepines and alley-cropped *Leucaena*-chiltepin polycultures; results are forthcoming.

E. Other Cultivation Requirements

Chiltepines may be started from seed or from cuttings, but because of desiccation and emergence problems, they cannot be planted directly into a field. The plants must be started in containers and later transplanted into the field. Seeds must be germinated under hot, moist conditions; a potassium nitrate or hot water pretreatment increases percent germination. Cuttings can be easily propagated with a rooting hormone and frequent misting. Consequently, water needs to be available to start the seeds or to root the cuttings, and to keep the plants turgid until they can be transplanted. Transplanting should be timed with the advent of the rainy season and after a period of acclimatization; however, in most situations, an irrigation source will be necessary to keep the transplants alive.

Being a drought-deciduous perennial crop, chiltepines would be able to survive drought years once they are established. However, without irrigation, dessication has caused over 70% mortality of plants grown in rainfall microcatchments on the desert edge in San Luis Potosi. As a perennial, it is also unknown how long the chiltepines will live under cultivation, although individual wild plants have persisted 30 years.[18] In addition, pest and disease problems may be aggravated through time. Discerning trends in productivity and susceptibility over time will help to determine how often the chiltepines will need to be replanted, and ultimately, if the resources required to start the chiltepines are worth the investment. These questions will only be answered by long-term trials and studies of chiltepin cropping systems.

F. Economic Considerations

Given that the plant can be propagated successfully, are there economic advantages to doing so? Morales-Cuen et al.[18] suggest that in the sierras of Sonora, a harverster can collect 4 to 5 l of fresh red chiltepines per day, equal to a dry weight for retail sale of 1.1 to 1.4 kg/d. Juan Trejo[29] suggests a 3 kg/d harvest for active, semi-skilled teenagers. In contrast, under the higher densities of his cultivated plots, Manuel Encina Morales estimates that harvesters can easily collect 10 to 20 kg/d each over a 2 month season of fruit ripening. Alfred Noriega projected a metric ton fruit yield per hectare from irrigated plants spaced 1 m apart, but has had difficulty mobilizing the labor to obtain a complete harvest. Greater harvesting efficiency and yield per unit area may compensate for increased production costs. The resulting economic benefits may not be shared with harvesters unless the crop development is planned collectively. Furthermore, neither production costs nor market stability can be adequately estimated at the present time. The only estimation of the potential market for chiltepines has been a preliminary unpublished report by Wittkowski.[20] The following evaluation is drawn largely from his report.

Wittkowski saw potential for the expansion of the chiltepin market based on the rapidly expanding consumption of chiles and spices in general in the U.S. He also cited the popularity of chiltepines among Mexican Americans and Indians in the American Southwest, and shortfall of the wild crop to meet their dried spice demand in certain years.

The unpredictability and occasional failure of the wild crop has limited the use of chiltepines by dried spice packagers and by salsa (hot sauce) canners. Few of the companies contacted were willing to print up labels, and package and market a product if they could not be guaranteed a somewhat stable supply from one year to the next. On the basis of economic theory, Wittkowski argues that a steady supply of chiltepines may not only create a stable market for them, but could increase rather than decrease average prices. Currently, prices fluctuate as much as twofold between years, and wild harvesters are paid a pittance ($1.20 to $1.40/kg) in bumper-crop years when buyers are unsure of how much surplus they can sell.

If more than a few growers each produce a ton or more of chiltepines, this added supply should add to competition in the marketplace. The harvest from cultivated fields has already eclipsed the wild harvest in its yield during a drought year. In contrast to the Wittkowski model, the California cultivation of wild rice *(Zizania)* has eclipsed the wild harvest in the U.S. and Canada, but has generally driven down prices.[30] However, Indians who still hand harvest wild rice from natural stands continue to ask higher prices for their quality product. In addition, they are opening additional markets for high cash value wild rice products in health food stores, food coops, airports, and regionally oriented gift shops.

By analogy, there are possibilities for both expanded markets and specialty marketing for chiltepines. A stable supply could enable production of additional hot sauces labeled as *chiltepin*-based; few companies in the U.S. do this today, and some of them actually use peppers other than chiltepines as their main ingredients. There are also the possibilities of selling them in vinegar or brine as a green fruit, mixed with garlic and Sonoran oregano (*Lippia palmeri* S. Wats); or as fresh, green fruit; or in preparations of *carne machaca, carne con chile, tamales,* or *salsa casera.*

Furthermore, the Noriega agribusiness experience has demonstrated that domestication can change product quality negatively for those who are accustomed to wild chiltepines, as has happened with wild rice. Consumers in both Sonora and Arizona have complained about Noriega's marketing of marble-sized, semi-cultivated chiles as *chiltepines;* they no longer resemble their wild counterparts. The consumer indignation over this selected chiltepin is a possible indicator that the *chiltepin de la region* remains highly valued for taste and cultural identity, even if other similar chiles are available. The chile industry promoters in the U.S. also see unique advantages to marketing chiltepines as "wild, original and untamed" to given them a distinctive niche.

With retail prices in the U.S. running up to $48 per pound ($22/kg) in 1988, and as low as $7.00/lb ($3/kg) at the same time in Mexican border towns, there will inevitably be a drive to sell as much of the product as possible in the U.S. Southwest. The gains from this price differential, however, are poorly shared between harvester and grower, local buyer exporter, packager, or retailer. It is obvious, however, that grower-harvesters stand to gain considerably if they begin to process or package, and export their own distinctive local products. Villages, ejidos, or collectives which could produce their own distinctive salsa with a label attractive to American buyers probably have the most to gain.

However, if the current market system used with the wild harvest is simply transferred to the farmers' market, chiltepin production would very likely not be profitable for the small farmer as he would not have control over the prices of chiltepin products. The small farmer would remain poor while the spice company benefits. This familiar scenario touches upon the all-too-common need for change in social structures.

A significant part of the rural population of eastern Sonora is comprised of people who live in small communities known as *ejidos*. The ejido, one form of rural corporate communities in Mexico in which certain resources are managed in common, was established by the ruling institutional revolutionary party government in order to redistribute land among the poor. Since its inception, however, the Mexican government has failed to adequately support the ejidos in their efforts to manage the land or themselves toward sustainable self-sufficiency. In addition, through outright neglect or through paternalistic programs which offer ejidos with inappropriate development schemes, ejiditarios over the last 3 to 4 generations since the revolution have become frustrated and skeptical of the federal and state government and its plans for them.

As a result, some ejidos in eastern Sonora may be blessed with many potential land resources close at hand, but they lack the infrastructure, the capital, or the market knowledge needed to take advantage of them. While certain economic wild plants are now either underutilized or overexploited, wealthy landholders nearby take advantage of their additional resources, and hire ejiditarios to work on their large, often state-of-the-art poultry, swine, and cash crop farms. As generations have passed, rural dwellers in the eastern Sonora uplands have conceded that this menial labor employment is their most reliable option. Renting their land to the wealthy, they sometimes hire on as peon labor to work the same land, or become migrant workers in the Rio Mayo, Rio Yaqui, or Costa de Hermosillo agricultural areas.

In order to utilize the rich store of resources available and to gain directly from his or her own investments, the small farmer must cautiously evaluate and develop confidence in any new development scheme. Chiltepin cultivation may be too risky for certain small farmers to undertake, whereas others may be opportunistic enough to make it work.

New markets for chiltepines have begun to develop but there is no guarantee that expanded production will not saturate the existing markets. Like any new crop or product, the inherent risks are perhaps greater than those for a conventional business. We must admit that this financial risk taking may make chiltepines inappropriate for many farmers in marginal lands, even though chiltepines are a relatively low-input crop. The poorest farmers must carefully consider use of all their time and productive land, because they have little to fall back on if a crop fails.

Nearby wild populations of chiltepines also influence the poorer farmers' decision not to grow the chiles. Why should he invest limited resources in a product that can be readily gathered in the wild? If the wild populations continue to decline from harvesting pressures.[24,25] perhaps there will then be more incentive to cultivate the chiltepines or to better manage the wild populations. Reforestation efforts with chiltepines and their nurses may prove to be just as viable as field cultivation, but with less land, time, and capital investments.

In other areas where the people are not strictly subsistence farmers, there may be room for risk taking. For those who choose to diversify their income-generating activities, adding chiltepines to their mixed cropping systems may be a lucrative option. However, even small risks can be cause for hunger in a poor community if not approached and managed carefully.

Before adopting a new crop or management technology the farmer must carefully weigh the probability of failure against the benefits it would give to his existing cropping system. If the probability of failure is high, this may discourage the farmer from adopting the measure. To the farmer, failure may not only mean the loss of a crop, but the loss of optimism and respect as well.[31] This is the situation with the domestication of chiltepines in ejidos near Escuela La Argentina. The chiltepin at this time is commonly considered a high-risk crop. Difficult and sporadic seed germination, together with the fact that chiltepines have not been domesticated are cited as reasons for it as being considered high risk. In addition, where water is limited in a community, an even greater risk is created when transplanting young moisture-loving chiltepin plants to the field. The primary production advantage is that once established, it is very hardy in the arid and semi-arid climates of the region.

On the other hand, one development theory suggests that certain people in poorer communities watch for and depend upon others to take risks for them in advance of their own trials.[31] Local people frequent the chiltepin fields at Escuela La Argentina to see how the chiltepines are coming along under the favorable conditions of a well-tended experiment. These people may be awaiting the outcome of the chiltepin production at the school before deciding to take their own risk.

Through research presently underway, we see that the reforestation or domestication of the chiltepin has considerable economic possibilities for the people of rural Sonora. There exist opportunities to form cooperatives to produce, process, and market the chiltepin and chiltepin products on both domestic and international levels, so that local people capture a larger percentage of the expanding market.

The first steps have been taken towards realizing that goal. Only when other small farmers decide that there are enough benefits from chiltepin production that a few of their neighbors have begun, will it ever become a part of economic improvement for communities as a whole. Without concommitant changes in socioeconomic structures, prospects of utilizing resources such as chiltepin peppers to improve the lives of the people within a rural community remain but a distant dream.

III. CONCLUSION

Returning to the criteria outlined in the introduction, a chiltepin/nurse tree agroforestry design, combined with hand-harvesting and a cottage-industry level of product development and marketing, may meet the following criteria:

1. It has already been embraced as a production possibility by small farmers in desert and arid subtropic areas with few other production alternatives; it is too early to tell if the subjective well being of workers or growers is improving.
2. It *can* be a low input production system. Although some large-scale farmers have added generous quantities of irrigation and agrochemicals attempting to increase their profit margins; there are considerable problems with growing dense monocultural stands of chiltepins.
3. It is labor dependent since mechanical harvesting cannot be accomplished on a small-fruited, irregularly maturing perennial crop without significant plant damage. Harvesters should be able to make more in the field than they do in the wild if wages are based on

weight harvested per day, not a standard fee. However, harvesters now obtain only a fraction of the crop's retail value.

4. The new cultivation can take pressure off the wild populations, and is occurring in areas where wild harvesters can be recruited for field harvesting. Yet, access to wild populations may deter cultivation efforts, especially if wild harvesters develop plans and establish rules to manage these populations. Encouraging reforestation of wild populations should also be considered.
5. It can be grown as a multiple product agroforestry system, intercropped with trees valuable for lumber, firewood, or food.
6. It has not yet replaced any subsistence crops, and probably will not do so on any scale. It is well suited to cottage-scale canning, pickling, or packaging.
7. Current growers have a strong cultural tie with this plant and its lore; some learn from its wild ecology more than others.
8. Some growers are using local seedstock, while others are actively selecting for higher yielding, large-fruited plants. However, there are indications that Sonorans do not accept the selected chiles as chiltepines. Some change in the balance between genetic variability and selection of high yielders will continue to occur.

Although chiltepines meet many of the above criteria, the unknowns of production costs, plant longevity under cultivation, and market expansion still make them a risk. They are not recommended for the most marginally economic farmers who cannot afford to make a risky investment. Nevertheless, they offer a model of how to evaluate any crop to see if it meets such needs.

REFERENCES

1. National Research Council, *Underexploited Tropical Plants with Promising Economic Value*, National Academy of Sciences, Washington D.C., 1976.
2. **Ritchie, G. A., Ed.**, Introduction, in *New Agricultural Crops*, Westview Press, Boulder, CO, 1979, 1.
3. **Aldon, E. F., Gonzalez-Vincente C. E., and Moir, W. H., Eds.**, *Strategies for Classification and Management of Native Vegetation for Food Production in Arid Zones*, U.S. Forest Service Rocky Mountain Region, Albuquerque, NM, 1987.
4. **Theisen, A. A., Knox, E. G., and Mann, F. L.**, *Feasibility of Introducing Food Crops Better Adapted to Environmental Stress*, Vol. I, Contract No. AER 77-19462, National Science Foundation, RA-780289, Washington, D.C., 1978.
5. **Alcorn, S. M., Mihail, J. D., and Orum, T. V.**, What? Our new crops are sick?, *J. Arid Environ.*, 11, 111, 1986.
6. **Downing, T. and Restrepo, I.**, New technology and dry land agriculture, *Cult. Agric.*, 7, 2, 1980.
7. **Cleveland, D. and Feldman, W.**, A methodology for assessing the appropriateness of arid lands agriculture: Jojoba and the Tonja of Zimbabwe, University of Arizona, Tucson, 1983.
8. **Worster, D.**, Thinking like a river, in *Meeting the Expectations of the Land*, Jackson, W., Berry, W., and Colman, B., Eds., North Point Press, San Francisco, 1984, 56.
9. **Nabhan, G. P.**, Replenishing desert agriculture with native plants and their symbionts, in *Meeting the Expectations of the Land*, Jackson, W., Berry W., and Colman, B., Eds., North Point Press, San Francisco, 1984, 172.
10. **McLaughlin, S.**, Economic prospects for new crops in the Southwestern United States, *Econ. Bot.*, 39, 473, 1985.
11. **Jackson, W.**, Introduction, in *Meeting the Expectations of the Land*, Jackson, W., Berry, W., and Colman, B., Eds., North Point Press, San Francisco, 1984, xiii.
12. **Ehrenfeld, D.**, Implementing the transition to a sustainable agriculture, and opportunity for ecology, *Bull. Ecol. Soc. Am.*, 17, 5, 1987.
13. **Nabhan, G. P. and Felger, R. S.**, Wild desert relatives of crops: their direct uses as food, in *Plants for Arid Lands*, Wickens, G. E., Goodin, J. R., and Field, D. V., Eds., George Allen and Unwin, London, 1985, 19.
14. **Mooney, P. and Fowler, C., Eds.**, Los Bancos de Semillas Locales; un Material de Apoyo, Rural Advancement Fund International, Brandon, 1986.
15. **Chapin, M.**, The seduction of models: are chinampas really an old answer to the future?, in *Traditional Cultures and the Conservation of Biological Resources*, Oldfield, M. and Alcorn, J., Eds., Westview Press, Boulder, CO, in press.
16. **Andrews, J.**, *Peppers: The Domesticated Capsicums*, University of Texas Press, Austin, 1984.
17. **Loaiza-Figueroa, F.**, Evolution of the Genus *Capsicum* (Solanaceae) in Mexico, M.S. thesis, New Mexico State University, Las Cruces, 1985.
18. **Morales-Cuen, A., Duarte-Ramirez, J. J., Gomez-Lorence, F.**, Ecologia y Productividad del Chiltepin, *Capsicum baccatum* L. Bajo Condiciones Silvestres en la Region del Rio Sonora, Mexico, Tesis profesional, Universidad Autonoma de Chapingo, Chapingo, 1986.
19. **Nabhan, G. P.**, Chiltepines! Wild spice of the American Southwest, *El Palacio*, 84, 30, 1978.
20. **Wittkowski, R. D.**, The Market for Chiltepines in the United States, Paper for Agribusiness: The Relationship Between Business and Government, Harvard Business School, Cambridge, January 30, 1987.
21. **Cotter, D. J.**, The scientific contribution of New Mexico to the chile pepper, in *Southwestern Agriculture: Pre-Columbian to Modern*, Dethloff, H. C. and May, Jr., I. M., Eds., Texas A & M University Press, College Station, 1982, 16.
22. **DeWitt, D. and Gerlach, N.**, *The Whole Chile Pepper Catalog*, Out West Publishing, Albuquerque, NM, 1987, 13.
23. **Duran, J.**, personal communication, COMPAS, Sonora chile broker, February 1988.
24. **Nabhan, G. P.**, *Gathering the Desert*, University of Arizona Press, Tucson, 1985, 123.
25. **Gentry, H. S.**, *Rio Mayo Plants*, Carnegie Inst. Washington Publ. 527, The Lord Baltimore Press, Baltimore, 1942, 234.
26. **Felger, R. S. and Nabhan, G. P.**, Agroecosystem diversity: A model from the Sonoran Desert, in *Social and Technical Management in Dry Lands*, Gonzales, N. L., Ed., Westview Press, Boulder, CO, 1978, 129.
27. **Nabhan, G. P.**, Nurse plant ecology of threatened desert plants, in *Conservation and Management of Rare and Endangered Plants*, Elias, T. S., Ed., The California Native Plant Society, Sacramento, 1987, 377.
28. **Farrell, J. G.**, Agroforestry systems, in *Agroecology, the Scientific Basis of Alternative Agriculture*, Altieri, M. A., Ed., Westview Press, Boulder, CO, 1987, 149.
29. **Trejo, J.**, personal communication, Mazocahui, Sonora chile broker, September, 1988.

30. **Vennum, T., Jr.,** *Wild Rice and the Ojibway People*, Minnesota Historical Society Press, St. Paul, 1988.
31. **Harwood, R.,** *Small Farm Development*, Westview Press, Boulder, CO, 1979.

4 THE SOCIAL AND CULTURAL DYNAMICS OF TRADITIONAL AGRICULTURAL COMMUNITIES

Cornelia Butler Flora

I. INTRODUCTION

Human communities, in a manner quite different from biological ecosystems, develop a variety of internal mechanisms to adapt to external forces and internal pressures. The community is a central social unit in traditional agricultural societies, since the community is more often the location where actual and potential conflicts over internal resource allocations are resolved. The community can either facilitate or resist extraction of resources by outside forces.

A. Definition of Traditional Communities

The term "traditional" in reference to agricultural communities is a relative one. In this case, I do not mean only groups that are totally self-sufficient and engage only in non-market, subsistence production. Few such groups currently exist, and there have been few communities which have been totally self-sufficient over the last thousand years. I define traditional agricultural communities as those that have been only partially penetrated by capitalist relations: although labor may be bought and sold by many in the community, there are still major areas of production where the household labor predominates and the household has the right to directly market or consume at least a portion of that production.[1] Traditional agricultural communities maintain a number of exchange relationships that exist outside the market that allow the household to reproduce itself. Part of the social and cultural dynamics of such communities often involves resistance to capitalist penetration and the resulting complete dependence of households on wage labor to survive. The process of reducing the factors of production solely to their monetary value and transforming them into articles of exchange is incomplete in traditional societies. Land, animals, tools, and even human labor have values that are not determined by the market, and those values can take precedence over exchange value in determining how individuals in those communities act toward them and toward each other. In particular, norms of exchange, reinforced by religious and social sanctions, block village relationships from becoming commercialized.[2] As a result, traditional agricultural communities tend to view the factors of production differently than do modern agricultural communities. Chevalier points out the need to link material practice to its cultural meaning.[3] Traditional agricultural communities attach different values to things than do commercial agricultural communities.

Traditional communities vary from each other in the degree to which the various factors of production are commodified.[4] As a result, almost every characteristic defining traditional agricultural communities can be challenged by a counter example. Nevertheless it is argued in this chapter that comparatively low commodification (reducing to only monetary value and market exchange) of at least some of the factors of production allow us to distinguish traditional agricultural communities from more modern ones.

B. Agriculture and Community

The thesis of this paper is that whereas the modern farmer can exist without community through a series of market linkages to the world economy, community is required for effective access to land, labor, and capital among traditional agriculturalists. By examining the social and cultural dynamics of the communities, we can understand these traditional agricultural systems' strengths and ability to resist the complete proletarianization of their residents.

II. THE FACTORS OF PRODUCTION

A. Land

In the prototypical traditional community, land is valued for its use as well as ownership. Assurance of long-term land access is socially important. Land may be in the hands of a few, as in South Asia and Latin America, but in traditional agrarian communities in those areas, usufruct rights provide the basis for relatively long-term social relations. Access to land provides an important alternative to the dependency and exploitation engendered by having to sell one's labor. Traditional societies vary in formal land tenure patterns. One of the thrusts of modernization has been to alienate land by setting up mechanisms for it to be bought and sold, making land that was once an organic unit divisible and commercialized. Under feudal-type arrangements, despite land concentration, peasants had access to land for their own use and stable communities resulted. Land concentration through purchase (or the pressure of agrarian reform), increases the need to make profit from it. Use rights are no longer honored, as maximization of production is sought. The majority of the peasantry in such circumstances is forced to seek wage employment. The land itself deteriorates under the shift from traditional communities to modern ones, as the need for immediate profit encourages land overuse and short term production gains.

1. Mechanisms of Land Access

Traditional agricultural communities, in contrast to modern ones, maintain a large number of social and cultural mechanisms that ensure that land is available to those who need to cultivate it. Tenure forms in traditional communities are extremely diverse, from communal lands to *latitundio-mini-fundio* complexes. The common element is assured land *access*. Mechanisms of land redistributive rites and festivals that include everything from the year of jubilee to the baptism of the peso all attempts to keep land distribution relatively equal and to protect the relative security of community members, thus assuring the maintenance of the social fabric of the community.

a. Year of Jubilee

The year of jubilee, mentioned in the Bible, was said to occur when land and other property that had been traded and conquered and thus concentrated for a period of seven sabbaticals plus one (50 years) was redistributed to its original owners to maintain equality and community systems of mutual support that allowed agriculture to prosper. This redistribution was divinely sanctioned, to keep the rich from exploiting the poor, thereby assuring each kin group continued access to the means of production. Those in bondage were also freed at that time, another mechanism to ensure that labor was not forced. The time period related to the contemporary Jewish perception of agricultural cycles and the time land needed to regain fertility.

b. Baptism of the Peso

Baptism of the peso presents itself in the cacao growing regions of the Cauca Valley in Colombia, where a whole series of social norms and belief systems imposed negative sanctions on those who would sell their small plots to the sugar interests encroaching at the entire valley floor. As with the year of jubilee, divine will was associated with blocking land consolidation, which would have forced the population to depend on selling their labor. The local community perceived that it was only through a pact with the devil that the descendants of African slaves were able to accumulate wealth at the expense of others.[5]

c. Political and Land Rights

Such communities can become political units as well. Indigenous agricultural communities, made official in such countries as Peru and Ecuador by government decree, are institutions of land tenure that regulate access to communal land for growing tubers.[6] Plots are allocated on a yearly basis (*laymi*) according to need[7] or to enrich the more powerful members of the community.[8] Maize plots are individually owned, and community dynamics encourage multiple marriages among sibling groups to combat land fragmentation.

B. Labor

Traditional agrarian communities, despite the degree to which households may depend on off-farm labor for capital, develop mechanisms to help ensure labor availability for key agricultural production throughout the agricultural cycle. Even when temporary migration for wage work may remove a key family member from the area, in traditional communities non-commoditized extra-household social relations maintain the individual household's options to maximize sources of income and subsistence. As Smith points out, "... the community provides the hearth from which the individual production units can go out to work. It is a source of 'unpaid' labor though a myriad variety of labour exchange institutions."[9] Mechanisms of mobilizing labor for the peaks of the agricultural production cycle and providing alternative uses of labor in the slack periods often develop within traditional agricultural communities.

1. Labor and Migration

Community/family groups then share labor during high labor demand periods. Labor sharing, either to work together or to substitute for absent members, is particularly important, because of male absence for lengthy periods throughout the year to harvest cane. For example, in highland Peru before the land reform, migratory work was done for the hacienda and was reimbursed both by a cash wage and by a grant of land access by the hacienda, which had acquired the upland subsistence land as a mechanism to capture peasant labor.[10]

2. Labor Exchanges

Labor exchanges vary in degree of formality. In the system that Skar and Alberti and Mayer[11] identify as *ayuda*, aid is given without record keeping or careful balancing, based on a sense of reciprocity rather than exchange. In contrast, *ayni* is a reciprocal labor exchange based on carefully calculated length and type of work. These ties, like those of *ayuda*, are very permanent. Less permanent mechanisms include the *minka*, where work is exchanged for food, liquor, and coca. Direct wages are never paid for labor on subsistence crops. Communal labor is mobilized for public works projects (*faena*), also without cash payment, but with community accounting.

3. Collective Labor and Collective Goods

Further, labor is mobilized from within the community to provide collective goods, from schools to roads to irrigation systems. These are collective goods to which all have access, and thus traditional communities develop norms and institutions to ensure that all contribute. These are necessary, as time and expense devoted to community improvement could have been devoted to household or individual production. When individual household units accumulate enough capital to make the labor within their domestic unit a great deal more productive than other units, the norms to enforce labor exchanges and collective labor enterprises decline and the institutions for such labor begin to crumble.

In modern agrarian communities, collective goods, when they are provided at all are provided by non-local government institutions, which may provide the capital to pay the labor (as well as to construct infrastructure beyond the means of community residents), but control the placement and rules surrounding access to that infrastructure. The individual and household has more *time* to devote to individual production, but must devote an increasing proportion of that production to taxes, furthering dependence on the cash economy, especially wage labor.

4. Labor for Wages and Non-Farm Production

Semi-proletarianization, particularly seasonal wage labor in the harvest of export crops, now dominates most traditional agricultural communities. Those communities that survive must mobilize labor to aid in household production. Further, traditional agrarian communities tend to develop patterns to rationalize the labor of all household members throughout the year, balancing agricultural labor with gathering and processing activities.[12] Often the marketing, if not the processing, is informally coordinated within the community.

C. Capital

1. Alternatives to Dependence on Capital

Traditional agricultural communities use capital. However, they seek to conserve it rather than to maximize capital accumulation. Decision making tends to be oriented to the long-term, with risk-reduction strategies given priority over profit-maximizing strategies. Traditional agricultural systems tend to be labor intensive. Labor intensification may reduce return to labor, but it also decreases dependence on outside sources of capital as a substitute for labor. Such strategies often increase dependence on other community members.

2. Low-Capital Use and Development

Wharton[13] has pointed out some criteria for subsistence farmers that focus on the relatively low level of capital use and capital investment. He derives these characteristics developmentally — as something to be overcome for development to take place. They are not theoretically derived, yet have strong ideological implications for applied development practice, which is devoted to changing traditional agriculture. These have major implications for the community:

1. Low sale of farm products. The subsistence farmer consumes a majority of his or her production, particularly crops. Thus agriculture is not a major source of capital within the community.
2. Low hired labor or purchased input factor ratio. Little capital is expended for crop or animal production.
3. Low level of technology. Traditional farmers use simple technological inputs, requiring little capital investment. Oasa and Swanson[14] point out that, since most of the surplus value produced by increased technology is absorbed by the market, cultural mechanisms in support of low-input technology help maintain the community.
4. Low income and level of living. Traditional farmers tend to be poor, particularly in terms of monetary income.

Wharton's focus on production decisions allowed him to ignore that traditional agriculture, by collectively maintaining alternative sources of income and sustenance for a large number of individuals, actually decreases the probability that a majority of the population must depend completely on wage labor for income and survival.

3. Low Capital Use and Dependence Avoidance

As Painter points out when looking at traditional agricultural communities in the highlands of Peru among the Aymara speakers, "The purpose of maintaining subsistence activity is to avoid becoming dependent upon purchased foodstuffs and to maintain control over their own food supply. Heavy reliance upon capital inputs and export crop production would foster precisely the sort of dependence that the maintenance of the subsistence sphere seeks to avoid."[15]

4. Traditional Sources of Capital

The high risk in agricultural production in many traditional communities means that some sort of credit mechanism to cover the bad times and emergencies must be available, if the members are to survive. Credit may also imply the future availability of labor that the store owner/landlord will require at harvest time. By offering credit, the landlord is assured of a labor supply. Income and market risk is decreased on the part of the peasant/worker/borrower. Creditors/buyers, employers are interested in such arrangements because they are able to manipulate prices and assure themselves of a profit. On the other hand, the agriculturalists find these arrangements advantageous because they have easy access to credit for production and family emergencies and to someone who will take their product at harvest time when there is a surplus to sell.

In traditional agricultural communities, credit is a constraint to increasing production efficiency. Credit in traditional agricultural systems tends to be more for consumption (survival in emergencies) than for investment in production. As a result there is low pay-off for modern sector lenders and high risk for the borrowers.

D. Summary of the Meaning of the Factors of Production in Traditional Agricultural Communities

The factors of production in traditional agricultural communities have different meanings as well as different relations than in modern agricultural communities. However, traditional agricultural communities are not stagnant nor are their members unwilling to innovate. In fact, ethnographic studies show constant experimentation on the part of traditional farmers with both cash crops and new potential garden and subsistence crops.[16] Traditionalism in agricultural communities is not simply constantly using particular production technologies with the same agricultural crops and livestock. Instead, it is based on reliance on highly complex social and economic interactions to ensure survival. In fact, a component of traditional agriculture is intense diversification of production including crops, livestock, forest products, value added, and labor trading and selling activities as a survival strategy. Each enterprise may have a high risk associated with it. But in combination, the various economic activities are expected to balance one another, thereby reducing risk.

III. RISK AVOIDANCE MECHANISMS

A. Low Risk as Survival Strategy and Value

In traditional farming systems, the community attempts to ensure the dependability and stability of farm productivity and to minimize risk through a variety of methods. As a result, traditional agricultural communities place primary importance on maximizing security rather than maximizing income.[17] Strategies of gaining access to and utilizing land, labor and capital all have security maximization as a guiding principle. They also preserve relative economic autonomy.

B. Agricultural Diversification to Avoid Risk

Community norms and institutions support diversified agriculture in order to (1) reduce the risk of complete crop failures; (2) distribute labor, equipment, and other production inputs as evenly as possible throughout the year; (3) reduce dependency on the outside by providing intermediate products or by-products to a major activity, such as stover for animal feed; (4) aid in the control of weeds, insect infestation, and plant disease; and (5) improve soil productivity; and distribute money income throughout the year.[18] Diversified farming systems often mean that the volume needed for efficient market integration is not achieved. This can be viewed as a barrier to progress — or a mechanism to maintain community autonomy.

1. Diversification and Long–Term Sustainability

Farming strategies legitimated in traditional agricultural communities often decrease risk because of an orientation to long-term sustainability of agriculture. For example, soil erosion is decreased because multiple cropping strategies and intercropping keep ground cover on the soil almost year round. A wide genetic pool for each crop is maintained, because farmers seek to have crops of the same species that mature at different times during the year in order to maximize food availability rather than marketing or harvesting convenience. A wider variety of crops is also maintained, including a large number of medicinal plants, often known to and gathered by women, which further increase the plant gene pool. Community traditions of health care, food storage, and food preparation all facilitate long-term agricultural sustainability.

2. Diversification and Family Well Being

Maintaining an autonomous subsistence sector through community exchanges has definite welfare implications. The importance of subsistence agriculture in providing the family food supply is shown by a study carried out in Tabasco, Mexico by Dewey.[19] She found that dietary diversity, dietary quality, and nutritional status of pre-school children are all negatively associated with lower crop diversity and increased dependence on purchased food. The traditional agricultural communities' emphasis on subsistence production as a way to stop outside dependence on purchased food clearly has wide-reaching implications in terms of family well being. The study further indicates that improving income does not necessarily improve family well being. Other studies suggest that raising income level of families within traditional communities tends to direct money disproportionately to the male members of the household. Men have different income and outgo streams than do women. A smaller proportion of male income is likely to remain in the family for child welfare purposes.

C. Mechanisms of Avoidance of Capitalist Penetration

Many of the mechanisms that traditional communities employ can be viewed as barriers to the encroachment of capitalist expansion. By protecting crucial areas of production from market intrusion, traditional communities can maintain their relative autonomy and equality. These social and cultural mechanisms work best when reinforced by geographic distance from the center and distance in time caused by lack of access infrastructure, such as roads, railroads, telephone, electricity, etc. Mechanisms of isolation, physical and social, do not mean that the economy in traditional agricultural communities is totally subsistence oriented. Rather, only certain parts of the economy — often those most central to family survival — are so protected. The capitalist and subsistence sectors of the economy involve the same people and are generally complementary, not contradictory.

1. Physical Barriers to Capitalist Penetration

The work of Sarah Lund Skar[20] regarding a relatively isolated Quechua-speaking community in the southern Andes of Peru illustrates a number of such mechanisms. Not only is the community physically isolated in a high mountain valley with poor roads and no rail connections, but language also isolates the community, particularly the women, from the larger Peruvian context.

2. Social Barriers to Capitalist Penetration

Social and cultural survival mechanisms in the Andes involve gaining access to production resources in as many ecological environments (generally determined by altitude) as possible.[21] To do so requires complex cooperation among multi-family units. While kinship relations may reinforce these obligations, community cultural forms are the basis for them.

Risk reduction strategies of small-farm owners stem from their desire to decrease risk in their production system and decrease risk when they enter market economies. Traditional peasants have often found themselves exploited by new market relationships. For example, Poffenburger[22] reports that when they began growing cloves, Indonesian farmers were exchanging 1 kg of cloves for 10 kg of rice, about one third the actual market value of cloves. The advantage traders have taken of farmers because of the farmers' isolation and need for foodstuffs has made farmers mistrustful. Time after time the exploitive nature of new market relationships make peasants hesitant to move fully into new production systems. That individual hesitancy is institutionalized through community mechanisms discouraging increased dependence on the outside. The stability of the whole community is placed at risk if such exploitation continues for long periods of time. Communities establish a variety of norms against getting highly involved in such risky trading deals.

D. Risk and Dependence on World Markets

Transition to cash crops that are traded internationally means susceptibility to very wide price swings in the international economy. The change from subsistence crops, with low expenditures for inputs, to cash crops which might increase profit but also increase dependence on purchased foodstuffs and inputs, is often discouraged by traditional agricultural community leaders and customs. Poffenburger illustrates the potential risk of export-oriented change. In the 1970s in Indonesia, a rapid fall in the international price of rubber was accompanied by a great increase in the price of rice. At that time, government policies stressed export agriculture. In Brazil, after a large push toward export crops, the world price of coffee dropped, while the price of basic grains increased. Such negative experiences employing profit maximization strategies through cash cropping have reinforced traditional strategies of diversification that emphasize subsistence production on the part of traditional agricultural communities.

IV. EXCHANGE RELATIONSHIPS: INTERNAL RESOURCE ALLOCATION

A. Production Exchange

Wolf[23] has pointed out the problem of maximizing self-interest in situations where it is impossible to be self-sufficient in traditional agricultural communities. These communities have devised a variety of equalizing or leveling arrangements that maximize collective surplus and minimize individual surplus accumulation. Periodic redistribution of land resources also acts as a restraint to capital investment in land, except when it recurs for the community at large, as with community-controlled irrigation systems. Such mechanisms maintain low input, extensive techniques used on annual crops. These responses of collective leveling and cooperation are not universal in traditional agricultural communities. However, their absence appears to be associated with more rapid differentiation of the peasantry and demise or transformation of the traditional agricultural communities.

Subsistence crops are not sold for cash in some traditional communities. Instead, community members with surplus of one crop and need for another exchange produce — such as maize for potatoes. This non-monetary exchange (*trueque*) is carried out in a ritual atmosphere and helps to more evenly redistribute essential products throughout the community.[24] While capitalistic linkages lead to increased class differentiation, community cooperation leads to maintaining equality — and maintaining surplus value within the community.

B. Labor Exchange

Exchange relationships involving labor in cash crops, based on social reciprocity rather than strict economic accounting, have been described in South Asia, Latin America, Africa, etc. The South Asian example used here is in a village more socially differentiated and more integrated into the national culture and economy than is the isolated Andean village. Nevertheless, there are times when non-market labor relations serve to both mobilize labor at key periods, carry out labor that cannot be done individually, affirm status, and reinforce community identity.[25] These non-market exchanges are of two types — *payra*, collective plowing using bullocks and labor, between equals (leaving out the poorest, who have no bullocks, and the richest, who have no need for extra bullocks), and *irjik*, between a male of relative wealth and his friendship group. Unlike the Andean case, these exchanges appear to involve only men.

1. Mutual Exchange

In the *payra*, a compatible group of bullock-owning individuals get together at the beginning of the plowing season to share bullocks and labor in order to plow each other's land. The land is so hard prior to the monsoons that teams of eight bullocks are usually necessary to pull the plow through the ground. While number of bullocks (amount contributed) and area to be plowed (amount collected) are correlated, there is no strict accounting: the men and bullocks keep working until all land is plowed. During the plowing, food is shared and social ties are rebuilt. The groups are relatively fluid, but participation is expected and necessary.

2. Status Difference and Labor Exchange

An *irjik* is a 1 day affair when a male landowner has a specific task that a large work crew can perform in a day. While the *payra* is the usual way of getting plowing done, the *irjik* is relatively rare. It will occur only several times a year. Friends and neighbors are summoned and work hard all day, receiving a noon-day meal. In the evening, a lavish feast is served. The more opulent and unusual the food, the more status accrues to the host/land owner, who has been working with the *irjik* all day. Schlesinger does not tell us if the preparation of the feast involves cooperative exchange labor or purchased labor on the part of the host's wife.

While labor is bought and sold in a variety of ways within the village, the exchange arrangements protect scarce cash and ensure a labor supply beyond the market. And, particularly in the case of the *payra,* it mitigates against accumulation within the village, since a *payra* member who unexpectedly loses a bullock often still is included in working and having his land plowed, so his family can produce a crop, repurchase a bullock, and not lose its access to the means of production.

3. Norms of Reciprocity

Labor exchanges have very clear norms of reciprocity built into them. It is recognized that there is value attached to that labor, although degree of exact accounting may vary. For example, in Chaingmai, Thailand, if a family is not able to fulfill its commitment to labor exchange, it must pay in kind, i.e., the equivalent of a day's wages, or it must hire someone to take its place. If these commitments are not fulfilled, eventually the family will be removed from the labor exchanges. Labor exchanges tend to occur among the middle strata of traditional villages. The very poorest members of the village who own no land cannot participate in labor exchange but only sell their labor. The richest members who do not farm their own land but rent it out also do not participate in the labor exchanges. For farmers who do participate, the exchange of labor often includes animals such as water buffalo or caribou.[26]

Non-market labor exchanges have been described for the Bantu Africans as well.[27] One type of exchange involves the host summoning community members, who work at a specific task, such as threshing, while consuming beer and food provided by the host (who is often a woman). As Randles points out, "Far from being strictly co-operative, the institution makes the rich even richer."[28] In the highly hierarchical, pre-colonial Bantu cultures, there was also a tradition of annual collective labor service for the chief, in return for a feast (*corvee*) and the assurance of the chief's benedictions to ensure the fertility of workers' fields. As the chief's power was undercut, so too was the *corvee.*

C. Irrigation and Formal Labor Exchange

A wide variety of formal traditional farmer organizations have been identified that provide a powerful mechanism for survival in traditional farming systems through labor exchange and resource distribution. The most well developed surround irrigation-based agriculture and the "hydraulic" social organization that results. In these cases access to water depends on collective labor efforts. The level of community organization is generally much more complex. For example, wet rice irrigation associations (*subaks*) in Indonesia allow for interchange of information among farmers, establishment of credit mechanisms, and putting pressure on governmental organizations to influence their policies.[29] Geertz explains that the *subak* is an irrigation society. It is "...a technological unit, marked out by collectively owned dam and canal; a physical unit, an expanse of terraced land with a defined border around it;... a social unit, a corporation consisting of people owning land in that expanse, serviced by the dam and canal"; and a religious unit.[30]

1. The Subak

Individual members of the *subak* carry out their own farming activities. The *subak* provides powerful collective control of the timing of agricultural activity. When to plant is not decided by the individual, because planting depends on water provided by the irrigation system. Everyone in the *subak* must plant at the same time. Upkeep of the canal system is carried out by work groups of *subak* members. Because of the high interdependence of this water use, the *subak* organization is extremely formal. There is a chief and a council with the power to regulate and to finr those that break the regulations. The *subak* has the power to tax in order to raise funds necessary for support of the canal system and buys the inputs necessary. The *subak* has a variety of shrines and appoints priests to conduct rituals at key points during the year at these shrines. Geertz compares the *subak* to a technically specialized, cooperatively owned public utility as opposed to a collective farm. Community mechanisms usually develop in response to the most all encompassing resource, whether land, water, or labor.

Ritual religious activities of each *subak* establish the pace and rhythm of individuals' cultivation decisions. Geertz shows how the ritual practices are correlated among *subaks,* so that there is coordination within a given drainage system. Intricate religious systems stabilize the demand for water over a crop year. They also assure staggered performance of the various cultural practices necessary for rice production in a given drainage region. In the *subak* system, water is a collective good to be managed but not owned. Geertz contrasts Bali to Morocco, where water is an individual property right, individually owned and controlled. Cropping systems differences are both results and causes of different types of community organization.

Irrigation helps spread out labor demand. Community organization of labor is encouraged in areas of irrigation of food crops. In dryland agriculture, especially where rainfall is highly unpredictable, as in certain areas of Africa, planting must be done as soon as possible after the rains. Labor groups are not possible, because by planting one family's field first, another family would risk crop failure.

D. Patron-Client Relationships

Traditional agricultural communities are not necessarily egalitarian. In many situations, high levels of inequality and substantial class hierarchy exist. However, there is also interdependence between the different classes.

As traditional communities become more tightly linked to the larger economy, the relationships of communities with the outside are often mediated by patron-client types of relationships. These relations protect the peasant household from losing its independence in the face of market forces and increasing state control, while at the same time it may serve to promote the interests of the richest peasant class.[31]

1. Traditional Communities in the Modern Economy

The logic of a modern economy[32] mitigates against peasant agriculturalists maintaining even their small access to resources. A modern economy is based on contract rather than custom. Universalistic criteria that are applicable in all circumstances are supposed to operate in modern economic relations, rather than particularistic criteria which take into account that a good person can have a run of bad luck. Further, a modern economy requires distinct role separation, instead of role diffuseness, which allows for consideration of multiple factors when making economic decisions. (My debtor is my son's godmother and provides medicinal herbs when my family is ill.) Role diffuseness also facilitates a single individual engaging in different forms of production, wage labor, and simple commodity production. Since integration into the modern economy leads to higher risks of exploitation, it is advantageous to build alliances with a patron as broker to the outsider, rather than to risk playing a game whose rules are stacked against the poor. Such patron-client relationships develop because risk is high and, therefore, the chance of not repaying on a contract schedule is great. A complexity of interrelationships develops between the buyer of labor (employer), owner of land (landlord), and provider of the capital (creditor), so that a single person may perform all functions. These multiple relationships are solidified through a communal mechanism of enforcement such as fictive kinship. As Platteau points out:

> ...in a world where uncertainties of all kinds are very high and where there are no written contracts or other securities to guarantee the fulfillment of agreed transactions, people have an interest in entering into multiple deals that cover several aspects of their economic (and social) life. In this way, they become enmeshed in a series of commitments which cannot be ended easily since they are part of an indivisible package that may be rescinded only *in toto*... As a consequence, interpersonal relationships are stabilized — an imperative objective in small village societies that must aim at maximum social cohesion — and risks of economic activities being interrupted or of various kinds of obligations being defaulted are minimized.[33]

2. When Traditional Relations Serve Best

In traditional agricultural communities, such relationships have built-in advantages for both the patron and the client, although the advantages have different dimensions for each and are weighed in favor of the patron. The patron can accumulate capital relatively easily, whereas the client is guaranteed relative security. A merchant can ensure himself or herself the availability of product at market time by providing credit during the year to someone who shops at her/his establishment. That "forward contracting" is a kind of futures marketing. Reformers often denounce patrons as exploitive intermediaries. In fact, in the tenuous situation where lack of access to resources means high yield risk and lack of negotiating power in the larger economy means high price risk, patrons provide a stable base for credit, marketing, and seasonal employment through a highly interpenetrated set of social relationships that maintain stability, although not enrichment, in traditional agricultural communities. Surplus value is extracted from the traditional agricultural community, and impoverishment may occur, but not at as fast a rate as would occur under total commercialization of all relations of production.

V. CONCLUSION

I have discussed social and cultural dynamics in traditional agricultural communities that serve to maximize the options of their residents in order not to become completely proletarianized. Such options mean that selling one's labor is one survival strategy a household may employ, but is only one strategy among a diverse array of community structures provided. It is hypothesized that in the absence of such community cultural mechanisms of exchange, reciprocity, and wealth leveling, total market integration is hastened and peasant differentiation proceeds apace. The presence of such social and cultural mechanisms protects the subsistence sector and the ability of the farming household to partially control at least in part both labor and product. These mechanisms often are attacked as sources of low productivity by outside change agents. The cultural and social dynamics of traditional agricultural communities should be viewed as barriers to increased dependency and exploitation. The traditional agricultural community lacks a wide range of economic alternatives for individuals in terms of paid wage labor, but provides security and identity not always available in more commercial, modernized economic relations. The choice is not which paid job to get, but whether or not to sell one's labor at all.

The dynamics of traditional agricultural communities can provide a barrier to market penetration of certain key aspects of production, thereby increasing economic security but sacrificing accumulation.

REFERENCES

1. **Smith, C. A.,** Forms of production in practice: fresh approaches to simple commodity production, *Journal of Peasant Studies*, 11, 201, 1984; **Bernstein, H.,** African peasantries: a theoretical framework, *J. Peasant Stud.*, 6, 421, 1979; **Friedmann, H.,** Household production and the national economy: concepts for the analysis of agrarian formations, *J. Peasant Stud.*, 7, 158, 1980; **Smith, G.,** Reflections on the social relations of simple commodity production, *J. Peasant Stud.*, 13, 99, 1985.
2. **Long, N., Ed.,** Introduction, in *Family and Work in Rural Societies: Perspectives on Non-Wage Labor*, Tavistock Publications, London, 1984, 1.
3. **Chevalier, J. M.,** There is nothing simple about simple commodity production, *J. Peasant Stud.*, 10, 153, 1983.
4. **Kahn, J.,** From peasant to petty commodity producers in Southeast Asia, *Bull. Concerned Asian Scholars*, 14, 3, 1982.
5. **Taussig, M.,** *The Devil and Commodity Fetishism in South America*, University of North Carolina Press, Chapel Hill, 1980.
6. **Lehmann, D., Ed.,** Introduction: Andean societies and the theory of peasant economy, in *Ecology and Exchange in the Andes*, Cambridge University Press, Cambridge, 1982.
7. **Skar, S. L.,** Interhousehold cooperation in Peru's southern Andes: a case of multiple sibling-group marriage, in *Family and Work in Rural Societies: Perspectives on Non-Wage Labor*, Long, N., Ed., Tavistock Publications, London, 1984, 83.
8. **Long, N. and Roberts, B., Eds.,** *Peasant Cooperation and Capitalist Expansion in Central Peru*, Institute of Latin American Studies, University of Texas, Austin, 1978.
9. **Smith, G.,** Reflections on the social relations of simple commodity production, *J. Peasant Stud.*, 13, 101, 1985.
10. **Skar, S. L.,** Interhousehold cooperation in Peru's southern Andes: a case of multiple sibling-group marriage, in *Family and Work in Rural Societies: Perspectives on Non-Wage Labor*, Long, N., Ed., Tavistock Publications, London, 1984, 83.
11. **Alberti, G. and Mayer, E., Eds.,** *Reciprocidad e Intercambio en los Andes Peruanos*, Instituto de Estudios Peruanos, Lima, 1974.
12. **Hecht, S. B., Anderson, A. B., and May, P.,** The subsidy from nature: shifting cultivation, successional palm forests, and rural development, *Hum. Organ.*, 47, 25, 1988; **Hudgen, R. E.,** Diagnostic Survey of Female-headed households in the Central Province of Zambia, in *Gender Issues in Farming Systems Research and Extension*, Poats, S. V., Schmink, M. and Spring, A., Eds., Westview Press, Boulder, CO, 1988, 373.
13. **Wharton, W. R., Jr., Ed.,** Subsistence agriculture: concepts and scope, in *Subsistence Agriculture and Economic Development*, Aldine, Chicago, 1969, chap. 2.
14. **Oasa, E. K. and Swanson, L.,** The limits of farming systems research and development: should development administrators be interested?, *Agricultural Administration*, 23, 1, 1986.
15. **Painter, M.,** Changing relations of production and rural underdevelopment, *J. Anthropol. Res.*, 40, 271, 1981.
16. **Poffenberger, M.,** Changing dryland agriculture in eastern Bali, *Hum. Ecol.*, 11, 123, 1983.
17. **Scott, J. C.,** *The Moral Economy of the Peasant: Rebellion and Subsistence in Southeast Asia*, Yale University Press, New Haven, CT, 1976.
18. **Lawas, J. M.,** Agricultural diversification and development: the Philippine viewpoint, paper presented at the Rural Development Panel Seminar, Manila, Philippines, January, 1971.
19. **Dewey, K. G.,** Nutritional consequences of the transformation from subsistence to commercial agriculture in Tabasco, Mexico, *Hum. Ecol.*, 9, 151, 1981.
20. **Skar, S. L.,** Interhousehold cooperation in Peru's southern Andes: a case of multiple sibling-group marriage, in *Family and Work in Rural Societies: Perspectives on Non-Wage Labor*, Long, N., Ed., Tavistock Publications, London, 1984, 83.
21. **Murra, J. V.,** *Formaciones Economicas y Politicas del Mundo Andino*, Instituto de Estudios Andinos, Lima, 1975.
22. **Poffenberger, M.,** Changing dryland agriculture in eastern Bali, *Hum. Ecol.*, 11, 23, 1983.
23. **Wolf, E. R.,** *Peasants*, Prentice-Hall, Englewood Cliffs, NJ, 1967.
24. **Alberti, G. and Mayer, E., Eds.,** *Reciprocidad e Intercambio en los Andes Peruanos*, Instituto de Estudios Peruanos, Lima, 1974.
25. **Schlesinger, L. I.,** Agriculture and community in Maharashtra, India, in *Research in Economic Anthropology*, Vol. 4, Dalton, G., Ed., JAI Press, Greenwich, CT, 1981, 233.
26. **Potter, J. M.,** *Thai Peasant Social Structure*, University of Chicago Press, Chicago, 1976.
27. **Goodfellow, D. M.,** *Principles of Economic Sociology, As Illustrated from the Bantu Peoples*, Routledge and Sons, London, 1939.
28. **Randles, W. G. L.,** Reciprocity in Bantu Africa, in *Research in Economic Anthropology*, Vol. 4, Dalton, G., Ed., JAI Press, Greenwich, CT, 1981, 12.
29. **Birkelbach, A. W., Jr.,** The Subak association, *Indonesia*, 16, 153, 1973.
30. **Geertz, C.,** The wet and the dry: traditional irrigation in Bali and Morocco, *Hum. Ecol.*, 1, 23, 1972.
31. **Long, N., Ed.,** Introduction, in *Family and Work in Rural Societies: Perspectives on Non-Wage Labor*, Tavistock Publications, London, 1984, 1.
32. **Weber, M.,** *Economy and Society*, University of California Press, Berkeley, 1978.
33. **Platteau, J.-P.,** The drive toward mechanization of small-scale fisheries in Kerala: a study of the transformation process of traditional village societies, *Dev. Change*, 15, 65, 1982.

5 WOMEN'S ROLE IN SMALL FARM AGRICULTURE

Louise Fortmann

I. INTRODUCTION

> I then explained that I wanted to join the fishing party. They were quite willing and said the ladies were just finishing planting their farms before the tornado season came on...[1]

In 1899, the English explorer, Mary Kingsley, went fishing with women who not only farmed and fished, but also made their own implements. A generation later, Baumann[2] credited women with the invention of hoe agriculture in Africa. There is ample additional evidence to the effect that women in all parts of the world have always been farmers. But the gender ideology of colonial powers frequently made women's roles in agriculture invisible to policy makers and agricultural staff. And the designers and implementers of agricultural policies, programs, and projects still tend to proceed as if all farmers were male.

This chapter is intended to make women farmers and their agricultural spaces visible. It discusses the elements of women's participation in agricultural production — the spatial distribution of women's agricultural activities, the magnitude of their involvement, their intellectual and physical contributions, their access to the means of agricultural production, and the factors which influence the nature of women's participation in agriculture.

In the course of this discussion the following will become apparent. First, without women's physical and intellectual work, the continuation of agricultural production would require substantial changes in social organization. Second, narrow definitions of agriculture have reinforced the influence of gender ideology in making women's contributions to agriculture invisible. Third, women often farm with fewer resources than do men and with less personal benefit.

II. DISAGGREGATING SMALL FARM AGRICULTURAL SYSTEMS BY GENDER

Small farms are for the most part run by kin-based social units, commonly referred to as households. Household, however, is a troublesome concept. For example, definitions such as those who eat from the same cooking pot fail to capture households whose livelihood strategy involves sending some members to urban areas as wage laborers. Given the problems of finding a satisfactory definition for a single culture,[3] trying to find a definition which would fit the broad sweep of cultures discussed here would be futile. Instead, I begin with the assumption that the women who participate in agriculture are members of various kinds of kin-based social units which, for the sake of convenience, shall be called households.

In many households there is more than one production system — male systems, female systems, and joint systems. Further, men and women may have conflicting interests over the use of household income.[4] Members of a single household might be involved in the production of a cash crop or livestock by the man, joint production of food crops by the man and woman, and production of food crops, poultry, and small stock by the woman.

The meaning of crops and resources will differ in different production systems. Agricultural researchers must face this reality when they try to improve crops such as cowpeas or amaranthus that can be utilized as a grain or a leafy vegetable. The strategy followed often depends on the production system the researcher recognizes. The kinds of clashes which can arise from differing production systems is illustrated by Palmer:[5]

> On a famine patrol in the New Guinea highlands, the men stated that the new sweet potato crop would take another nine to thirteen months to mature, but the women (who cultivated the crop) assured the patrol it would take only another four to five months. The sexes' interests were different: the men wished the famine rations to continue as they were a means of exercising local patronage and power, while the women saw the next sweet potato harvest as the point when their own role in society could be resumed.

The costs of not paying attention to these parallel but separate systems were revealed in a Tanzanian cotton project. Women's work was required for carrying water at the same time women's own maize crops needed attention. The women let the cotton die and carried on with their own maize farming. The project failed.[6]

III. AN OVERVIEW OF WOMEN'S ROLES IN AGRICULTURE

Whatever the production system, whatever the task — every facet of agricultural production in virtually all of the Third World involves or relies on women. Women produce food and cash crops on their own fields and in the confines of the homestead. They assist their husband in the production of his cash and food crops. They work as wage laborers in the fields and homesteads of other small farmers.[7] They join with other women in cooperative agricultural labor on their own field, the fields of others, or in joint enterprises. They own and care for animals of all sizes. They raise and catch fish. They gather and nurture wild plants. They hire and supervise wage laborers.

Agricultural field work is predicated on help from other women household members with domestic work. As an Indian woman said, the only reason she could go out and work in the field was because her mother-in-law was "in the house to do all the housework, attend to the childrens' needs, and cook for the workers."[7] Women maintain and reproduce the household, that is, women bear and care for children, cook, clean, nurse the sick, repair the house, fetch water, gather firewood — all of which frees and enables other household members to do directly productive work. Women may work to provide the resources of production including those cases where rent includes the tenant's wife's unremunerated domestic work in the home of the landowner.[5]

Women produce goods and services necessary for the performance of directly productive agricultural tasks. They cook for wage laborers and work parties. In Bangladesh they make rope to tie goats and cows.[8] In parts of India they are responsible for ensuring that the necessary equipment is available for laborers, paying laborers, and keeping some kind of record.[7] Women's assumption of these facilitating and managerial roles allows men to give their full attention to physical tasks.

IV. THE SPATIAL DISTRIBUTION OF WOMEN'S AGRICULTURAL ACTIVITIES

Farming takes place not just on fixed fields outside the homestead but also on plots and pots inside the homestead, and on shifting sites outside the homestead. With the exception of the influence of the institution of female seclusion, little attention has been given to the factors which lead women to farm in one place or another or to the agroecology of these separate places. These are ample areas for future research.

A. In the Field

Women may have an entire field for their production. Or they may, as is common in some parts of West Africa, intercrop their plants in the fields where their husband grows trees or cash crops.[9-11] McMillan[12] found in Burkina Faso that "wives' private fields tend to be planted in areas alongside the cooperatively worked fields in order to reduce the amount of time involved in commuting from one field to another".

Even in societies that practice female seclusion women work in the agricultural fields. In Bangladesh, Wallace et al.[8] found that 75% and 25% of women (mostly the landless and to a lesser extent small farm women) in two villages work in the fields. Likewise, Mencher[7] notes that in India, especially if they are not of high caste, women from small land-owning and tenant households work on their own lands, sometimes work for others, and supervise field operations. Taylor[13] reports that in the south of Egypt women from half the households work in the field. This women's labor involves virtually every type of agricultural activity.

Although it is often considered to be heavy men's work especially if it is done with a tractor or animal-drawn plow, women are involved in *land preparation.* In Bangladesh, women prepare seedbeds for rice.[8,14] Women plow in Sierra Leone[15] and clear land in Senegal.[16] Henn[11] found that women in the Cameroons spent more than twice as much time as men clearing the major food crop fields. Women in New Guinea cut shrubs and bushes and prepare the ground for burning while men fell the trees.[17] And women prepare land in the Philippines[18] and Niger.[19] Some women participate in preparing plots in Colombia and Peru.[24] Sen[21] reports that Indian women in the poorer and hilly districts of the old state of Punjab plowed the fields when men were absent. Taylor[13] reports UN data showing that women in the south of Egypt plow and level the land in half the households.

Planting Crops — Women plant crops all over the world — alone, with help from spouses or hired labor, or with other women in work groups. Women's role in planting crops has been documented all over the world. For example, women plant crops in Bangladesh,[8] The Gambia,[10,22] Sierra Leone,[15] Senegal,[16] Burkina Faso,[16] Trinidad (where they also spread canes for planting),[23] Colombia, Peru,[20] India (where they also transplanted rice seedlings),[21] Egypt,[13] and the Philippines.[18] Among the Mundurucú Indians of Brazil, the women plant manioc shoots after the men have dug the holes and plant all minor crops.[24] Despite the norms of seclusion, Moslem women are involved in planting field crops. Both secluded and non-secluded Moslem Hausa women plant crops in the field.[25] In Turkey, women plant wheat and other crops.[26]

Crop maintenance — Crop maintenance tasks are labor intensive, time consuming, must be repeated throughout the cultivation cycle, and require considerable skill in many cases. Women are frequently involved in the tedious jobs of weeding, thinning, mulching, fertilizing, spraying pesticides, hand watering and bird scaring. In Bangladesh, women weed the paddy field, pull rice seedlings from the field, spread cow dung, harrow, and irrigate the field with bucket and pond water.[8] In The Gambia, women scare birds from millet, inspect corn for pests and hand water during the dry season, guard ripening millet and generally look after the minor crops.[10] In Sierra Leone, women weed, bund, and scare birds for rice production.[15] Niger women collect manure in front of their compound and spread it on their fields.[19] Women are involved in all crop maintenance tasks for millet in Senegal and millet, groundnuts, sorghum, rice, greens, and vegetables in Burkina Faso.[16] In Trinidad[23] they fertilize, mold, and deflower tobacco, fertilize and weed rice, weed vegetables, tie them to stakes, and irrigate them by hand. Women weed in the Philippines,[18] Colombia and Peru,[20] and among the Mundurucú Indians of Brazil.[24] In Colombia, Bari women cultivate forest gardens.[28] Women in Nepal are heavily involved in spreading manure.[29] In India, women traditionally loaded, unloaded, and spread manure, hoed and weeded.[21] Moslem women also participate in the maintenance of field crops. In Morocco they weed the crops.[30] In sections of Turkey they do all the work associated with the production of hazelnuts, tea, and wheat and most of the work associated with producing sugar beets and tobacco.[26] Egyptian women help their husbands in the fields although it is only the men's labor which is considered "prestigious and skillful."[31]

Harvesting — Harvesting must be carried out in a timely fashion if potentially catastrophic crop losses are to be avoided. Women's labor is crucial for harvesting in many parts of the world. In The Gambia, women harvest garden products, minor crops,[27] millet, cotton,[10] and rice.[22] In Sierra Leone, women harvest rice.[15] In Cuba, women traditionally picked tubers.[32] In Trinidad they harvest tobacco, rice, and vegetables and gather cocoa fruits.[23] In southern Brazil, women comprise about 30% of the wage laborers who harvest crops.[33] In Cameroon, women harvest cocoa although it is a men's crop.[61] In the Philippines,[18] Senegal, Burkina Faso and Ivory Coast,[16] Colombia and Peru,[20,28] and Nepal[29] women harvest crops. Women in New Guinea participate in harvesting almonds.[17] Moslem women also participate in harvesting. In Morocco they harvest beans, peas, maize, cut hay and in some areas pick fruit.[30] In Turkey they harvest wheat, cotton, sugar beets and tobacco and all hazelnuts and tea.[26] In southern Egypt women in 35 to 40% of the households are involved in harvesting.[13] Both secluded and non-secluded Moslem Hausa women harvest beans, groundnuts, and cotton.[25] Moslem women also pick cotton on household plots on the Gezira Scheme in the Sudan.[34]

B. In the Homestead

In areas where women's outside activities are restricted by female seclusion, the importance of women to agricultural production is dramatized by structures that bring agriculture to her since she can not go to it. Even in cultures with no female seclusion, significant amounts of agricultural production are located in the homestead.[35,36]

In Bangladesh women grow potatoes, squash, pumpkins, beans, spinach, tomatoes, eggplants, cauliflower, beans, banana, papaya, jackfruit, date nut palm (grown for juice), and other vegetables and fruits in the homestead. In establishing and maintaining their homestead production women dig mud and carry and pile it up to level the plot, typically at the rear of the hut along the slope of the homestead mound. They fence their garden with cut bamboo, hoe it to prepare the soil, plant, construct trellises for beans and pumpkins or allow them to grow up the side of the hut and over the roof, weed, water, and harvest.[8,14,37,38] Women in New Guinea and India are also the primary workers in home gardens.[39] Moslem women in Baluchistan, Pakistan harvest windfall dates near the home.[40] In Latin America, "dooryard gardens", in many cases a complex mix of horticultural and tree crops and animals, are planted and managed by women.[33]

Seed selection and storage — This is a women's speciality in Bangladesh, India, and Nepal. Seed must be carefully dried, cleaned, sorted, and stored sometimes in special baskets sealed with a mixture of cow dung and mud.[7,14,38] In Bangladesh women are responsible for testing the germination rate of seed.[41] Nepali women made 60% of the decisions concerning seed use in studies summarized by Acharya and Bennett.[29]

Crop processing — Crop processing is often women's work. In Bangladesh, women clean, thresh, dry, and store mustard.[8,14] They remove seeds from jute plants, clean jute seed, extract jute fibers from jute stems, dry jute stems, and store jute.[8] They process jute fiber into rope and macrame wall hangers. Both men and women process jute stalks into fencing or siding.[14] Women sun, stir, and thresh peas with animal power, and sun and dry hay.[8] They pick straw from paddy, husk, dry, winnow, and parboil paddy, and store it in house.[8] When rice is threshed women prepare the hard mud threshing floor, unbind the rice bundles, spread the stalks and turn them as they are threshed, gather and stack the straw after threshing, dry the straw, and clear the threshing floor.[14] They sun and stir wheat, thresh wheat and pulses using animal power, sun pulses, winnow pulses and wheat.[8] Women thresh, dry and store millet grain.[14] In India women take paddy to the rice mill, and clean and measure the dehusked paddy.[7] In The Gambia women clean cotton, pound grain, and help men winnow groundnuts.[27] Cuban women grind corn and string tobacco up to dry.[32] In Trinidad, women process tobacco, rice, and vegetables and remove cocoa seeds.[23] In Cameroon they process cassava.[11] Moslem women in Baluchistan, Pakistan process dates.[4] Moslem women in Morocco winnow wheat, a task which at one time would have been prohibited for fear of "mystical contamination".[30] In South America the processing of bitter manioc into farina is almost universally done by women.[33]

Wallace et al.[8] and Chen[14] report that women also store the processed crops — paddy, hay, jute, and seed in Bangladesh. Women prepare the storage bins and supervise the storage of grain. In India Mencher[7] notes, "Often it is farm wives with little help from their men who are responsible for the prevention of post-harvest losses."

Poultry production is often done in or near the compound

by women. In Bangladesh women are responsible for poultry production including hunting for snails to feed ducks and killing chickens.[8] Women are also noted as taking care of chickens in India[7] and Cuba.[32]

C. The Spaces in Between

Rocheleau[42] has pointed out that women depend on wild plants growing in uncultivated spaces to sustain their households.[43] Women's use and nurturing of these plants is another area for research. Although not cultivation, this management of natural plant resources is an integral part of the ecology and livelihood system that form the context for agriculture as traditionally defined.

D. Livestock Production

Livestock and poultry production is also distributed over all three spaces. Information on women's involvement in livestock production is relatively hard to come by, probably because it is even more firmly believed that livestock, particularly among nomadic pastoralists, is a male sector. But such data as do exist indicate that even among nomadic pastoralists, women play an important role in livestock production.

In Bangladesh and India, women take care of goats and cattle. This involves removing dung from the cow shed and sweeping the shed, milking, feeding and watering the cows (which itself involves fetching water, pulling grass and mulching and stacking hay, and mixing wheat and pulse chaff into fodder), washing calves, and, in the case of older women and young children, grazing the livestock.[7,8,14]

In The Gambia, in addition to livestock they bought with their own money, Wolof women traditionally received female kids from their husbands to raise. When the kids are born, the woman kept the females for herself and the husband got the males.[27] McMillan[12] describes a Burkina Faso settlement scheme where during the late 1970s women began to acquire herds of 20 to 30 goats and two senior wives had acquired cattle. In Malawi, although men signed up for a stall-feeder program, it was the women who actually hauled the water and fed the animals.[44] Small-stock raising is a traditional women's activity in Burkina Faso.[16] Peters[45] found women to be as knowledgeable as the men about the state of the family herd. Women raise goats in Nigeria.[46] In Somalia, women own livestock, they perform almost all the tasks of men including handling camels although they are unlikely to herd larger numbers of camels away from the village for extended periods of time and women household heads tend to keep their animals near the village.[47] In Niger, women own animals which they may pen at some inconvenience outside the compound (owned by the man) so as to retain ownership of their manure.[19]

Deere and Leon de Leal[20] showed that 45 and 88% of the women in sample households in two regions in Colombia were involved in the care of large farm animals while 95% of the women in a region in Peru cared for large farm animals as well as "house animals" such as chickens, guinea pigs, and rabbits. Similar findings were reported for Peru by other researchers who also found that women had an important decision-making role in livestock production.[48] Hecht's[49] Brazil study found that "virtually all the women raised animals and sold them or their products in the town's local market". Stubbs[32] describes Cuban women feeding pigs. Wollenberg[18] found that upland women in the Philippines care for livestock. Moslem women in Morocco care for livestock.[30] Moslem girls in Turkey watch the cattle and gather their droppings and the women are in charge of all dairy production.[26]

E. Intellectual Tasks

Women make intellectual contributions to agriculture as inventors, decision makers, holders of specialized knowledge, and managers. These too are distributed across all three spaces although perhaps a larger proportion of them are located within the homestead.

Hoe agriculture may have been one of the earliest of women's inventions; it was hardly the last. When women lack the resources to fulfill their responsibilities, the responsibilities do not disappear. So women invent new techniques, new social forms. For example, Chen[14] reports that although wheat in Bangladesh is usually threshed using bullocks, many households do not have access to a bullock for threshing. "Women in those households have had to devise methods of threshing wheat." Reboussin[50] argues that disproportionate outmigration of Diola women can be explained by the loss of male labor to clear and prepare rice fields and maintain dikes when men turned to groundnut cultivation. In order to meet their provisioning activities in the face of declining production, women invented new forms of livelihood and sent their daughters off to be housekeepers in urban areas. Senegalese women rice farmers exchange varieties and experiment with new kinds of rice.[10] Home gardens, primarily the preserve of women, often serve as mini-experiment stations.[36]

Women are independent decision makers on their own fields and in the production arenas they control, they are consulted in joint decisions, and they influence the effect of decisions made by others by the way they decide to implement them.

Safilios-Rothschild[15] reported that in Lesotho women in about half of the *de jure* female-headed households made all agricultural decisions. In *de facto* female-headed households if the woman is over 25 she makes the day-to-day agricultural decisions, although long-term decisions over agriculture or cattle may be made by the husband or jointly. Wealthier Botswana women may hire labor.[45] Mencher[7] describes a widow in Tamil Nadu as saying, "I am having a radio. I will listen to the farm news and from that I will tell my son what sort of fertilizers we should use and what kind of pesticides to buy." Halim and Ali[41] report that even in patriarchal Bangladesh, women decide about hiring women laborers, distribution of work among the children, and livestock raising. In addition, a survey in two villages found that

20% and 37% of women in two villages felt they had considerable influence on persuading their husbands to adopt high yielding rice varieties; 8 and 21%, respectively, for the adoption of irrigation; 12 and 35%, respectively, for the adoption of fertilizer; and 21 and 31%, respectively, for the adoption of pesticide.[8] Half the farmers in a Caribbean study indicated that women were involved in agricultural decision making.[51]

In Nepal, Acharya and Bennett[29] found that the wealthier the household, the less say women had in farm management decisions, but the greater a woman's personal income, the greater her say. Widows and postmenopausal Moslem women in Morocco often take care of agricultural land.[30]

Women influence decision making because they have specialized knowledge necessary to sustaining the livelihood. In Ghana, Owusu-Bempah[52] reported that women farmers could list an average of 57 forest products that could be harvested or produced from the nearest forest reserve while men farmers could, on average, list only 14. In Bangladesh, it is women who can judge rice, paddy, and seed for its quality and moisture before and during storage.[14,38]

Women organize the production process itself by hiring and supervising laborers, keeping records, ensuring that tools are ready, and preparing areas (such as winnowing floors) for productive activity. They also manage the household ensuring through their organizational skills that other workers in the household can rely on the availability of food, water, heat, light, clothes, and salubrious surroundings.

V. THE MAGNITUDE OF WOMEN'S INVOLVEMENT IN AGRICULTURAL PRODUCTION

Cataloging the diversity of women's participation in agricultural activities does not indicate how widespread or how important such participation might be. Unfortunately, as Dixon-Mueller[53] has noted, the data base which would allow quantifying and comparing women's involvement in agriculture does not exist. The trends indicated by available empirical data suggest that in many societies women spend more time on agricultural production than men. When the double day (much of which facilitates directly productive agricultural activities) is considered, women's workload is staggering. A sample of available quantitative data follows.

Chen[14] recorded Bangladeshi women who during the harvest season rise at 3 a.m. to prepare breakfast and finally serve dinner at 9 p.m. — over an 18-h day. In Nepal, women were found to work an average of 10.8 h/d compared to 7.5 for men.[29]

African figures are similar. In the Gambia during the 1947 rice harvest, Haswell[22] found men and boys averaged 110 working days on the farm while women and girls averaged 155. That year rice produced by women constituted 80% of the food supply. In the Ivory Coast, women do 66% of agricultural work.[54] In Cameroon, Henn[11] reports that rural men spend 12 h/week working in agriculture compared to women's 26 h.

In Colombia, Castro and Velz[55] found men did 5.78 h/d of "productive work" while women worked 6.56 h. Taylor[13] reports UN data showing that in certain areas of Peru women in 86% of the households did almost all the farming.

In Taiwan, Gleason[56] found that women provided 54% of all agricultural labor. Families who lacked female family labor could not grow certain crops since the production of labor intensive crops relied heavily on female family labor. In two sites, families without female labor planted 12 crops of which 6 accounted for 95% of the acreage. Farms with female family labor planted 22 crops with only 41% concentrated in two crops. In a third site, families without female labor planted 5 crops, with 73% of the land area under sugar cane. Families with female labor planted 15 crops with only 32% of the land under cane.

Acharya and Bennett[29] found that women provided 57% of the adult time to the family farm enterprise (agricultural and animal husbandry activities, food processing, collecting wild food, fuel and water, and construction or repair of one's own dwelling) and contributed 50% of the total household income.

These fragments of quantitative data are of limited value in providing a wider picture. For this purpose, qualitative data are both more accessible and provide us with a better understanding.

Answering the question, "what happens when a man loses his wife?" provides clear evidence of the importance of women's labor. Peters[45] notes that in Botswana:

> Where a wife's death or illness had removed her from the typical worker-manager role, the men and their households either did not plow or they experienced a series of poor harvests which caused some to withdraw from planting. Only very old widowed men remained long without acquiring another wife. One man whose wife had become frail complained that he would have to sell a beast for cash to hire workers to replace her labor.

While female-headed households are common in rural areas, in many places the femaleless male-headed household is an anomaly. That is, men seem to find farming without women far more difficult than women find farming without men.

The ubiquitous nature of strategies used to obtain and control women's labor is a second qualitative indicator of women's importance in agriculture. These strategies include paying women wages,[4,25] seclusion,[38,57] and beatings.[4]

Thus, while the role of women in agriculture does vary, both quantitative and qualitative data show that on the whole women's contribution to agricultural production is substantial. This suggests that investments in research and providing resources to women farmers, women's crops, and women's spaces are likely to have a significant impact on agriculture and perhaps more return than investment in mainstream research and outreach programs.

VI. WHO ARE WOMEN FARMERS? FACTORS DIFFERENTIATING WOMEN'S PARTICIPATION IN AGRICULTURAL PRODUCTION

The preceeding text has made generalizations about the agricultural activities of "women" in a given society or ethnic group. This oversimplification is necessary because the available sources often treat rural women as a homogeneous mass. But not all women farm and by no means are women farmers all alike. The extent and nature of women's participation in agriculture in a country varies within a country by ethnic group, class, over time, within households, and for a given woman over the span of her lifetime.

A. Historical Variations

The nature and magnitude of women's involvement in agriculture is in continual flux. This may take the form of the feminization of agriculture. In some places economic pressure (including colonial poll taxes) and economic opportunity have pushed and pulled men into off-farm employment leaving farming to their female relatives.[7] Economic and population changes have also forced women into field agriculture who otherwise would have practiced only household agriculture. The shift from subsistence agriculture to landlord-managed capitalist farming has increased landlessness and near-landlessness, forcing women into the fields.[7,21,58] The need to buy inputs has similarly increased the cost of agriculture so that often either women must undertake the field work while men work to accumulate capital[7] or women are sent to work as wage labor on the fields of others while the men farm at home.[21]In other areas specific tasks may become the purview of women as has bookkeeping on farms in the U.S.[59] Conversely, in other places traditional women's agricultural activities are taken over by men as production is commercialized and mechanized technology is adopted.[58]

B. Societal Variations

Every class and culture has a gender ideology which among other things specifies the tasks and responsibilities which are considered suitable for women. This affects the practice of agriculture. For example, women in societies which practice seclusion are more likely to do agricultural work within the compound if they do it at all although this often varies by class.[60] In other cultures specific tasks such as plowing, herding cattle, driving machinery, or planting trees may be considered unsuitable for women.[61]

Even within a single ideology, the presence of opportunity leads to different patterns of participation. Henn[11] found that women with market access worked significantly longer in their fields than women with no market access. On the other hand, women who have the option of trading, marketing, or cottage industry (all of which pay better) may shun agricultural work.

The nature of the production system also may affect women's participation. Boserup[63] noted that women's participation was less in plow agriculture than in hoe agriculture.

C. Tenure Variations

Where a woman fits in the system of land ownership will also affect her production decisions or whether she will be able to farm at all. Is a woman farming on her own land, her husband's land, leased land, or community land of one sort or another? The production possibilities open to a land owner are quite different from those open to a tenant or tenant's wife. If the land is community land, certain activities such as tree planting may be forbidden.[64] If the land is her husband's, she may lose the rights to permanent tree crops as well as the land itself if he divorces her. For example, in Cameroon, if a woman is divorced and leaves the area, all her personal possessions, food supplies, and unharvested plants including trees are forfeited to the husband.[65] A factor crucial for widows is whether they retain rights to their husbands' fields.

D. Class Variations

The class of the household may often determine both the nature of women's participation in agriculture and how much she benefits from it. Wallace and his colleagues[8] found different work patterns for women in households with different sizes of land in Bangladesh. Likewise, Deere[66] found the participation of Peruvian women in agriculture varied by class. The farming practices of rich women may be more similar to those of rich men than those of poor women.[67] Indeed, poor women may work as hired labor in the fields of rich women. Historically, female slave labor was important in decreasing the agricultural labor burden of rich women.[68] And, particularly where it is a sign of status, wealth may be used to relieve women from the necessity of field work.

E. Individual Variations

The interrelated factors of a woman's relationship to the head of household and her marital status have very important effects on her participation in agriculture. A woman can be the head herself or she can be the daughter, daughter-in-law, wife, or mother of the head. As Cain et. al.[165] noted for Bangladeshi women, "different patterns of work, hours worked, and authority relations with other women in the household" are characteristic of each status. The work of a wife in a polygamous household depends on her childbearing responsiblities, age, and health; the number, sex, and age of her older resident children; and the status of her co-wives.[5,12]

Female household heads are increasingly prevalent. It is estimated that they account for one fifth of all households and 40 to 45% of rural households in Subsaharan Africa.[69] Female-headed households are found in every country and every culture including pastoral societies and societies practicing seclusion. For example, researchers have found 60 to 70% of rural households were headed by women in Lesotho,[32] 20% in Sierra Leone,[32] 25% in a Ghana sample,[52] 15% in a Cameroon sample,[11] 8% in a sample of Somali pastoralists,[47] 20 to 43% in Botswana,[67] and 17% in a Colombia sample.[55] Female household heads may be widows, divorced, deserted, separated, or never married. In addition, many women whose husbands are absent for long periods of time are de facto heads of household. Thus, a female household head may do work that in a male-headed household is done by men. Women with resident sons, daughters, or daughters-in-law may be able to pass the workload on to someone else. Old women may do different work from younger women and so on.

Women's work also varies with training, age, health, pregnancy, region, caste, or community.[7] Acharya and Bennett[29] found that Nepali women's work burden increases with age and that women whose husbands spend a lot of time out of the village and who have female children aged 10 to 14 spend more time on subsistence agricultural activities. Henderson[10] reports for The Gambia that "although all rural women are involved in productive activities, young women make up the majority of working women and they are growing resentful of the heavy labor burden left to them by migrating men".

In summary, women can be differentiated along a number of dimensions which affect if, where, and how they practice agriculture. This includes what decisions they make and what are made for them, their contact with sources of service and information, their access to and control over resources, the amount of time available for directly productive agricultural activities, and the nature and amount of benefits they get from their labor.

VII. WOMEN'S ACCESS TO THE MEANS OF AGRICULTURAL PRODUCTION

Female household heads are often poor and have less access to resources than men. Castro and Velez[55] found that 44% of male-headed Colombian households were landless; while 60% of the female-headed households were landless. Further, no female household head in their sample owned more than 20 ha while a small percentage of men did. Henn[11] found in two Cameroons villages that the average male-headed household held 4.7 ha of agricultural land compared to 2.2 ha for female-headed households. Female-headed households tend to own fewer animals and have less access to draft power than men.[47,61,67,70-72] Female-headed households also have less available labor than men.[67,72-74] As Nyhus and Massey[47] point out:

> When a man wants to improve his lot, the most direct means is through his increasing the number of wives and children he has. A female household head is precluded from taking this strategy, thus she is cut off from reproducing her labor in the way that is available to men.

Women in male-headed households often must depend on the male head for access to resources. Thus, while female-headed households are particularly disadvantaged, women in general have less access to resources than men. Women are less likely than men to own land in their own right,[75-79] and they have smaller acreages than men.[73,80,81] Women have less contact with extension workers[16,19,67,82-84] and less access to credit.[85] Since women comprise only 8.5% of agricultural advisors in Latin America, 2.9% in Africa and 0.7% in Asia, the latter findings are hardly surprising.[13]

Women, in short, often farm with less.

VIII. CONCLUSIONS

While farming systems research has taught us to look at the system of directly productive agricultural activities, it has often proceeded as if these were field activities taking place in a social vacuum. Instead we must look at the household livelihood system which provides the context within which directly productive agricultural activities take place and that often shapes the form, timing, and extent of those activities. A clear example is the way in which women's household responsibilities constrain the time she has for directly productive agricultural activities and/or lead to a steady decline in her own physical well being. A second example is the demand for cash which drains off both male and female labor into off-farm employment limiting the potential for agricultural production. A third example is the dependence of agricultural production on women's indirectly productive activities and ability to manage both the home and the logistics of agricultural production.

Successful incorporation of women into agricultural development requires the recognition that agricultural production is not found only in the field; does not end with the harvest; is not accomplished by physical activity alone; and is not encompassed by one production system per household. It requires legislation and other social arrangements which establish and enforce property rights for women in their own right. It requires policies and projects (such as domestic water projects) which reduce the labor required for household responsibilities. It requires providing training and resources to women and ensuring that there are women in technical professions to work with women and look out for women's interests. It requires crop, forestry, and livestock research to increase the productivity of women's crops, women's animals, and women's spaces. If the study of agricultural production systems is to reflect the household livelihood system, the definition of on-farm must be broadened to include non-field farming — the kitchen garden, homestead agro-forestry, and so on. It may also include what might be termed "mobile farm sites" — the pots in which urban or landless women may grow plants for household consumption or tree seedlings for sale.[86-88] Finally, agricultural scientists must learn to seek out the specialized knowledge and skills that women farming experts possess to expand and buttress their own more traditional and narrow understanding of agriculture.

ACKNOWLEDGMENTS

The comments of R. Dixon-Mueller, S. Feldman, S. Hecht, F. McCarthy, and E. Roe are gratefully acknowledged. Tanis Toland undertook the enormous task of converting a standard bibliography to the format used here.

REFERENCES

1. **Kingsley, M. H.,** *West African Studies,* Macmillan, London, 1899, 91.
2. **Baumann, H.,** The division of work according to sex in African hoe culture, *Africa,* 1(3), 289, 1928.
3. **Guyer, J. I.,** Intra-household processes and farming systems research: perspectives from anthropology, in *Understanding Africa's Rural Households and Farming Systems,* Moock, J. L., Ed., Praeger, Boulder, 1986, 92.
4. **Jones, C. W.,** Intra-household bargaining in response to the introduction of new crops: a case study from North Cameroon, in *Understanding Africa's Rural Households and Farm Systems,* Moock, J. L., Ed., Westview Press, Boulder, 1986, 109.
5. **Palmer, I.,** Women's issues and project appraisal, *Sussex Inst. Dev. Stud. Bull.,* 12(4), 37, 1981.
6. **Shapiro, K. H.,** Water, Women and Development in Tanzania, presented at Third Annu. Conf. on the International Water Resources Association, Sao Paulo, Brazil, 1978, cited in Jiggins, 1983.
7. **Mencher, J. P.,** South Indian Female Cultivators and Agricultural Laborers: who are they and what do they do?, presented at Conf. on Gender Issues in Farming Systems Research and Extension, Gainesville, FL, February 26 to March 1, 1986, 4.
8. **Wallace, B. J., Ahsan, R. M., and Hussain, S. H.,** Women's Work in Rural Bangladesh, unpublished, 1986, 99.
9. **Bradbury, R. E.,** The Benin Kingdom and the Edo-speaking peoples of Southwestern Nigeria, in *Ethnographic Survey of Africa,* Forde, D., Ed., International African Institute, London, 1957, 23.
10. **Henderson, H. K.,** Women in Development Project Advisement Report for The Gambia Mixed Farming and Resource Management Project (635-0202), paper prepared for Consortium for International Development/Women in Development Project AID/OTR-G-1871, n.d., 1.
11. **Henn, J. K.,** Intra-household Dynamics and State Policies as Constraints on Food Production: Results of a 1985 Agroeconomic Survey in Cameroon, presented at Conf. on Gender Issues in Farming Systems Research and Extension, Gainsville, FL, February 26 to March 1, 1986, 5.
12. **McMillan, D.,** Distribution of resources and products in Mossi households, in *Food in Sub-Saharan Africa,* Hansen, A. and McMillan, D. E., Eds., Lynne Reinner, Boulder, 1986, 267.
13. **Taylor, D.,** Part 1 — Women: an analysis, in *Women: A World Report,* Oxford University Press, 1985, 17.
14. **Chen, M. A.,** *A Quiet Revolution: Women in Transition in Rural Bangladesh,* Dhaka, BRAC Prokashana, 1986, 47.
15. **Safilios-Rothschild, C.,** The persistence of women's invisibility in agriculture: theoretical and policy lessons from Lesotho and Sierra Leone, *Econ. Dev. Cult. Change,* 308, 1985.
16. **Savané, M. G.,** The effects of social and economic changes on the role and status of women in Sub-Saharan Africa, in *Understanding Africa's Rural Households and Farming Systems,* Moock, J. L., Ed., Praeger, Boulder, 1986, 126.
17. **Hogbin, I.,** Tillage and collection in Wogio, in *Studies in New Guinea Land Tenure,* Hogbin, I. and Lawrence, P., Eds., Sydney University Press, Sydney, 1967, 60.
18. **Wollenberg, E.,** An Evaluation of Methodologies Used in Time Allocation Research, presented at Conf. on Gender Issues in Farming Systems Research and Extension, Gainesville, FL, February 26 to March 1, 1986a, 12.
19. **Barres, V., Brigatti, P., Correze, A., Debourg, M., Ducet, M., Gentil, V., Malachowski, S., Pieressa, F., and Snoy, O.,** *The Participation of Rural Women in Development: A Project of Rural Women's Animation in Niger, 1966—1975,* IRAM, Paris, 1976, 34.
20. **Deere, C. D. and León de Leal, M.,** Peasant production, proletarianization, and the sexual division of labor in the Andes, in *Women and Development: The Sexual Division of Labor in Rural Societies,* Beneria, L., Ed., Praeger, New York, 1985, 74, 75.
21. **Sen, G.,** Women workers and the Green Revolution, in *Women and Development: The Sexual Division of Labor in Rural Societies,* Beneria, L., Ed., Praeger, New York, 1985, 35.
22. **Haswell, M. R.,** *The Nature of Poverty. A Case History of the First Quarter Century after World War II,* St. Martin's Press, New York, 1975, 120.
23. **Harry, I. S.,** Labour Patterns in Agriculture in Trinidad, presented at Conf. on Gender Issues in Farming Systems Research and Extension, Gainesville, FL, February 26 to March 1, 1986, n.p.
24. **Murphy, Y. and Murphy, R. F.,** *Women of the Forest,* Columbia University Press, New York, 1985, 153.
25. **Longhurst, R.,** Resource allocation and the sexual division of labor: a case study of a Moslem Hausa village in nothern Nigeria, in *Women and Development: The Sexual Division of Labor in Rural Societies,* Beneria, L, Ed., Praeger, New York, 1985, 102.
26. **Cosar, F. M.,** Women in Turkish society, in *Women in the Muslim World,* Beck, L. and Keddie, N., Eds., Harvard University Press, Cambridge, 1978, 130.
27. **Gamble, D. P.,** The Wolof of Senegambia, in *Ethnographic Survey of Africa, Western Africa Part XIV,* Forde, D. Ed., International African Institute, London, 1957, 34, 35.
28. **Pinton, F.,** The tropical garden as a sustainable food system: a comparison of Indians and settlers in Northern Colombia, in *Household Food Production: Comparative Perspectives,* Niñez, V. K., Ed., International Potato Center, Lima, 1985, 26.
29. **Acharya, M. and Bennett, L.,** Women and the Subsistence Sector. Economic Participation and Household Decision Making in Nepal, World Bank Staff Working Papers No. 526, World Bank, Washington, D.C., 1983, 42.
30. **Maher, V.,** Women and social change in Morocco, in *Women in the Muslim World,* Beck, L. and Keddie, N., Eds., Harvard University Press, Cambridge, 1978, 110.
31. **Morsy, S.,** Sex differences and folk illness in an Egyptian village, in *Women in the Muslim World,* Beck, L. and Keddie, N., Eds., Harvard University Press, Cambridge, 1978, 605.
32. **Stubbs, J.,** Gender Issues in Contemporary Cuban Tobacco Farming, presented at Conf. on Gender Issues in Farming Systems Research and Extension, Gainesville, FL, February 26 to March 1, 1986, 11.
33. **Hecht, S.,** personal communication, 1987.
34. **Isikdag, F.,** Women as the captive labor force: the Gezira scheme, Sudan, in *Women Creating Wealth: Transforming Economic Development,* Gallin, R. S. and Spring, A., Eds., Association for Women in Development, Washington, D.C., 1985, 126.
35. **Niñez, V.,** Household Gardens: Theoretical Considerations on an Old Survival Strategy, Potatoes in Food Systems Research Series, Report No. 1, International Potato Center, Lima, 1984.
36. **Niñez, V.,** Food production for home consumption: nature and function of gardens in household economies, *Arch. Latinam. Nutr.* XXXV, 1, 9, 1985.
37. **Cain, M., Khanam, S. R., and Nahar, S.,** Class, patriarchy and women's work in Bangladesh, *Popul. Dev. Rev.,* 5(3), 420, 1979.
38. **Feldman, S. and McCarthy, F.,** Purdah and changing patterns of social control among rural women in Bangladesh, *J. Marriage Family,* November, 951, 1983.
39. **Vasey, D. E.,** Household gardens and their niche in Port Moresby, Papua, New Guinea, in *Household Food Production: Comparative Perspectives,* Niñez, V. K., Ed., International Potato Center, Lima, 1985, 41.
40. **Pastner, C. McC.,** The status of women and property on a Baluchistan oasis in Pakistan, in *Women in the Muslim World,* Beck, L. and Keddie, N., Eds., Harvard University Press, Cambridge, 1978, 439.
41. **Halim, A. and Ali, M. H.,** The Effects of Farming System Research and Extension on Farm Families, presented at Conf. on Gender Issues in Farming Systems Research and Extension, Gainesville, FL, February 26 to March 1, 1986, 3.

selecting the characteristics of grain cultivars, they rarely have considered the characteristics required for optimal domestic milling. Among the varieties available locally to producers, characteristics have been selected which are compatible with the domestic technologies used in processing, preserving, and preparing food and non-edible products. Changing varietal characteristics without changing post-harvest domestic technologies places undue burdens on local food systems and is one source of so-called resistance to change.[22]

Ay and Nweke[23] have demonstrated the close relationship between varieties, processing technology and women's livelihoods with respect to a new cassava variety, 30572, released by IITA to villages north of Ibadan in Nigeria. The cultivar has short maturation, high yield, and vigorous growth but also a higher water content than comparable local varieties. Women cassava processing specialists found that, using the usual flour processing sequence, 30572 required a longer drying period after fermentation, the flour had a greyish appearance which lowered its market price, and the flavor suffered. Through experimentation with techniques borrowed from *gari* production, the women found a way to express the excess water but at the cost of a higher labor input, which raised production costs significantly and lowered their profits.[24] During a recent on-farm evaluation of further new varieties, the women complained to IRRI scientists: "The cassava roots are too big. It is much harder to peel such big roots, it takes longer to 'fry' them into gari because of their higher water content, not to speak of the problem of carrying such heavy roots..."[25]

F. Focus on Single Utilities

The models of production favored by agricultural researchers tend to assume the goal to be maximization of single utilities, such as edible yield per unit area. In reality, the majority of smallholders continue to seek multiple utilities from production and to optimize returns to a mix of factors. Certainly those living in areas of low potential, whether climatically, geographically, agronomically, or economically so defined, cannot anticipate in the near future the provision of social and economic infrastructures, or the development of incomes and markets, which would supply the multiple values they presently derive from their environments.

The question of multiple utility is of particular concern to the poor, whose prospects of obtaining these values through the market are low as long as their alternative income potential remains low. A Bangladesh study, for example, has pointed to the interdependence between rice varieties and poor families' livelihoods. Analysis of data from different survey sites pointed to the importance of the rice varieties grown as an index of the availability to women of harvesting wages, gleaning, and post-harvest threshing employment.

> The long strawed broadcast *aman* rice grown in Comilla was less uniform in length. Consequently, some crops remained unharvested in the fields. On all land but that belonging to the poorest households, it was a prerogative of the women and children from poor landless households to glean the fields. They would then obtain acess to a neighbour's *dheki* to dehusk the rice.[26]

The transition to short-straw, uniform maturity varieties in areas where women's gleaning income may form up to 20 to 30% of incomes for the poorest, or even the sole seasonal income source, has had serious livelihood consequences in Bangladesh.[27] Alternative livelihoods, such as those offered by Food for Work programs, are essential but need complementary measures if the longer term development impacts of agricultural modernization are to be positive overall for both men and women.

III. DIAGNOSIS, MEASUREMENT, AND VALUATION

It is not reasonable to expect agricultural research and extension to provide all the solutions to livelihood maintenance and development. It is reasonable to expect diagnostic procedures, more adequately than at present, to identify at an early stage of technology development both the interventions agricultural research can make and those areas in which other development institutions need to make complementary efforts.

This in turn requires greater conceptual sophistication on the part of agricultural technology developers, as noted in Section II. It also requires greater sophistication in the use of diagnostic procedures that capture the *interdependent* relations of smallholder production systems. The following examples illustrate these intrahousehold and domestic domain - public domain interactions.

A number of studies show that choice of crop mix, plant characteristics, domestic processing, preservation, and food preparation technology can be linked to the seasonal availability of different types of fuel for cooking and space heating and the different fuels available to women in different income strata through the seasons.[28] The switch to short-straw varieties and commercial rice milling and the introduction of shorter stem jute varieties in Bangladesh, for example, has reduced the availability as fuel of straw and chaff to poor households. Food intake and processing of foodstuffs for sale is limited not merely by deficient food availability but also by deficient biomass or alternative fuels.

Vidyarthi records data from one Indian village on the use of dung and firewood by women in bullock-owning households and the increasing reliance by poorer women on crop residues for fuel.[29] The poorer women use spiky millet stems for fuel through the end of the *Kharif* season in November, pigeon pea stems through the end of *Rabi* in April (which gives the best sustained heat of all residues), and then a weed, *Ipomea fitulosa*, which gives smoky heat and must be gathered, cut, and dried for a month before use, and gathered wood. He estimates that agricultural residues may form around 40% of all fuels used by poorer women.

The point is not that pigeon pea or other residues are desirable fuels per se but that introduction of new varieties which might displace fuel utilities has to be accompanied by action to make alternative fuels accessible to poor women. Research decisions, such as that made by the International Center for Research in the Semi-Arid Tropics (ICRISAT) not to record backyard pigeon pea production in its long time series village studies, and its selection of short stem varieties with high grain yield for breeding trials, are based on assumptions blind to multiple biomass use.[22]

Huss-Ashmore has shown how the type of fuel used in Mokhotlong in highland Lesotho, and the time spent getting it, vary according to the seasonal availability of dung.[30] Slow-cooking protein sources are not used up equally through the year but are depleted during the cold season when slow-burning compacted dung is available. "During the summer the population relies heavily on wild vegetable protein sources, which require more time to locate and gather but which can be rapidly cooked, using the horse and cattle dung picked up on high pastures and kindled with quick-burning resinous and woody shrubs."[30] It is fuel seasonalities and not crop availability which determine which foods are eaten.

There is increasing interest in the development and field-testing of methodologies which can pick up such interdependencies, cost effectively, early on in the determination of research agendas. Feldstein et al.[31] have suggested a three-fold diagnostic focus, using checklists:

A. ACTIVITIES: who does what (by time, site, crop(stock) variety), for both production and post-harvest operations
B. ACCESS TO AND CONTROL OF RESOURCES: by household member with respect both to domestic domain and public domain resources
C. ACCESS TO AND CONTROL OF BENEFITS: stakeholder analysis with respect to end uses and incentives

Other methodologies which have proven useful and cost effective for addressing such issues include rigorous screening for gender bias and for potentially negative impacts on women, at each stage of research programming,[31] time-allocation studies,[32] and peer group workshops.[33,34] There remain, however, some problematic methodological issues of measurement and valuation of labor and time inputs and output.[35] Values for female labor, time, and energy expenditures cannot with any accuracy be assumed to be a standard fraction of or equivalent to male values. Energy coefficients over time, peak labor stress, multiple work performance, and site- as well as crop-specific patterns vary systematically by gender and it is important that adequate empirical data be collected before "rules of thumb" are drawn.

The volume and value of official market trade are usually one of the elements considered in prioritization between crops in research programming. Better instruments therefore are needed for recording the volume and value of products which are consumed in the household and traded outside official channels. They are also needed for assigning weights to the relative contribution of outputs to household survival and (particularly low income) consumption. A low value crop which is nonetheless a main food source for a critical part of the year or which, for example, is an essential snack food at a time when labor or fuel constraints restrict cooking, has in terms of livelihood analysis a higher value than crude official market analysis might suggest. Moreover, in areas where women's domestic "agribusiness" is important, the value added to commodity production through post-harvest transformation needs to be considered.

IV. ADAPTATION AND INNOVATION

The pro-innovation bias in diffusion studies has overshadowed analysis of the rationality of non-innovation. More recent work indicates that not only must producers want to, know how to, and have the resources to innovate: the technology offered must be appropriate and profitable.[36] The notion of "appropriateness" implies that there is a *context* to innovation which is bound in time and space. Fitting the often long time horizons of plant breeding work to the contextual dynamic of smallholder production is the subject of much recent procedural and methodological experimentation, notably in Farming Systems Research and Extension (FSR/E).

But it is curious how strongly the implicit normative predispositions outlined in Section II. above continue to support the neglect of women's roles in the contextual dynamic. The non-adoption of dent-type maize in Malawi illustrates the irrationality and the waste of research and extension effort which result.

Both men and women in Malawi grow maize, but their objectives are somewhat different. Access to labor, credit, and land resources is also different. The nature and extent of women's participation in maize growing in Malawi varies according to both class and community variables as well as agro-climatic characteristics. Overall the position is as follows:

- The percentage of *de jure* female heads of household is high and rising; FHH now form over 18% of all rural households; in the majority of these, women are full-time farmers on their own account.
- Part-time off-farm work for male family members is common and wives then become wholly responsible for cropping 'family fields' and looking after their husbands' plots during their spouses' absence.
- In most though not all parts of Malawi, women are wholly responsible for supplying household food needs, which means they farm both for direct consumption and to generate cash to buy relishes and other household items.

In sum, women farmers form around 70% of all full-time

TABLE 1
Male and Female Percentage Labor Contribution (1975)

Crop	Activity				
Maize	Plant	Weed	Harvest	Shell	
Men	32	21	27	16	
Women	50	51	60	66	
Sprayed cotton	Plant	Weed	Thin	Harvest	Grade
Men	35	29	39	21	26
Women	41	32	32	36	42
Tobacco	Plant nursery	Weed	Harvest	Cure	
Men	58	44	24	36	
Women	39	39	64	47	
Unsprayed cotton	Plant	Weed	Thin	Harvest	Grade/ Market
Men	70	94	90	27	33
Women	24	6	20	61	58

Note: Percentages do not add to 100 because of labor contributed by children and hired workers.

Source: Clark, B. C., The work done by rural women in Malawi, *East Afr. J. Rural Dev.*, 8 (1 and 2), 89, 1975.

food crop farmers and have substantial interests in non-food cropping. Their participation is reflected in percentage labor contribution. Aggregated data from 1975 for all farm sub-enterprises on farms in which both men and women cultivate are shown in Table 1.

Notwithstanding the fact that women's participation on this scale has been visible in official statistics, as well as in academic studies and project reports, for at least the last 15 years, service provision does not function equally or 'neutrally' for male and female farmers. In 1985, there were approximately 1680 Technical Assistants whose main responsibility was farm improvement, with 238 TAs responsible mainly for credit and the farmers' clubs, and 188 Farm and Home Assistants (all the HAs were women).[37] Within the Ministry of Agriculture's Women's Programme, in total there were around 3 professional female staff at the Department of Agriculture and 254 assigned to the Agricultural Development Districts (ADD). Although some male TAs and FAs do work with women farmers, in the majority of cases they work with the larger-scale male farmers growing hybrid maize and a range of non-food crops. Female heads of household are the least likely to receive any kind of extension contact.[38]

Access to credit is channelled through membership of farmers' clubs. Membership is open to both men and women and both husband and wife can be members of the same club. Neither the percentage of women farmers who are club members nor the percentage of members who are female has been documented but only about 4% of all farmers get credit through the clubs in any case. Of those who do receive credit, microstudies suggest that women form between a fourth and a fifth of all members taking credit. The Women in Development Agricultural Production Programme/United States Agency for International Development (WIDAPP/USAID) project found that in 1981/82, 23% of all farmers receiving credit in the project area were women. Around a fifth to a quarter of these received CAN fertilizer packs, 19% received improved seeds for Faya rice and 15% received S/A fertilizer packs. Women received only about 6 to 7% of the credit in the form of hybrid maize seeds, cotton insecticide packs, and 20:20 fertilizer.[38]

The persons chiefly responsible for guiding the clubs are the 238 male (credit) Technical Assistants. Female Home Assistants do encourage women to join but they are not directly responsible for club activities. The criteria for joining are not clearly spelled out but in practice, many TAs take adoption of hybrid maize as one of the main eligibility criteria.

Within the generally low levels of attendance at farmer training courses, women farmers' participation, particularly of female heads of household, is extremely low, as Table 2, based on 1980—81 data, indicates.

Furthermore, differences in resource endowments, access, and production objectives lead to somewhat different maize technology requirements between men and women. Highly productive maize hybrids, mostly of the dent type, have been available in Malawi for many years. Presently available hybrids have a yield potential on farmers' fields of 3.5 to 4.5 tons/ha but by 1980—81, only 3.4% of maize plots were planted to hybrids. Three composites were available but in restricted supply. Their on-farm yield potential is around 2.5 tons/ha; local varieties average around 1.2 tons/ha.

It is clear that restricted access to credit and fertilizer continues to block the further spread of hybrids. It is also

TABLE 2
Male and Female Attendance at Farmer Training Courses (1980—81)

	Farmer attendance at a training course (%)			
ADD[a]	Male heads	Wives	Female heads	Male heads attending a residential center (%)
Karonga	34	11	6	11.9
Mzuzu	30	15	19	15.5
Kasungu	36	16	12	8.3
Salima	3	13	13	6.1
Lilongwe	25	15	6	3.3
Liwonde	10	5	6	2.7
Blantyre	12	5	8	3.3
Ngabu	42	19	16	5.0
Malawi	23	11	8	5.2

[a] ADD: Agricultural Development District.

Source: National Statistical Office 1984; National Sample Survey of Agriculture 1980—1981, Zombe.

clear from sample survey data that women have a strong preference for the local flint-type varieties because of their superior domestic processing qualities, better on-farm storage qualities, and because their very white flour commands a higher price on local markets. Moreover, local varieties in variable and uncertain rainfall areas give a more certain yield than the hybrid dent-types.[39]

One flint-type hybrid, widely grown in South Africa, was introduced in 1984 by the national seed company (as NSCM41). Since it had not been approved for release as a Malawi hybrid, the parastatal agricultural marketing organization (ADMARC) did not stock it. Hybrid varieties are produced for sale to ADMARC and are regarded by women essentially as a cash crop. They therefore compete on relative price and returns to female labor with other cash crops, rather than on yield with local maize. The relationship is very clearly indicated by the sharp decline in groundnut production (a crop grown mainly by women) in areas in which a World Bank project encouraged the production of hybrid maize for export.[40]

Research-extension workshops have been introduced in the last few years to improve feedback from farmers to researchers on these kinds of issues but they will not serve as an effective channel for female farmers' evaluations of, and preferences for, crop characteristics as long as the main extension service fails to reach women farmers. The new Adaptive Research program will need to make special efforts to ensure that female farmers are included in adaptive research procedures.

Some experience as to how this might be done in Malawi has been provided by two special women's agricultural programs, one conducted under the Phalombe Rural Development Project[41] and the other as a national program within the Ministry of Agriculture.[42,43] Both were successful, partly through collecting and making visible gender-sensitive production data. Other measures contributing to success were advocacy at policy levels, targeted in-service training, and careful adaptation of incentives for changed work practices. All proved helpful in encouraging the service system and adaptive research cadres to pay greater attention to women farmers and respond to their needs. However, without sustained support and continued funding for these kinds of special initiatives, it is doubtful if agricultural development bureaucracies and personnel will *of themselves* adopt a more rational approach to agricultural realities.

V. CONCLUSIONS

The record of the last 15 or 20 years is mixed. On the one hand, deep-seated normative and methodological problems continue to block adequate data collection, conceptualization, and model building in agricultural research and extension. On the other, there is increasing understanding that gender issues are a legitimate and important area of concern for agricultural research and extension staff.

The need for conceptual and methodological revision is the more urgent because of the speed at which the forces of change are modifying family structures and undermining customary relations between men and women in rural communities.[44-46] In many areas, the interplay of structural change in the organization of national economies and the process of capital-intensive agricultural modernization, is forcing rural women to take increasing responsibility for family food security, both through direct production and agro-based income generation. At the same time, the environmental conditions in which women take on new roles are often under severe stress, as more of the best land is reserved for export cropping, fuelwood supplies diminish, and populations increase.

At the practical level, experience is also mixed. Few research and extension services have translated awareness into action. By far the majority of women are not involved in the technology development process and are not reached by production and marketing services. On the other hand, there is an expanding body of experience concerning how to develop understanding and competence so that the issues may be addressed early on in the process of technology development and innovation.

REFERENCES

1. **Acharaya, M. and Bennett, L.,** Women and the Subsistence Sector, Economic Participation and Household Decision-making in Nepal, Staff Working Paper No. 256, World Bank, Washington, D.C., 1982.
2. **Burfisher, M.E. and Horenstein, N.,** Sex Roles in the Nigerian Tiv Farm Household, a Framework for Analysing Impacts of Women and Men, Economic Research Service, USDA, Washington, D.C., 1985.
3. **Gupta, A.K.,** Survival under stress — A note on the case study method, BARI, Joydepur, 1986.
4. **Jones, C.,** Women's Labour Allocations and Irrigated Rice Production in North Cameroon, presented at Int. Assoc. of Agricultural Economics Meeting, Jakarta, August 30, 1982.
5. **Palmer, I., Subhadira, S., and Grisanaputi, W.,** The Northeast Rainfed Agricultural Development Project in Thailand: A Baseline Survey of Women's Roles and Household Resource Allocation for a Farming Systems Approach, Study No. 3, Case Studies of the Impact of Large-scale Development Projects on Women, A Series for Planners, The Population Council, New York, 1983.
6. **Waters-Bayer, A.,** Women as Decision-makers in Dairying, presented at Workshop on Women in West Africa, Ibadan, May 7 to 9, 1984.
7. **Deere, C.D. and Leon de Leal, M.,** Women in Andean Agriculture, Women, Work and Development Series No. 4, ILO, Geneva, 1982.
8. **Jiggins, J.,** Problems of Understanding and Communication at the Interface of Knowledge Systems, presented at Conference on Gender Issues in Farming Systems Research and Extension, University of Florida, Gainesville, Feb. 26 to March 1, 1986.
9. **Stoler, A.,** Garden use and household economy and future development, *Culture Agric.*, Bulletin of the Agricultural Study Group on Agrarian Systems 5, 1, 1978.
10. **Brierley, J. S.,** Kitchen Gardens in the West Indies with a contemporary study from Grenada, *J. Trop. Geogr.*, 43, 30, 1976.
11. **Callear, D.,** The Social and Cultural Factors involved in Production by Small Farmers in Wedza Communal Area, Zimbabwe, of Maize and its Marketing, Report RRD 17, UNESCO, Paris, 1982.

12. **Colfer, C. J., Pierce, Evensen, C., Evensen, S., Agus, F., Gill, D., Wade, A., and Chapman, B.**, Transmigrants' Gardens: A Neglected Research Opportunity, in Proc. Centre for Soil Research Technical Meetings, Bogor, 1985.
13. **van Eijnatten, C. L., M.**, Home gardens: principles and experiences, presented at Seminar on Agricultural Research in West Africa, University of Ibadan, 1971.
14. **Ninez, V. K.**, Household Gardens in Peru: ecological, nutritional and economic perspectives on indigenous small-scale food production, International Potato Center (CIP), Lima, 1984.
15. **Watson, G. A.**, Women's roles in the improvement of rice farming systems in coastal swamp lands, in *Women in Rice Farming Systems, Proc. Conf. on Women in Rice Farming Systems*, International Rice Research Institute (IRRI), Gower, 1985, chap. 11.
16. **Fortmann, L. and Rocheleau, D.**, Women and agroforestry: four myths and three case studies, *Agrofor. Syst.*, 2, 253, 1985.
17. **Gay, J.**, Liberia, Lesotho and Tanzania: students of survival, Reports, *RRDC Bull.*, Reading, 1982.
18. **Owusu-Bempah, K.**, The Role of Women Farmers in Choosing Tree Species for Agroforestry Systems in Ghana, presented at Conf. on Gender Issues in Farming Systems Research and Extension, University of Florida, Gainesville, Feb. 26 to March 1, 1986.
19. **Hoskins, M.**, Rural Women, Forest Products and Forestry Outputs, Community Forestry Office, Food and Agricultural Organization (FAO), Rome, 1986.
20. **Chambers, R. and Jiggins, J.**, Agricultural Research for Resource-Poor Farmers, Part I, Transfer-of-Technology and Farming Systems Research, *Agric. Admin. Exten.*, 27, 1, 35, 1987.
21. **Farrington, J. and Martin, A.**, Farmer Participatory Research: a review of concepts and practices, Discussion paper 19, Agricultural Administration Network, ODI, London, June, 1987.
22. **Jiggins, J.**, Gender-Related Impacts and the Work of the International Agricultural Research Centres, Study Paper No. 17, CGIAR, Washington, D.C., 1986.
23. **Ay, P. and Nweke, F.**, Women in Farm Production and Research, Discussion Material, presented at Workshop on Women in Agriculture in West Africa, International Livestock Center for Africa (ILCA) at International Institute for Tropical Agriculture (IITA), Ibadan, May 7 to 9, 1984.
24. **Ay, P., Odediran, O., and Ogunsakan, L. A.**, IITA Cassava now part of Local Farming System, draft report, International Institute for Tropical Agriculture (IITA), Ibadan, November, 1983.
25. **Lundborg, G.**, Meet the African farmer — and her husband, *Afr. Farm.*, March/April, 43, 1987.
26. **Begum, S.**, Women and technology: rice processing in Bangladesh, in *Women in Rice Farming Systems, Proc. Conf. on Women in Rice Farming Systems*, International Rice Research Institute (IRRI), Gower, 1985, chap. 13.
27. **Jiggins, J.**, Women in Bangladesh: an economic strategy paper, prepared for Office of Women in Development, World Bank, Washington, D.C., January, 1987.
28. **Foley, G., Moss, P., and Timberlake, L.**, *Stoves and Trees*, International Institute for Environment and Development (IIED), London and Washington, 1984.
29. **Vidyarthi, V.**, Energy and the poor in an Indian village, *World Dev.*, 12, 8, 821, 1984.
30. **Huss-Ashmore, R.**, Seasonality in Rural Highland Lesotho: Method and Policy, in a Report on the Regional Workshop on Seasonal Variations in the Provisioning, Nutrition and Health of Rural Families, African Medical and Research Foundation (AMREF), Nairobi, 1982.
31. **Feldstein, H., Poats, S. V., and Rocheleau, D. E.**, Intrahousehold Analysis and On-Farm Research and Experimentation, presented at International Centre for Maize and Wheat (CIMMYT) Network-shop on Intrahousehold Dynamics and Farming Systems, Lusaka, April 25 to 30, 1987.
32. **Colfer, C. J., Pierce, Yost, R., Kasmini, V., and Kurdiana, A.**, Time Allocation Studies: A Methodology for Food Production Systems, keynote address presented at World Food Day Symposium on Food Production Systems, Cornell University, Ithaca, Oct. 16, 1984.
33. **Food and Agriculture Organization**, *Learning from Rural Women*, Agriculture Services Division, Rome, 1985.
34. **Hahn, N.**, personal communication, 1986.
35. **Jiggins, J.**, Research and methodology notes: collection of labour data for OFR/FSP, *Farm. Syst. Newsl.*, Nairobi, 25, April—June, 1986.
36. **Röling, N. G.**, *Extension Science: Information Systems in Agricultural Development*, Cambridge University Press, Cambridge, 1988.
37. **Ian MacDonald and Associates Ltd.**, Manpower Development and Training Needs Study, U.K., 1985.
38. **Spring, A., Smith, C., and Kayuni, F.**, Women Farmers in Malawi, their Contribution to Agriculture and Participation in Development Projects, Women in Development Office, United States Agency for International Development (USAID), Washington, D.C. and Ministry of Agriculture, Lilongwe, 1983.
39. **Hansen, A., Mwango, E. N., and Phiri, B. S. C.**, Farming Systems Research in Phalombe Project, Malawi, Another Approach to Smallholder Development, Center for Tropical Agriculture, University of Florida and Farming Systems Analysis Section, Dept. of Agricultural Research, Ministry of Agriculture, Lilongwe, 1983.
40. **United Nations**, World Survey...1985, Paper A/Conf. 116/4, presented at World Conf. Women: Equality, Development and Peace, to Review and Appraise the Achievements of the United Nations Decade for Women, Nairobi, July 1985.
41. **Evans, J.**, Malawi: Reaching Women Farmers, Reports, *RRDC Bull.*, Reading, September, 1986.
42. **Spring, A.**, The Women in Agricultural Development Project in Malawi: Making gender free development work, in *Women Creating Wealth: Transforming Economic Development*, Gallin, R. and Spring, A., Eds., Association of Women in Development, Washington, D.C., 1985, 71.
43. **Spring, A.**, Using Male Research and Extension Personnel to Target Women Farmers, presented to Conf. on Gender Issues in Farming Systems Research and Extension, University of Florida, Gainesville, February 26 to March 1, 1986.
44. **Leon, M., Ed.**, *Las Trabajadores del Agro*, ACEP, Bogata, 1982.
45. **Hafkin, N. and Bay, E. G., Eds.**, *Women in Africa, Studies in Social and Economic Change*, Stanford University Press, Stanford, CA, 1976.
46. **Jain, D.**, *Women's Quest for Power*, Vikas Publishing House Pvt. Ltd., Ghaziabad, Uttar Pradesh, 1980.

7 THE ECOLOGICAL RATIONALITY OF PEASANT PRODUCTION

Victor M. Toledo

I. INTRODUCTION

In the last decade there has been a renewed interest in the ecological knowledge and strategies of traditional rural cultures.[1,2] Using information from a growing number of case studies, various authors have stressed the importance of traditional cultures in the design of alternative, ecologically sound, agricultural systems,[3] and for the conservation of both biotic and genetic resources.[4-8] In addition, traditional ecological knowledge and strategies have been proposed as the basis for new styles of rural development in such different areas of the world as West Africa,[9] Southeast Asia,[10] México,[11] and the Amazonian region.[12]

This research has been noteworthy for one reason. Although the investigators represent different disciplines and their works are essentially descriptive accounts of various ecological aspects of traditional cultures, the investigations are explicitly or implicitly based on one central assumption: *In contrast to most modern systems of rural production, traditional cultures tend to implement and manage ecologically correct systems for appropriation of natural resources.* This assumption, which can be considered potentially as a new scientific paradigm, encloses a second thesis: A certain *ecological rationality* to traditional production exists, although it has not yet been carefully analyzed.

This paper explores this ecological rationality, which is inherent to traditional production. The paper also focuses on peasant phenomena, because the somewhat vague terms "traditional" and "indigenous" cultures (which are commonly used by scholars to denote non-literate peoples inhabiting the rural areas of developing countries) encompass a complex array of human societies. While tribal groups (including nomadic and agricultural peoples) generally inhabit villages geographically and economically isolated from the rest of their national societies, peasants live in communities connected to regional, national, and even international markets. Peasants represent the majority of the population of so-called traditional cultures, and their proportion steadily increases as the social tendencies of the world transform tribal societies into peasant groups. Nevertheless, this discussion is also applicable to tribal societies.

II. CHARACTERISTICS OF PEASANT PRODUCTION

Beyond the theoretical and political discussion of whether or not peasant production constitutes a specific *mode of production* or simply a *class* or a fraction of a class within different modes, there is a set of characteristics that define the peasant economy. (1) A main feature of peasant production is its relatively high self-sufficiency. Peasant families (the peasant unit of production) consume a substantial portion of their own production and, concomitantly, produce almost all the goods they need. To use Marx' terminology, in peasant production there is a relative predominance of *use values* (goods consumed by the unit of production) over *exchange values* (goods not self-consumed but circulated as commodities outside the unit of production). (2) Peasants are engaged in a production process predominantly based on family labor with minimal external inputs. Human and animal power, rather than fossil fuels, are the main sources of energy. The household consequently functions, simultaneously, as a unit of production, consumption, and reproduction. (3) The combined production of use values and commodities is not for profit-making but for simple reproduction of peasant households. (4) Peasants, generally, are small-scale land owners, due to technological reasons, and often due to scarcity and/or unequal distribution of land. (5) Although agriculture tends to be the main activity of peasant households, *peasant subsistence is based on a combination of practices,* which include plant gathering, domestic cattle raising, handicrafts, fishing, hunting, and off-farm labor on a part-time, seasonal, or intermittent basis.

III. AN ECOLOGICAL-ECONOMIC APPROACH TO RURAL PRODUCTION

How peasants produce goods is a subject of considerable interest to economists and other social scientists, particularly after the rediscovery of Chayanov's seminal works.[13] These studies, however, examine the phenomena separately from its environmental context.[14,15] In this narrow view the natural factors are simply eliminated from analysis or are taken as constants, usually called raw materials,[16] such that peasant production becomes a process realized in an ecological vacuum, an obvious consequence of the compartmentalized fashion of modern science. In this context the main components of the phenomena are separated and dissected to be analyzed as the subjects of specific disciplines. By contrast, an interdisciplinary ecological-economic approach can resolve certain fundamental contradictions and obscure aspects of the peasant productive process, inherent to Chayanov's rule, the "ecological reductionism" of the energetic analysis by anthropologists, and the actual views on the potential role of the peasantry.[17]

Before constructing a coherent model of peasant production, it is necessary to recognize that the peasant economy is, in the last analysis, a particular form of rural or agrarian production, and that in these areas producers utilize natural resources as the basic and irreplaceable means of production. Essentially, there are two specific sets to be analyzed and integrated. Since rural inhabitants are *primary producers* facing both natural and social forces, they are economic actors within an ecological and economic context. Therefore, any analysis of rural and peasant production must include both ecological and economic variables that affect this process. Peasant production always implies the combination of use values and exchange values; it is the result of both natural processes and market forces acting upon the peasant as producer and consumer. This two-sided nature of rural production must then be considered during the construction of an appropriate conceptual framework.

Production is both a theoretical and a practical category. Therefore, rural production can be empirically reduced to flows of materials, energy, labor, commodities, and information.[18] The key to understanding and explaining the productive process of rural societies, then, is to describe the ways in which these flows are arranged and integrated into the concrete reality where they take place. All this implies, finally, the *"spatialization"* of the phenomena. By this I mean that in attempting an operational, ecologically oriented approach to rural production, we need to elaborate a *topology* of the productive process, as suggested by Godelier.[19]

Human society, as an artificial, humanized "second nature" that was constructed upon the foundation of nature,[20] can be visualized as an island in the waters of a natural sea (Figure 1). Any *rural unit of production* (P) is a cell on the periphery of the island, carrying out two basic types of material exchanges: with nature (N), and with other sectors of the social organism (S). In the first case P interchanges

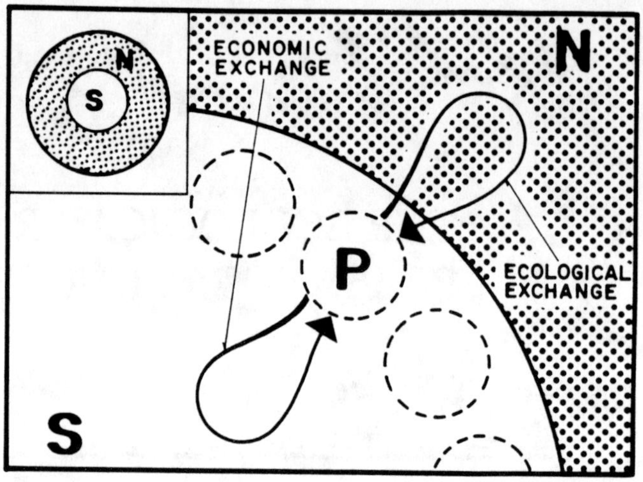

FIGURE 1. The double material exchange between a rural unit of production (P), nature (N), and society (S).

materials outside the social organism, while in the latter case P exchanges materials with the inner portions of the island of the social organism. During the interchange of materials with Nature, P makes its particular contribution to the general metabolism that exists between Nature and Society, an eternal, natural, and pre-social condition.[20] On the other hand, when P carries out interchanges of materials with other sectors of the society, it functions in an historical process, subject to the specific social and historical conditions under which it is performed. Therefore, the rural productive process can be analyzed in terms of an ecological exchange and an economic exchange.[21,22]

We must, however, make another distinction in the case of the ecological exchange. The appropriation of nature is, basically, an appropriation of ecosystems, which are the basic units of nature. During production, P deliberately channels materials and/or energy resources out of the ecosystem, and towards the social organism. We can distinguish two principal levels of human intervention in the ecosystems. (1) At the first level, natural resources are obtained and transformed without provoking substantial changes in the structure, dynamics and architecture of natural ecosystems. The first level includes most known examples of hunting, gathering, fishing, forest-product extraction, and certain types of cattle breeding or pastorialism. (2) On the second level, the natural ecosystems are partially or completely disarticulated and partially or totally replaced by assemblages of plant and animal species in the process of domestication. Examples of the second level are plant agriculture, forest plantations, cattle ranching, and aquaculture. As several scholars have stressed,[23] the main difference between natural ecosystems and human-manipulated ecosystems is that the former have the capacity for *self-maintenance, self-repair, and self-reproduction*. Transformed ecosystems, however, are intrinsically unstable systems, which necessarily require external energy (be it human, animal, or fossil fuels) for self-maintenance.

Based on the previous discussion, we can conclude that P exchanges materials in a three-dimensional form. These exchanges are realized by P with concrete entities, which have particular places in space. P performs in three arenas: (1) *the natural environment* (NET), which includes the set of natural ecosystems and their successional stages existing in P's territory; (2) *the transformed environment* (TET), represented by the set of artificial ecosystems or agro-ecosystems; (3) *the social environment* (SET), which is defined as the social space where P realizes its economic exchange (Figure 2). While NET and SET can be situated relatively easily in concrete natural space, SET is harder to locate in concrete terms. NET and TET can be defined by delimiting discontinuities in the natural landscape, using vegetational, topographical, and pedological criteria. SET, on the other hand, can be defined only by tracing the relationships between P and other local units of production, and regional, national, and international markets.

Having defined P, SET, NET, and TET, it is necessary to describe the different relationships that may exist among them. As a point of departure, we can use the key concept of *labor power*,[24] which is defined as the material and intellectual means used by the members of P to extract their means of existence from nature. This requires human effort (F^o), the force, strength, and energy necessary to overcome the resistance to transformation inherent in any ecosystem. Since P in reality acts upon two different types of ecosystems (NET and TET), the labor power is channelled in two ways, towards the NET (F^{oa}) and towards the TET (F^{ob}).

As a result, P obtains two flows of materials from nature. The first comes from the natural ecosystems (F^1), and the second from the transformed ecosystems (F^2). Materials from these two flows can be utilized by P in two ways, self-consumption (F^{1a} and F^{2a}), and in exchange with SET (F^{1b} and F^{2b}). In the first case, P generates materials that are retained as use values for domestic consumption. In the second case, P produces goods (generally as raw materials) that are circulated as commodities. Finally, P consumes goods provided by SET (F^3), and generates transformed materials in a small-scale, artisan-like style, including crafts, tools, textiles, etc. (F^4) (Figure 3).

IV. PEASANT PRODUCTION AS A SUBSISTENCE ECONOMY

The conceptual framework of the previous section identifies in space and time the key variables and main processes to be measured and analyzed in order to accurately characterize rural production, and recognizes the dual (ecological and economic) character of the process. However, the model is only a synchronic and historically abstract representation of the rural production process. Animating the model allows us to observe the various specific forms that rural production may take, revealing at the same time the historical character of the process. There is a continuous spectrum of combinations between two extreme types of social organization, *production for use* and *production for exchange*. Each of the potential combinations to be found in the model corresponds to and arises from historically determined specific forms of society. For instance, in the simplest forms of production for use, the process is reduced to an ecological exchange between P and NET (Figure 3a) as exemplified by hunting and gathering societies, or between P, NET, and TET, in the case of sedentary agricultural societies (Figure 3b). In both cases, P acts as a "species" within the ecosystem, and the rural productive process of these "natural economies" is basically an ecological process. By contrast, in a completely commodity-oriented case, the rural productive process is simplified by the abolition of use value flows (F^{1a} and F^{2a}). In this case, P becomes a specialized entity, the rural productive process is completely integrated to market networks, and the ecological exchanges are subordinated to economic dynamics, as in the example of commercial monoculture agriculture (Figure 3d).

In peasant production, the implicit objective of the productive process is the simple reproduction of the peasant household and, consequently, of the entire peasant community. In addition, however, peasants also produce goods that are externally circulated as commodities. In fact, in the peasant economy, ". . . production for livelihood, does not exclude producing a surplus; the enigma of production for use is not whether a surplus is produced but why the surplus which is produced does not accumulate and transform the system."[25] Whatever the causes provoking this situation, *the exchange sphere of peasant production remains subordinated to the aim of self sufficiency,* and this subsistence economy is fundamentally dependent on the exploitation of natural resources. In summary, despite the fact that peasants carry out both ecological exchanges as well as economic exchanges, the maintenance and reproduction of the producer and his family is based more on products coming from Nature (obtained from NET and TET) than on goods obtained from markets (SET) (Figure 3c). In the last analysis, *peasant production is a subsistence economy.*

V. WHY PEASANTS ADOPT A MULTI-USE STRATEGY

Since peasant production is based more on ecological exchanges than on economic exchanges, peasants are obliged to adopt survival mechanisms that guarantee an uninterrupted flow of goods, materials, and energy from the natural and transformed environments (NET and TET). They are then forced to carry out a non-specialized production based on the principle of diversity of resources and productive practices. This results in the utilization of more than one ecogeographical unit, the integration and combination of different practices, the recycling of materials, energy, water and wastes, and the diversification of the products obtained from ecosystems. Such a strategy operates at the household level as well as the community level and even regional level. This pattern is carried out through time and space. On the spatial axis, maximum utilization is sought for all the available ecosystems. In terms of time, the goal is to

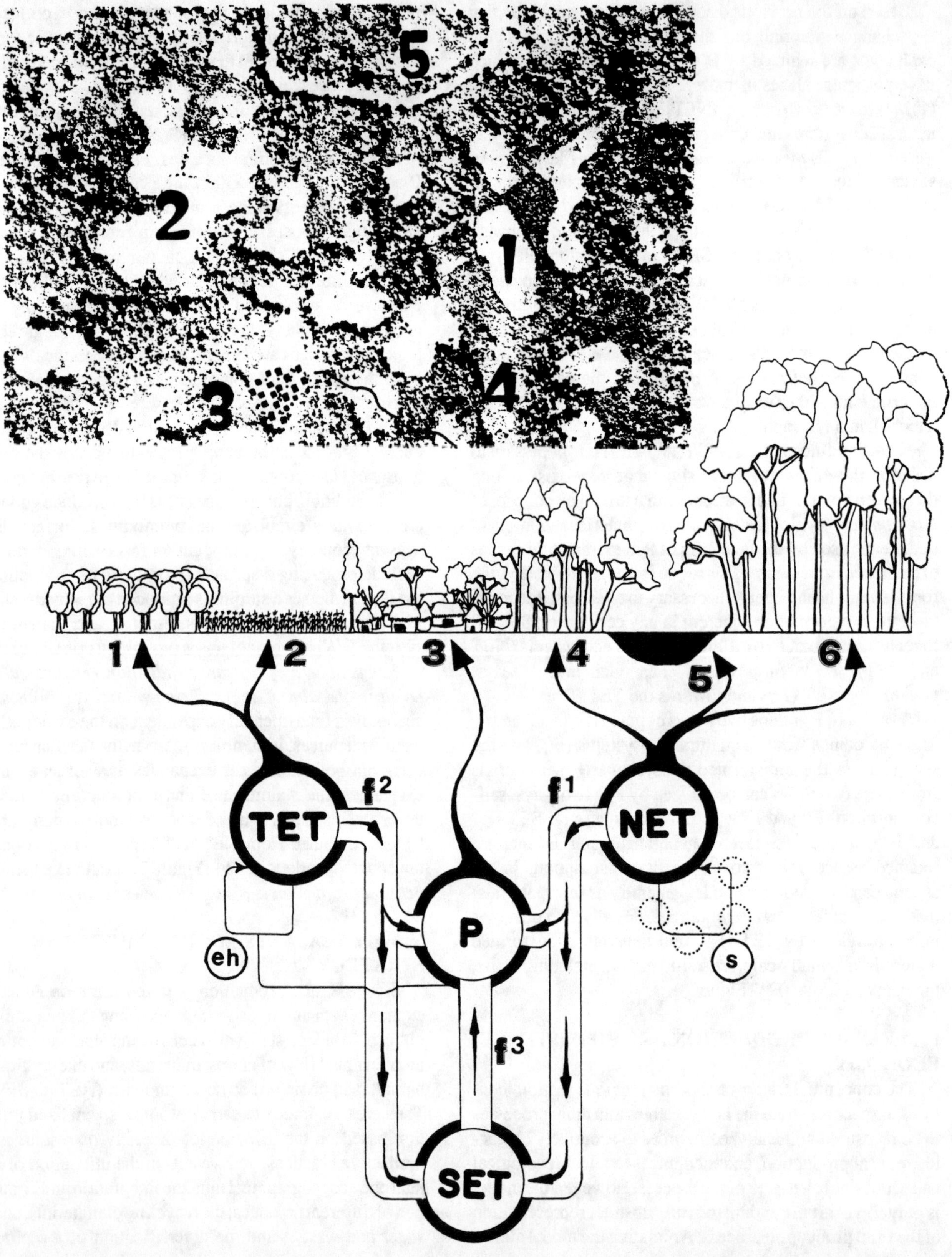

FIGURE 2. An idealized scheme of the material exchanges carried out by a rural unit of production (P) into a concrete (tropical) landscape. 1: Plantation area. 2: Maize fields. 3: Human settlement. 4: Secondary forests. 5: River. 6: Primary forests. NET: Natural environment. TET: Transformed environment. SET: Social environment. EH: Energy derived from human power. S: Ecological succession. F^1 and F^2: Ecological exchanges. F^3: Economic exchanges.

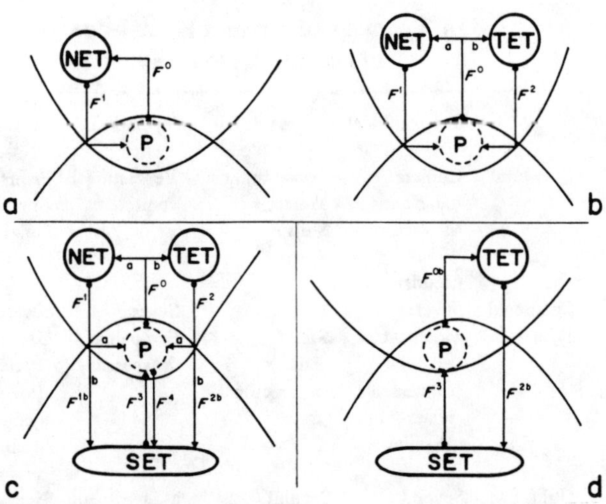

FIGURE 3. Theoretical representation of the ecological-economic exchanges carried out by a rural unit of production (P) in four different social (and historical) instances. a: Hunter and gathering band. b: Tribal agricultural village. c: Peasant community. d: Modern (commercial) farm. NET: Natural environment. TET: Transformed environment. SET: Social environment.

obtain a maximum number of necessary products that each ecosystem offers throughout the year.

A. Biotic and Non-Biotic Components of Ecosystems

Peasant households utilize biotic and non-biotic components of the ecosystems in order to satisfy their basic life requirements. Peasant production implies, then, the generation of a myriad of products, including foods, domestic and work instruments, housing materials, medicines, fuelwoods, fibers, animal forage, and such substances as gums, resins, dyes, drugs, and stimulants. Economic exchanges allow peasants to obtain manufactured and industrial goods and natural goods not produced by themselves from SET. From a theoretical viewpoint it is possible to predict that those peasant groups exploiting ecosystems with more limited resources (for example, non-forested or highly seasonal ecosystems) will be more fragile and vulnerable to economic, technological, and cultural exchanges than those living in resource-rich environments (for example, humid tropical areas or ecotones).

In the context of predominant use-value economic rationality, peasants are obliged to adopt a strategy that maximizes the variety of products produced, in order to provide basic household requirements throughout the year. This is a major feature of peasant units of production. Peasants manipulate the natural landscape in such a way that two environmental characteristics are maintained and favored: *spatial heterogeneity* and *biological diversity*. This multi-use strategy[26,27] allows peasants to manage different geographical units, with different biotic and physical components. Peasants seek to avoid the specialization of their natural spaces and their productive activities, a feature intrinsically in contradiction with the predominant tendencies of the majority of rural modernization projects. All this explains why *peasant producers are generally not solely agriculturalists.* Although plant agriculture tends to be the central productive activity of any peasant household, it is always complemented (and in some cases replaced) by such practices as plant-gathering, forest product extraction, fishing, hunting, cattle-raising, and handicrafts. The combination of all these practices buffers the peasant household against both market fluctuations and environmental changes and hazards.

As a result, on a typical peasant property the natural and transformed environments become a complex landscape mosaic in which agricultural fields, fallow areas, primary and secondary forests, household gardens, cattle-raising areas and water bodies are all segments of the entire production system. This mosaic represents the field upon which the peasant producer, as a multi-use strategist, plays the game of subsistence through the manipulation of ecological and geographical components (species, soils, topography, climate, water and space), and ecological processes (succession, life cycles and movement of materials). The same diversified arrangement tends to be reproduced at a micro-level, with multi-species, multi-story crops favored over monoculture.

In summary, *variety*, in geographical, ecological, biological, and even genetic terms, is the main feature of peasant production, because variety itself is a risk-reduction mechanism. This *multi-use strategy*, through which peasants maintain and reproduce their productive systems, constitutes an ecologically valuable characteristic that tends to conserve natural resources by maintaining environmental and biological diversity. The acclaimed and somewhat enigmatic ecological rationality of the peasant and traditional producer is just a subsistence strategy developed within a non-commodity system of production. It is a direct consequence of the process of appropriation of nature in an economy predominantly directed to production for use.

VI. ETHNOECOLOGY AND PRODUCTION: THE PEASANT COGNITIVE SYSTEM

As any producer, peasants need intellectual means in order to realize a correct appropriation of ecological systems during the process of production. In this context, the set of knowledges that peasant producers put in action to exploit natural resources became conclusive. This knowledge has a substantial value for clarifying the ways in which peasants perceive, conceive, and conceptualize the ecosystems on which they depend for a living. Moreover, in the context of a subsistence economy, this knowledge of nature became a decisive component in the implementation of the peasant survival strategy based on the multiple use of natural resources.

With very few exceptions[12,27-29] the predominant tendency in the studies of peasant knowledge of nature has been based on an approach where: (a) the peasant cognitive phenomenon appears separated from its practical purposes;

in other words, the intricate system formed by *corpus* and *praxis* is artificially segregated, and (b) the cognitive body is only partially studied, so that the researcher only studies "fractions" (plants, animals, soils, etc.) or "dimensions" (classificatory systems, utilitarian elements, and others) of the entire system. So, in general, ethnoscientists have treated the investigation of practical activities as secondary in importance to the investigation of peasant cognitive systems, perpetuating a tendency to regard *culture* as distinct and largely autonomous from *production*. Because of the above, research on resource management by peasants[30] has been generally conceived and carried out without connection to the body of knowledge studied by the ethnoscientists. On the contrary, as has been pointed out by Barahona,[31] it is very difficult to reach a coherent and complete understanding of these cognitive systems by separating them from the concrete and practical daily activities and behaviors of peasant producers.

A. Resource Knowledge

It seems clear that in relation to the concrete and practical problems to be solved in ecosystem management, peasant producers must possess knowledge of resources in at least four dimensions: *geographical* (including macrostructures and events such as climate, winds, mountains, etc.), *physical* (topography, minerals, soils, micro-climate, water, etc.), *vegetational* (the dominant forms of vegetation) and *biological* (plants, animals and fungi). In the same way, based on the extensive ethnoscientific literature it is possible to distinguish four kinds of knowledge: *structural* (which is relative to the natural elements or components), *dynamic* (which refers to processes or phenomena), *relational* (linked to the relationships between or among elements or events) and *utilitarian* (which is circumscribed to the usefulness of the natural resources). As a result of the above, it is possible to integrate a preliminary typology of peasants' knowledge of natural resources (Table 1), which may serve as a conceptual and methodological framework to ethnoecological studies. The problem is, however, how this cognitive body is connected and integrated to the production logic of the peasant systems: the *multiple-use strategy*. Although this is a topic to be solved through future research, Figure 4 shows a hypothetical scheme that integrates the different types of peasant knowledge of Nature in relation to production. This scheme, inspired by the results of numerous ethnoecological studies (see for example References 32—36), is based on the idea of practical "management units," through which peasant producers manipulate natural resources (components and processes). These units are derived from the peasants' recognition of eco-geographical units in the landscapes, which in turn are the result of the peasants' knowledge of vegetation, soils, and topography. In summary, it seems clear that the peasant cognitive system, which is permanently used by the producer during the management of ecosystems, underlies the ecological rationality of the peasant production.

TABLE 1
A Typology of Peasant Knowledge of Natural Resources

	Geographical	Physical	Vegetational	Biological
Structural	Climate Landforms Mountains Winds Clouds	Topography Minerals Soils Water	Vegetational units	Plants Animals Fungi
Relational	Several	Several	Several	Several
Dynamic	Moon cycles, material movements, water-table changes	Soil erosion, microclimatic events	Ecological succession	Life cycles, flowering periods, nesting seasons
Utilitarian	Several	Several	Management units	Several

VII. ECOLOGY, PEASANT PRODUCTION, AND THE MODERNIZATION PROCESS

With very few exceptions, the recent modernization process of the world rural areas has been an ecologically and culturally disruptive event, especially in Third World countries. During modernization both natural resources as well as peasant communities tend to be destroyed and replaced by "modern" forms of production based on spatial, productive and human specialization, and an exclusively market-oriented production. To achieve rural development without destroying natural resources and without transforming peasant households into specialized and wage-labor dependent units, it is necessary to abruptly change the main objectives of rural modernization. The recognition of an ecological rationality of the peasant production alone challenges the central paradigms of rural modernization, forcing a re-evaluation of the meaning and potentiality of peasant cultures normally considered as an archaic or traditional sector without importance.

In addition, peasant self-sufficiency, which is largely based on a permanent *symbiosis* with local natural resources, should constitute the starting point for any ecologically sound alternative development. In this perspective, the surplus-production strategy must be implemented only after a self-sufficiency stage. Local and regional subsistence and not commercial production should be the first productive goal in any rural development policy, especially in those areas characterized by high eco-geographical complexity and high biological and genetic richness. Finally, since the active role played by peasants in the maintenance of ecological and biological resources as well as in genetic conservation has been largely unappreciated, it is urgent to design a new conception where all peasant contributions will be emphasized. Needless to say, this new orientation of rural modernization is obliged to challenge and to overcome the economic and political mechanisms through which peasants' work is permanently exploited by a non-producing

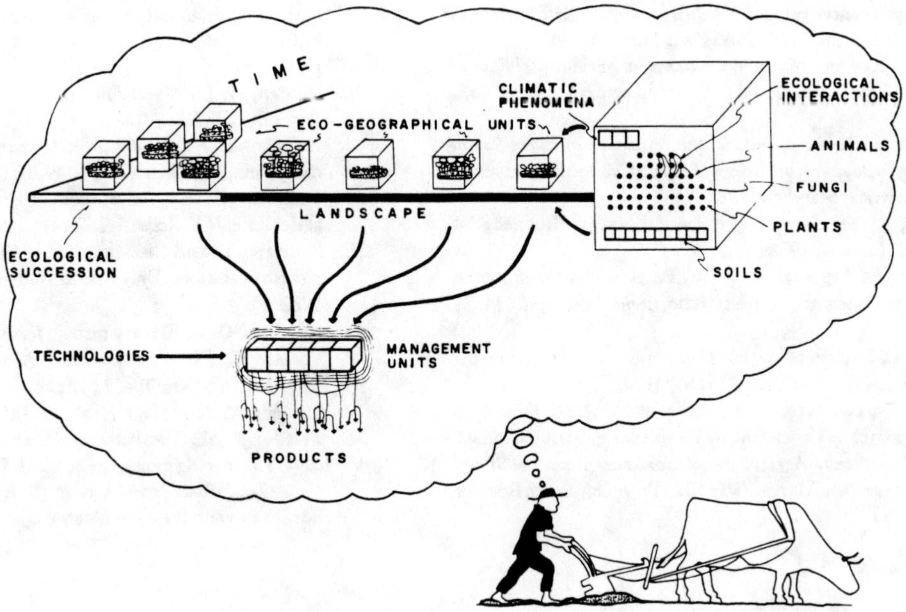

FIGURE 4. An integrative scheme of peasant knowledge of Nature.

sector (through various mechanisms of surplus extraction), through which peasants tend to be converted into a subordinated social sector.

ACKNOWLEDGMENTS

The major portion of this paper was written while the author was teaching as a Visiting Professor at the University of California, Berkeley (College of Natural Resources, Department of Geography) on sabbatical from the National University of México (UNAM). I want to express my cordial thanks to Dr. Miguel Altieri, who made possible my stay in Berkeley and who stimulated the creation of this paper. Thanks are also extended to Phil Daniels (México City) and John Cloud (Berkeley) for many hours of help in the review of the English text. The figures were done by the artistic hand of Felipe Villegas (UNAM).

REFERENCES

1. **Klee, G. A., Ed.,** *World Systems of Traditional Resource Management*, John Wiley & Sons, New York, 1980.
2. **Brokensha, D. W., Warren, D. M., and Werner, O., Eds.,** *Indigenous Knowledge Systems and Development*, University Press of America, Maryland, 1980.
3. **Altieri, M.,** *Agroecology: The Scientific Basis of Alternative Agriculture*, Westview Press, Boulder, CO, 1987.
4. **Brush, S. B.,** Genetic diversity and conservation in traditional farming systems, *J. Ethnobiol.*, 6, 151, 1986.
5. **Altieri, M. and Merrick, L. C.,** *In situ* conservation of crop genetic resources through maintenance of traditional farming systems, *Econ. Bot.*, 41, 86, 1987.
6. **Altieri, M., Merrick, L. C., and Anderson, M. K.,** Peasant agriculture and the conservation of crop and wild plant resources, *Conserv. Biol.*, 1, 49, 1987.
7. **Michon, G. and Bompard, J. M.,** Agroforesteries indonesiennes: contributions paysannes a la conservation des forest naturelles et de leur resources, *Rev. Ecol. (Terre et Vie)*, 42, 3, 1987.
8. **Oldfield, M. L. and Alcorn, J. B.,** Conservation of traditional agroecosystems, *Bioscience*, 37, 199, 1987.
9. **Richards, P.,** *Indigenous Agricultural Revolution*, Westview Press, Boulder, CO, 1985.
10. **Marten, G. G., Ed.,** *Traditional Agriculture in Southeast Asia: A Human Ecology Perspective*, Westview Press, Boulder, CO, 1986.
11. **Toledo, V. M., Carabias, J., Mapes, C., and Toledo, C.,** *Ecologia y Autosuficiencia Alimentaria*, Siglo XXI Eds, México, 1985.
12. **Posey, D. A., Frechione, J., Eddins, J., et al.,** Ethnoecology as applied anthropology in Amazonian development, *Hum. Organ.*, 43, 95, 1984.
13. **Durremberger, P. E., Ed.,** *Chayanov, Peasants and Economic Anthropology*, University of California, Berkeley, 1984.
14. **Perez, L.,** The human ecology of rural areas: an appraisal of a field of study with suggestions for a synthesis, *Rural Sociol.*, 44, 584, 1979.
15. **Dunlap, R. E. and Martin, K. E.,** Bringing environment in the study of agriculture, *Rural Sociol.*, 48, 201, 1983.
16. **Deere, C. D. and De Janvry, A.,** A conceptual framework for the empirical analysis of peasants, *Am. J. Agric. Econ.*, 601, 1979.
17. **Toledo, V. M.,** La sociedad rural, los campesinos y la cuestion ecologica, in *La Sociedad Rural Hoy*, Zepeda, J., Ed., El Colegio de Michoacán, México, in press.
18. **Cook, S.,** Production, ecology and economic anthropology: notes toward an integrated frame of reference, *Soc. Sci. Inform.*, 12, 25, 1973.
19. **Godelier, M.,** Infrastructures, societies and history, *Curr. Anthropol.*, 19, 763, 1978.
20. **Schmidt, A.,** *The Concept of Nature in Marx*, NLB, London, 1971.
21. **Toledo, V. M.,** La ecologia del modo campesino de produccion, *Anthropol. Marxismo*, 3, 35, 1980.

22. **Toledo, V. M.,** Intercambio ecologico e intercambio economico en el proceso productivo primario, in *Biosociologia y Articulacion de las Ciencias*, Leff, E., Ed., UNAM, México, 115, 1981.
23. **Odum, E.,** Properties of agro-ecosystems, in *Agricultural Ecosystems*, Lowrance, R., et al., Eds., Wiley Interscience, New York, 1984.
24. **Cook, S.,** *Zapotec Stoneworkers: the Dynamics of Rural Simple Commodity Production in Modern Mexican Capitalism*, University Press of America, Maryland, 1982.
25. **Gudeman, S.,** *The Demise of a Rural Economy*, Routledge & Kegan Paul, London, 1978.
26. **Toledo, V. M., Argueta, A., Rojas, P., et al.,** Uso multiple del ecosistema; estrategias de ecodesarrollo, *Cienc. Desarrollo*, 11, 33, 1976.
27. **Johnson, A.,** Ethnoecology and planting practices in a swidden agricultural system, *Am. Ethnol.*, 1, 87, 1974.
28. **Brosius, J. P., Lovelace, G. W., and Marten, G. G.,** Ethnoecology: an approach to understanding traditional agricultural knowledge, in *Traditional Agriculture in Southeast Asia: A Human Ecology Perspective*, Marten, G. G., Ed., Westview Press, Boulder, CO, 1986, 187.
29. **Alcorn, J. B.,** Process as resource: agricultural ideology in the humid tropics, in *Indigenous Resource Management in Amazonia*, Balee, R. and Posey, D., Eds., Westview Press, Boulder, CO, in press.
30. **Wilken, G. C.,** *Good Farmers*, University of California Press, Berkeley, 1987.
31. **Barahona, R.,** Conocimiento campesino y sujeto social campesino, *Rev. Mex. Sociol.*, 49, 167, 1987.
32. **Conklin, H. C.,** An ethnoecological approach to shifting agriculture, *N.Y. Acad. Sci. Trans.*, 17, 133, 1954.
33. **Zizumbo, D. and Colunga, P.,** Los Huaves: la apropiacion de los recursos naturales, Universidad Autonoma de Chapingo, México, 1982.
34. **Sanabria, O. L.,** El uso y manejo forestal en la comunidad de Xul, Yucatan, *Etnoflora Yucatanense (INIREB, México)*, 2, 1, 1986.
35. **Altieri, M. and Trujillo, J.,** The agroecology of corn production in Tlaxcala, México, *Hum. Ecol.*, 15, 189, 1988.
36. **Toledo, V. M.,** The Patzcuaro's lesson: nature, production and culture in an indigenous region of México, in *Biological Resources Under Traditional Management*, Oldfield, M. L. and Alcorn, J. B., Eds., Westview Press, Boulder, CO, in press.

8 THE DYNAMICS OF PEASANT AGRICULTURE IN LATIN AMERICA: IMPLICATIONS FOR RURAL DEVELOPMENT AND AGROECOLOGY

Alain de Janvry and Steven Helfand

I. THE PEASANT PUZZLE

Recent estimates of the relative rates of growth of employment in modern vs. peasant agriculture show that the peasant sector has been able to absorb economically active population at a much faster pace than the modern sector. Thus, while between 1960 and 1980, employment in modern agriculture only grew at the average annual rate of 0.76% for Latin America as a whole, employment in the peasant sector grew at the rate of 1.75%.[1] This observation raises the question of whether this differential growth indicates that the sector of family and sub-family farms that constitutes the "peasantry" is, indeed, able to compete successfully with commercial farming and, for that reason, increase in relative numbers. In this case, the economic status of peasants would likely be improving along with their successful competitive ability. An alternative explanation posits that the rising number of peasants indicates accumulation in that sector of a surplus population unable to obtain either sufficient access to land or sufficient access to employment in nonagricultural activities and is consequently unable to escape the status of peasant. In that case, the growing absolute and relative sizes of peasant employment would be accompanied by persistent poverty and would reflect the economic failure of a segment of the population trapped in a low productivity activity. A growing number of peasants would symptomize failure of the development model followed by the Latin American nations to create employment at a sufficiently rapid pace to match population growth and/or to provide access to additional land. Far from increasing in number as a result of successful competition, the peasant sector would do so because it has ceased to compete on the product market and has become a refuge for surplus population and a reservoir of cheap labor. It is this puzzle which we address in this paper.

The answer to this question has important implications for the design of rural development programs directed at the alleviation of rural poverty and for the potential contributions of agroecology to that goal. If peasants can successfully compete, rural poverty can be decreased through the implementation of production-oriented rural development programs. This is the approach that has been at the core of the international community's attack on rural poverty for the last 15 years, particularly under the World Bank's leadership. An agroecology approach to rural development would focus on improving the ability of peasant farming systems to generate higher monetary income for the households and to sustain these gains over time. If, by contrast, peasants are not in a position to compete, principally because their resource base is insufficient, then other approaches to rural development need to be followed. They include land reform to increase access to land, creation of employment opportunities in both commercial agriculture and in nonagricultural activities decentralized to the rural areas, introduction of complementary activities to farming at the level of the household such as handicrafts and trade, and an agroecology approach to raising productivity in subsistence-oriented home plots.

To test the proposition that a growing peasant sector is, in fact, a sign of competitive failure, we proceed in two steps. First, in Section II of this chapter, we analyze the dynamics of the evolution of the number of farms, average farm size, and share of production of different crops by farm size categories in the Latin American countries. This allows us to identify which social groups have been gaining from agricultural growth and which have been losing. In Section III, we turn to a characterization of the sources of income for peasant households. While there are no data over time, this approach enables us to see the extent of reliance on off-farm sources of income for a large majority of the rural population. Data on the evolution of rural poverty over time, however, help confirm the hypothesis that a large segment of the peasantry has the fate of a "cornered" population. By cornered we mean that they are a social sector trapped between a lack of access to land and to sources of productivity gains in agriculture on the one hand and a lack of access to urban activities allowing them to abandon farming on the other hand. This leads us, in Section IV, to contrast the dynamics of several subsets of the agricultural population and to identify different approaches to reduce rural poverty, including the role that agroecology can play in a strategy of rural development.

II. COMPARATIVE DYNAMICS BY FARM SIZE

An analysis of the comparative dynamics of different groups of producers requires creating a typology based on quantifiable indicators that allows us to contrast different forms of organization of production. In particular, we want to contrast the following four types of farms.

1. *Sub-family (SF) Farms* where the resource base is insufficient to absorb productively the totality of the household's labor. This results in the need to seek a variety of sources of income in addition to farming the home plot. They include activities such as handicrafts, trade, wage labor, and migration.
2. *Family (F) Farms* where the household's labor is the main source of work for the farm. While the net amount of annual labor used is close to the total availability of family labor, labor may be traded in and out at particular points in time to match the seasonality of labor needs.
3. *Medium-size Commercial (MC) Farms* which are run as business enterprises, use hired labor in excess of the sale of family labor, and where the return to the different factors of production can generally be accounted for separately.
4. *Large-scale Commercial (LC) Farms* where the contribution of family labor becomes insignificant relative to that of hired workers.

Creating such a typology, based on sources of labor, has been the ambition of decades of structural analysts of the Latin American farm sector, starting with the classic study of CIDA.[2] Systematic data are, however, missing to implement this typology. Based on years of accumulated experience in detailed case studies, a rough correspondence has been established between ranges of farm sizes and the four types of farms described above. Farms of less than 5 hectares (ha) are thus generally categorized as sub-family, between 5 and 20 ha as family, between 20 and 100 ha as medium-size commercial, and above 100 ha as large-scale commercial. In specific countries and regions, these boundaries are redefined. While this typology generally holds true for countries of the Altiplano, in Brazil, family farms are acceptedly in the range of 10 to 50 ha and in Uruguay of 20 to 100 ha. We will use these modifications when dealing with specific countries separately.

TABLE 1[3]
Average Annual Growth Rate of Number of Farms

Country	Years	Farm size (hectares) 0—5	5—20	20—100	100+	All sizes
		Annual growth rate				
Brazil	1950—1970	7.1	4.7	2.5	2.0	4.4
Costa Rica	1950—1973	3.8	1.4	2.2	4.5	2.8
Peru	1961—1972	3.7	7.3	8.1	3.8	4.4
El Salvador	1950—1971	3.4	0.4	0.0	−0.8	2.9
Colombia	1954—1971	1.9	0.7	1.1	0.5	1.5
Panama	1950—1971	1.7	−0.2	3.4	4.9	1.5
Ecuador	1954—1974	1.6	2.6	4.2	2.6	2.1
Honduras	1952—1974	1.6	0.0	0.2	1.1	1.0
Dominican Republic	1950—1971	0.5	0.1	1.0	2.6	0.5
Uruguay	1961—1970	−1.0	−1.3	−2.3	−0.5	−1.3
Venezuela	1961—1971	−2.4	−0.9	1.4	2.3	−1.0
Mexico	1950—1970	−2.5	0.6	−0.1	0.8	−1.5
Number of positive changes		9	8	9	10	9

A. Changes in the Number of Farms

Using agricultural censuses over time we see, in Table 1, that the number of farms has increased rapidly in a majority of countries. We also see that this rise in numbers has been fairly evenly distributed across farm types, with only a slightly higher frequency of rising numbers among the MC and LC farms. This tells us that, as far as the number of farms is concerned, we do not clearly see one type of farm becoming dominant. The very high degree of inequality in the distribution of farms across farm sizes thus tends to be reproduced over time and, if anything, increased in spite of decades of presumed efforts at redistributive land reform.

Relative constancy in the growth of the number of farms across categories hides, however, potentially very different reasons why the numbers increase. To identify these determinants, we need to look at the changes in average farm sizes and at the distribution of production across farm types.

B. Changes in Average Farm Size

Here we see a very different situation than with the number of farms. Average farm size nearly always declines among both the smallest (SF) and the largest (LC) farms while the frequency of changes in the two middle groups of farms is about 50% (Table 2). This indicates that farmers in the two middle groups have been able to effectively protect their access to land on an individual basis even though their numbers have been increasing. Decline in the average farm size in the other two groups responds to markedly different causes.

TABLE 2[3]
Average Annual Growth Rate of Average Farm Size

Country	Years	0—5	5—20	20—100	100+	All sizes
		Farm size (hectares)				
		Annual growth rate				
Mexico	1950—1970	0.3	0.1	0.3	−1.0	1.3
Peru	1961—1972	0.0	0.5	−0.3	−2.0	−1.7
Dominican Republic	1950—1971	−0.1	−0.2	−0.1	−1.4	0.3
Uruguay	1961—1970	−0.4	−0.2	0.3	0.4	1.1
Ecuador	1954—1974	−0.5	−0.1	0.1	−2.6	−0.7
Colombia	1954—1971	−0.7	0.0	0.0	0.0	−0.8
Venezuela	1961—1971	−0.7	0.1	0.4	−2.1	1.2
Honduras	1952—1974	−0.8	0.1	0.3	−1.1	−0.8
Brazil	1950—1970	−0.8	−0.3	−0.1	−1.2	−3.1
El Salvador	1950—1971	−1.4	−0.1	0.2	−0.7	−3.1
Costa Rica	1950—1973	−1.7	−0.1	0.0	−1.8	−0.4
Panama	1950—1971	−2.7	0.4	0.0	−0.7	1.4
Number of positive changes		1	5	6	1	5

For the SF, it reflects their inability to compensate for demographic pressures through either competition on the land market or through outmigration toward other economic activities. Accumulation of surplus population in SF farms thus leads to a decline in land availability per household. For the LC farms, rapidly rising numbers and falling average size reflect a process of restructuring toward smaller size operational units that are managerially more efficient than the old *latifundio* and hacienda system. This fall in average farm size among the LC did not, however, lead to a decline in the inequality in the distribution of land across farms. This is due to the fact that numbers increase and farm size declines at both extremes of the distribution of land. As a result, Gini coefficients of land distribution, if anything, tend to increase slightly for the whole of Latin America between the 1960s and the 1970—1980s. Gini coefficients for these two periods in Latin America are as follows:[4]

Country and years	Gini coefficient 1960	1970—1980
Countries with rising inequality		
Brazil, 1960—1980	0.83	0.85
Costa Rica, 1963—1973	0.78	0.81
Panama, 1960—1972	0.74	0.85
Countries with falling inequality		
Colombia, 1960—1970/71	0.87	0.86
Mexico, 1960—1970	0.95	0.93
Peru, 1960—1972	0.94	0.91
Uruguay, 1961—1979/80	0.82	0.80
Venezuela, 1961—1971	0.93	0.91
Latin America, weighted average	0.87	0.88

TABLE 3[5]
Post World War II Changes in Share of Crop Production for Selected Latin American Countries

Country	Period	Farm size (hectares)	Share of crop production (%) Period 1	Period 2	Average 10 year growth rate
Costa Rica[a]	1963—1973	0.5—4	7.6	7.6	−0.3
		4—20	21.3	19.2	−10.0
		20—100	31.4	27.8	−11.6
		100+	39.7	45.5	14.6
El Salvador	1950—1971	0—5	33.2	37.3	5.6
		5—20	18.6	14.4	−11.4
		20—100	20.9	18.8	−4.8
		100+	27.3	29.5	3.7
Honduras	1952—1974	0—5	34.8	32.7	−2.9
		5—20	35.3	27.2	−11.3
		20—100	21.0	21.3	0.8
		100+	8.9	18.8	40.7
Venezuela	1961—1971	0—5	18.0	12.0	−33.5
		5—20	31.6	27.2	−14.0
		20—100	25.0	26.1	4.5
		100+	25.4	34.7	36.8
Brazil	1970—1980	0—10	19.9	12.5	−37.2
		10—50	37.2	30.6	−17.7
		50—200	19.8	23.3	17.5
		200+	23.1	33.6	45.3
Uruguay	1970—1980	1—20	7.9	5.9	−24.3
		20—100	18.9	14.9	−21.2
		100—500	33.7	31.8	−5.7
		500+	39.5	47.4	19.8

[a] The 1960—1970 comparison for Costa Rica is not exact due to slightly different farm size categories used in each period.

C. Changes in Shares of Total Crop Production

The data in Table 3 show that the contribution of peasants (the two smallest farm size categories, corresponding to SF and F farms) to production is, indeed, very important. In the latest period for which census data are available, their share of total crop production reached 60% in Honduras, 52% in El Salvador, 43% in Brazil, 39% in Venezuela, 27% in Costa Rica, and 21% in Uruguay. On the average for these six countries, the contribution of peasant production is 40% for crops and livestock.[6] It is also significant in certain export crops as it reaches 41% for coffee and 33% for cocoa.

Latin American agriculture has grown rapidly during the post-WWII period. Different social groups have, however, not shared equally in this growth. In every country in Table 3 but El Salvador, the peasant sector has suffered considerable losses in its share of total crop production. The surprising observation is that this is true for family farms as well as for the sub-family farms. Thus while the sub-family farms lost less market share than the family farms in Central America, the opposite is observed in South America. The performance of medium-size commercial farms is mixed, with as many countries showing gains as countries showing losses. The systematic gainers are the largest commercial farms. The average gain over a decade can be very large,

TABLE 4[5]
Costa Rica 1963—1973: Change in Production Share By Farm Size

	Change in production share (%)[a]						
	Farm size (hectares)				Crop area (%)		
	0.5—4	4—20	20—100	100+	1963	1973	Average annual growth of prod. (%)
Rice	−2.4	−7.9	−12.5	22.8	14.8	19.5	10.0
Bananas	−1.2	−5.1	−15.1	21.4	7.5	10.8	—
Plantains[b]	−0.4	10.3	−7.0	−2.9	2.1	1.9	—
Maize	−0.2	−5.7	2.0	4.0	15.5	15.5	−0.7
Sugarcane	−0.2	−0.2	−1.7	2.1	10.2	11.6	7.5
Cassava	0.7	−0.7	−2.4	2.4	0.7	0.6	8.1
Edible dry beans	1.0	−2.4	1.9	−0.5	12.8	8.0	−3.7
Cocoa	1.3	11.9	13.5	−26.7	11.1	6.1	−11.7
Coffee	2.8	3.2	−0.7	−5.3	23.8	24.9	—
Total[c]	0.0	−2.1	−3.6	5.8	98.5[d]	98.9[d]	—

[a] Figures show percentage point difference not percentage change, i.e., 1973 share minus 1963 share. The 1963—1973 comparison is not exact due to slightly different farm size categories used in each period. Total arable land actually decreased in the 1963—1973 period by about 30% while the number of cattle grew by over 50%.
[b] Plantain refers to grown alone only.
[c] The total changes in production share by farm size are calculated from the ten most important crops weighted by area.
[d] Area totals sum to 100% for the ten most important crops.

TABLE 5
El Salvador 1950—1971: Change in Production Share By Farm Size

	Change in production share (%)[a]						
	Farm size (hectares)				Crop area (%)		
	0—5	5—20	20—100	100+	1950	1971	Average annual growth of prod. (%)
Rice	−8.7	−9.1	0.6	17.2	3.0	2.8	3.5
Cotton	−2.5	−1.8	−8.5	12.8	4.5	12.4	11.0
Coffee	−1.9	−1.0	2.6	0.2	30.5	28.4	2.6
Sugarcane	−1.6	0.1	5.6	−4.2	4.1	4.5	3.5
Maize[b]	10.4	−2.4	−1.0	−7.0	24.8	20.7	5.0
Beans[c]	16.4	−4.9	−5.9	−5.6	7.8	7.0	1.0
Total[d]	4.0	−4.2	−2.0	2.2	74.7[e]	75.7[e]	—

[a] Figures show percentage point difference not percentage change, i.e., 1971 share minus 1950 share.
[b] Maize: figures refer to cultivation alone only.
[c] Beans: figures refer to cultivation alone and with other crops.
[d] The total changes in production share by farm size are calculated from the nine most important crops weighted by area.
[e] Area totals sum to 100% for the nine most important crops.

reaching 20% in Uruguay, 37% in Venezuela, 45% in Brazil, and 41% in Honduras. These are enormous redistributions of market shares toward the larger farms who are thus the clear winners in the competition for markets across farms.

D. Changes in Shares by Crops

In Tables 4 through 9, we use the six countries for which census data on the production of specific crops by farm size are available over time. This allows us to test two hypotheses.

The first is that there are distinct crops in which production has increased in small versus large farms and, hence, that there exists a considerable degree of specialization in farming systems by farm sizes. On peasant farms, the crops that have gained are either traditional low income food crops (maize and beans in El Salvador, beans and sorghum in Honduras, plantains in Venezuela, and cassava in Brazil) or labor intensive export corps such as coffee and cocoa in Costa Rica. By contrast, the crops which have gained on the larger farms are either urban food crops (rice in Costa Rica, El Salvador, Honduras, Venezuela, and Brazil; bananas in

TABLE 6[5]
Honduras 1952—1974: Change in Production Share By Farm Size

	Change in production share (%)[a]						
	Farm size (hectares)				Crop area (%)		
	0—5	5—20	20—100	100+	1952	1974	Average annual growth of prod. (%)
Banana/plantain	−17.8	−25.6	−15.8	59.2	10.9	4.7	—
Rice	−7.9	−10.9	6.1	12.7	2.2	2.3	2.6
Coffee	−2.4	−7.0	1.3	8.0	9.9	19.6	4.2
Maize	1.3	−6.8	0.5	5.0	52.3	48.5	2.6
Beans	5.8	−7.7	−0.9	2.9	10.2	10.5	2.0
Sorghum	6.5	−7.5	−5.6	6.7	12.6	8.9	−1.0
Total[b]	−2.2	−8.2	0.4	9.9	98.2[c]	94.4[c]	—

[a] Figures show percentage point difference not percentage change, i.e., 1974 share minus 1952 share.
[b] The total changes in production share by farm size are calculated from the nine most important crops weighted by area.
[c] Area totals sum to 100% for the nine most important crops.

TABLE 7[5]
Venezuela 1961—1971: Change in Production Share By Farm Size

	Change in production share (%)[a]						
	Farm size (hectares)				Crop area (%)		
	0—5	5—20	20—100	100+	1961	1971	Average annual growth of prod. (%)
Maize	−9.5	−8.6	1.9	16.3	41.7	39.3	2.7
Cotton	−7.6	−2.0	3.9	5.7	3.1	3.6	5.9
Rice	−7.3	−5.2	−3.0	15.5	3.6	9.8	9.2
Beans[b]	−7.3	−1.2	3.6	4.9	6.7	6.2	−1.4
Sugarcane	−2.9	−1.4	0.1	4.1	5.4	6.2	3.9
Bananas	−2.2	5.0	5.6	−8.4	7.2	4.9	—
Cocoa	0.1	4.5	2.7	−7.4	4.9	4.8	3.1
Coffee	0.5	2.3	2.0	−4.7	21.8	20.9	0.6
Plantains	1.8	5.4	−2.3	−4.9	5.5	4.4	−2.0
Total[c]	−6.0	−4.4	1.1	9.3	100.0[d]	100.0[d]	—

[a] Figures show percentage point difference not percentage change, i.e., 1971 share minus 1961 share. Sesame excluded due to no prod. info. for 1971 (area is about 10% of total, prod. tripled, 88% produced by 20+ in 1961).
[b] Beans: black and kidney
[c] The total changes in production share by farm size are calculated from the nine most important crops weighted by area.
[d] Area totals sum to 100% for the nine most important crops.

Costa Rica and Honduras) or industrial inputs (cotton in El Salvador and Venezuela) and nontraditional export crops (soybeans in Brazil and oranges in Uruguay).

The second hypothesis is that the dynamics of growth of particular crops explain which group of farms will be able to gain in market shares. Small farms tend to gain in production in the crops whose effective demand is relatively stagnant. Overall, they lose market share as they are being displaced from the production of the more dynamic crops. The large farms, by contrast, gain in production in the crops that have substantial market dynamic. As a result, they capture the lion's share of the gains in market expansion.

This second hypothesis can be tested with the following regression equation using the pooled data from Tables 4 through 9:

Change in production share   Growth of production

Farm size	Regression coefficient	t-statistic	R^2
SF	−0.24	−1.39	0.05
F	−0.45	−2.53	0.14
MC	−0.15	−0.82	0.02
LC	0.28	2.89	0.17

We see that it is only the large farms that are able to systematically gain market shares by being associated with the more dynamic crops. For the others, the dynamism of crops is a factor of displacement as they are being outcompeted by the largest farms. The reverse implies that peasants (SF and F) attempt to protect their market shares by being associated with the production of the more stagnant crops that are being

TABLE 8[5]
Brazil 1970—1980: Change in Production Share By Farm Size

	Change in production share (%)[a]						
	Farm size (hectares)				Crop area (%)		
	0—10	10—50	50—200	200+	1970	1980	Average annual growth of prod. (%)
Soybeans	−10.5	−17.0	9.0	18.5	7.2	19.6	21.1
Rice	−6.2	−9.3	−2.8	18.4	14.2	14.4	4.4
Beans	−5.9	−1.1	4.5	2.6	13.4	11.0	1.3
Cotton[b]	−5.7	0.7	1.6	3.4	4.9	2.6	−0.7
Millet	−5.1	−3.2	3.0	5.3	35.1	26.0	2.1
Wheat	−3.5	−1.8	5.6	−0.3	6.8	6.6	2.4
Sugarcane	−2.4	−3.9	−0.6	7.0	5.6	6.5	7.5
Coffee	0.3	0.9	1.6	−2.8	5.4	6.2	6.4
Cassava	3.4	−3.9	0.3	0.2	5.7	2.9	−2.7
Total[c]	−7.4	−6.6	3.5	10.5	98.3[d]	95.8[d]	—

[a] Figures show percentage point difference not percentage change, i.e., 1980 share minus 1970 share.
[b] The 1980 data for cotton was taken directly from the Brazilian census due to some inconsistencies in the FAO report.
[c] The total changes in production share by farm size are calculated from the ten most important crops weighted by area.
[d] Area totals sum to 100% for the ten most important crops.

TABLE 9[5]
Uruguay 1970—1980: Change in Production Share By Farm Size

	Change in production share (%)[a]						
	Farm size (hectares)				Crop area (%)		
	1—20	20—100	100—500	500+	1970	1980	Average annual growth of prod. (%)
Oranges	−14.4	−5.0	11.1	8.4	—	—	11.7
Potatoes	−10.6	−5.2	1.4	14.4	2.6	3.3	−0.8
Maize	−2.6	−1.5	0.9	3.3	26.4	23.8	−4.0
Wheat	−1.0	−4.5	−2.3	7.8	52.4	52.5	−0.3
Rice	0.0	−2.1	5.0	−2.9	4.2	9.0	5.7
Sunflowers	−0.2	−1.2	−4.5	5.9	14.4	11.4	−6.3
Total[b]	−1.9	−4.0	−1.9	7.8	100.0[c]	100.0[c]	—

[a] Figures show percentage point difference not percentage change, i.e., 1980 share minus 1970 share.
[b] The total changes in production share by farm size are calculated from the five most important crops weighted by area.
[c] Area totals sum to 100% for the five most important crops.

abandoned by the largest farms. Yet, their relative gains in the stagnant crops are not enough to compensate for their relative losses in the dynamic crops and they, thus, lose their overall market shares over time.

E. Conclusion

Put together, these results reveal a sharply contrasted dynamic among the four farm types in the typology used here. Sub-family farm households have expanded in numbers like farms in the other three categories. They have, however, been squeezed on the land market as average farm size has fallen significantly. They have also been squeezed on the product market, as their production is not associated with the dynamic crops, and they have lost overall market share. Clearly, they have been the main losers as agricultural producers.

Family farm households have been able to protect their access to land as average farm size has remained relatively constant in spite of the rapidly rising number of households as in the other three farm categories. They have, however, lost significantly in market share, indicating lack of productivity gains and concentration in the relatively stagnant activities. This poor performance, in spite of protection of the assets base, indicates serious neglect in providing this social sector with sources of productivity gains and reveals biases in access to institutions such as credit, information, infrastructure, and technology that jeopardizes their competitive ability. In terms of rural development, it is conse-

TABLE 10[9]
Sources of Income

Region	Year	Farm size (hectare)	Share of farm households (%)	Shares of income derived from: (%) Farm	Wages	Other
Cajamarca (Peru)	1973	0—3.5	72	23	50	27
Puebla (Mexico)	1970	0—4	71	32	58	11
Garcia Rovira (Colombia)	1972	0—4	20	79	16	5
South Bolivia	1977	0—5	67	38	62[a]	
Region IV (Chile)	1976	0—5	59	47	40	13
Vertentes (Brazil)	1979	0—10	16	—[b]	56	—
Northwest Altiplano (Guatemala)	1978	0—3.5	85	29	59	12
El Salvador	1975	0—2	71	64	27	9
Sierra (Ecuador)	1974	0—5	77	37	44	19
Coast (Ecuador)	1974	0—5	77	48	41	11
Chamula (Mexico)	1973	—	—	11	89[a]	

[a] Sum of wages and other.
[b] Dashes indicate no data available.

quently the social category where the largest gains can potentially be achieved.

While the medium farms have rapidly increased in numbers, it is striking to observe that their gains in average farm size were not translated into more favorable market shares. This indicates that the dynamics of capital accumulation and productivity gains in Latin American agriculture does not have its origin in the emergence of a middle class of agriculture producers. It is, in contrast, the largest farm category that is at the source of the dynamics of accumulation. While the number of large farms has increased at par with the increase in the other categories and average farm size has fallen sharply, indicating an important reorganization toward managerially more efficient and politically less vulnerable operational units, they are the ones who are rapidly increasing their control of the market. Emergence of a dynamic class of "farmers" in Latin America is thus not a middle class phenomenon but a transformation of the old agrarian establishment into a class of modern large entrepreneurs which is able to monopolize the benefits of growth while peasants and middle farmers are increasingly marginalized from the market.

III. SOURCES OF INCOME FOR PEASANT HOUSEHOLDS

With the SF households losing in access to land and in market shares, while their numbers increase rapidly, their survival strategies must increasingly rely on off-farm sources of income as complements to farming on the home plot. There unfortunately do not exist data over time characterizing changes in the sources of income of farm households. From the cross-sectional data on sources of income among SF households in 11 regions of Latin America reported in Table 10, we see that, with the exception of two cases, farm income constitutes considerably less than 50% of their total income. Among nonfarm activities, wages are by far the most important source of income as they account, on the average, for some 43% of total household income. From this table, we can thus conclude that some 60% of the farm households in the areas surveyed derive nearly 60% of their income from off-farm activities.

A more detailed characterization of sources of income is available from a farm survey conducted in 1974 in the Sierra of Ecuador (Table 11). It shows a clear contrast in the private sources of income across farm sizes. Monetary income derived from agriculture is only important as a share of total income in farms larger than 2 and, even more so, 5 ha. It is thus in this category that productivity gains in farming can have a significant income effect. For the smaller farms, it is agricultural and nonagricultural wages that are the main sources of income. Consequently, these households crucially depend for their welfare upon access to sources of employment in activities located principally in the rural areas. On the very small farms, other activities such as handicrafts and trade also make significant contributions. Finally, the home plot is important for home production in all the SF and F farms, clearly indicating the importance of family labor to production. Since men are usually the main participants in the labor market, women tend to predominate in the home plot, handicraft, and petty trade activities.

Since we do not have data on changes in the sources of income over time—that would expectedly show the rising importance of off-farm sources of income as SF households need to compensate for the observed losses in access to land and in market share—we can turn to data on the evolution of rural poverty to test the hypothesis that these households have the character of a cornered segment of the labor force. Table 12 shows that the incidence of rural poverty has, indeed, increased between 1970 and 1980 in every country but Argentina. Rural poverty rose, therefore, in spite of the

TABLE 11[8]
Sources of Income in Ecuador Sierra by Farm Size

Farm size (hectares)	Home consumption	Net monetary agricultural income	Payments in kind	Trade (%)	Handicrafts	Agricultural wages	Nonagricultural wages	Total per capita income (U.S. $)
0—1	9.4	13.6	0.6	7.1	4.3	24.4	40.6	129
1—2	15.3	30.1	1.2	4.2	2.1	23.9	23.2	93
2—5	16.8	49.2	0.7	4.3	0.6	15.5	12.9	124
5—20	14.7	64.9	1.0	6.0	0.8	5.6	7.0	210
20—100	6.3	81.0	1.0	5.2	0.6	1.1	4.8	761
100+	1.1	88.0	3.9	1.9	0.0	0.3	4.8	5,433

TABLE 12[9]
Dynamics of Rural Poverty in Latin America

	1970		1980		Change, 1970—1980	
Country	Millions	Percent	Millions	Percent	Millions	Percent
Argentina	0.9	19	0.5	10	–0.4	–9
Brazil	27.6	73	28.8	73	1.2	—
Colombia	4.4	54	6.2	67	1.8	13
Costa Rica	0.3	30	0.4	34	0.1	4
Chile	0.6	25	1.2	56	0.6	31
Honduras	1.3	75	1.9	80	0.6	5
Mexico	10.6	49	15.9	68	5.3	19
Panama	0.4	52	0.6	67	0.2	12
Peru	4.2	68	3.9	68	–0.3	—
Venezuela	1.2	36	1.6	64	0.4	28
Total	51	62	61	65	9.5	4

fact that the 1970s was a decade of remarkably rapid growth of the urban economy—a consequence, in part, of the accumulation of debt in all countries and oil booms in several. Even in periods of rapid growth, structural features have thus constrained employment creation, leaving large segments of populations cornered in the SF sector of the peasantry. In the 1980s, with the dramatic slowdown of growth associated with the debt crisis and falling international commodity prices, rural poverty has certainly increased much more rapidly. In this context, SF farms harbor the social failures of the rest of the economic system. Household-oriented rural development and agroecology have important roles to play as survival strategies to improve home consumption derived from production on the home plot.

IV. RECOMMENDATIONS FOR RURAL DEVELOPMENT AND AGROECOLOGY

The contrasted social dynamics which we have observed in the rural sectors of Latin America lead us to identify two segments of the peasantry for which rural development and agroecology are important, but from very different perspectives. Family farms have been able to protect their access to land but have lost in market share. As a result, their resource base is still sufficiently large that they could benefit from production-oriented rural development programs that can have a significant impact in enhancing the welfare of family farm households. This requires, in addition to a macroeconomic and sectoral policy context favorable to overall agricultural development, a reduction in the anti-peasant biases that exist in the access to institutions and the organization of effective rural development programs.[10] Agroecology is an important subset of these programs. In this case, it must strive to enhance the productivity and sustainability of farming systems that have an important component of market orientation. The objective is to increase the ability of the family farm to generate monetary income. A clear identification of the monetary costs and returns involved is an economic exercise which is largely left to be done.

The sub-family farms mainly have the character of a refuge population which is inserted in the rest of the economy principally through the labor market and ancillary household-based activities and only secondarily through agricultural production. Nevertheless, the home plot is an important element of survival strategies, particularly in the context of the current economic crisis and of the corresponding rise in food prices and loss in employment opportunities in the nonagricultural activities. Agroecology should here strive to improve productivity of the home plot largely for home consumption. Since these households need to manage a highly diversified portfolio of activities to complement their lack of access to land, new technological alternatives and new activities must be carefully tailored to the specific opportunity costs of different members of the household, to the explicit role of women, to the limited ability to bear risk, to weak access to markets and institutions, and to lack of monetary resources (investable income or credit). Indeed, much of the agroecology work performed by nongovernmental organizations has been oriented to production in home plots for family consumption.[11] The nature of the agroecological advances required in this case are likely to be fundamentally different from those desirable for family farms. A clear identification of the survival strategies of peasants and of their specific resource and institutional constraints is thus a prerequisite for a meaningful research agenda for agroecologists.

REFERENCES

1. **International Labor Office,** *Mercado de Trabajo en Cifras, 1950—1980,* PREALC, Santiago, Chile, 1982.
2. **Barraclough, S., Ed.,** *Agrarian Structure in Latin America,* Lexington Books, Lexington, MA, 1973.
3. **United Nations, FAO,** *1970 World Census of Agriculture. Analysis and International Comparison of the Results,* Rome: Food and Agriculture Organization, 1981.
4. **Haggblade, S. and Hazell, P.,** Prospects for Equitable Growth in Rural Sub-Saharan Africa, Washington, D.C.: Agriculture and Rural Development Department, The World Bank, April, 1988.
5. **United Nations, FAO,** *Report on the World Census of Agriculture, Results by Countries.* Rome: Food and Agriculture Organization, various years. For Table 8, see also the *IX Recenseamento Geral do Brasil—1980,* Vol. 2, Tomo 3, Numero 1, Censo Agropecuario. Rio de Janeiro: IBGE, 1984.
6. **ECLAC/FAO,** *Peasant Agriculture in Latin America and the Caribbean,* Santiago, Chile: Joint ECLAC/FAO Agriculture Division, 1986.
7. **de Janvry, A., Sadoulet, E., and Wilcox, L.,** *Rural Labour in Latin America,* ILO Working Paper WEP 10-6/WP79, Geneva: International Labour Organization, June, 1986.
8. **Ministry of Agriculture of Ecuador and ORSTOM,** *Diagnostico Socioeconomico del Medio Rural Ecuatoriano,* Quito: PRONAREG-ORSTOM, November, 1978.
9. **United Nations, FAO,** *Potentials for Agricultural and Rural Development in Latin America and the Caribbean.* Rome: Food and Agriculture Organization, 1988.
10. **de Janvry, A., Marsh, R., Runsten, D., Sadoulet, E., and Zabin, C.,** *Rural Development in Latin America: An Evaluation and a Proposal,* San Jose, Costa Rica: Inter-American Institute for Agricultural Cooperation, 1988.
11. **Altieri, M.,** Potential Contributions of Agroecology to Rural Development in Latin America, Division of Biological Control, University of California, Berkeley. April, 1988 (mimeograph).

Small Farm Development Approaches

9 AGRICULTURAL DEVELOPMENT SYSTEMS IN LATIN AMERICA: A HISTORICAL PERSPECTIVE

Nicolo Gligo

I. INTRODUCTION

The history of Latin America is closely linked to the evolution of agriculture. The pre-Columbian empires were agricultural. Their territory was organized on the basis of exchange and agricultural complementarity and on forms of power which made possible the appropriation and redistribution of the surpluses generated by this sector.

Notwithstanding the fact that the settlement strategies of the colony (with the exception of Brazil) were predominantly mining oriented, agriculture (on the basis of the implantation of the European system which incorporated the pre-Columbian method) provided food for the new settlements. Agriculture gradually began to produce specific crops for the Crown and subsequently other products, which were sold on the international market under the aegis of Spain.

The large agricultural export potential of the New World was not allowed to develop under the Iberian domination, and this was one of the main causes of the independence of the American nations.

Once the barriers to international trade were removed, the new nations consolidated their development and their borders on the basis of the agricultural settlement of the land. At the same time, the foreign enterprise enclaves gathered force little by little until in some nations they became the most important factors in the power structure. The predominant form of agricultural development was the form used in Europe during the previous century (although the crops, i.e., cotton, sugarcane, etc., were different) and it coexisted side by side with pre-Columbian social forms and technologies in the peasant sector. Both forms became well established in the temperate and subtropical areas and highly productive enclaves entrenched themselves firmly in the tropical areas. The wet tropical hinterland remained virtually untouched.

Up to the time of World War II, the agricultural area continued to expand as large irrigated areas were added. After the war in the 1960s and 1970s the green revolution exploded with great force and proved to be the key technological factor in developing the capitalist productive forces.

Agricultural production grew apace, land and labor productivity increased and vast areas of land were settled; but at the same time, rural poverty increased, there was a massive emigration to the cities, which created enormous problems of urban marginality, and the environment deteriorated, with irreversible losses of the ecosystem's resources.

The financial crisis of this decade is exerting pressure for more foreign exchange to be generated, even at great ecological cost.

The principal aspects outlined here will be discussed further and an attempt will be made to provide a historical view of the most outstanding agricultural development methods adopted in Latin America.

II. AGRICULTURAL DEVELOPMENT DURING THE COLONIAL PERIOD

The colonial period in American history is characterized by the breakdown of the social and economic structure of the pre-Columbian cultures, the occupation of land by the conquerors and the use of new lands. This use imposed new forms of organization, introduced some crops and technologies, incorporated others, partially discarded traditional production systems, and established new productive structures.

The differences in the land settlement methods of the systems adopted can be explained by the divergent strategies established by the Europeans according to the mining and agricultural capacity of each region and in particular according to the reaction of the indigenous groups and cultures. In fact, the social and economic development of some pre-Columbian societies, especially their stratification and power structure, made possible the partial use of this organization and structure and especially the subjection of the sedentary Indian who was attached to the land. But the nomadic Indian who hunted and fished, who had low productivity and lacked any social organization or the tradition of transferring surpluses could not be assimilated into the sedentary system. The result was that he was either enslaved or clashes ensued between the overlords and the ethnic groups who revolted.This hypothesis seems to explain the attitude and position of the Europeans, although there are other factors to be taken into account such as the fact that the mentality that had developed in Spain during the wars to drive out the Arabs was transferred to America.

Although there were areas where social organization was structured around agricultural development, the main thrust was usually mining exploitation. The "visible" wealth was mining; the great empires had an exploitation structure already in operation. In order to take advantage of this structure, the conquerors inevitably triggered off a demographic catastrophe, which occurred in the sixteenth century.

The number of natives in Latin America was drastically reduced as rebellions were crushed, settlements displaced, food production was disorganized, and epidemics spread. The destruction and dislocation of the conquered cultures became genocidal.[1] Not only was there direct killing but a high percentage of the natives died with the introduction of malaria, measles, small pox, and yellow fever.[2] In their data, Denevan and Parsons[3,4] refer to some 150 million inhabitants in Latin America. It is estimated that the Mexican population, which had allegedly reached 16 million during the period of the conquest, had been reduced to about one tenth of that number one century later.

A consequence of the demographic catastrophe was the destruction of the productive activities which had been pursued in harmony with the environment. The priestly class and others which possessed the most highly developed empirical knowledge disappeared and with it, the techniques and methods adapted to the ecology.

The method used in carrying out this settlement is most important as it spawned embryonic ways of using the resources. The production methods established slavishly served the aims of the metropolis, as they had under the Incas and Aztecs but in another framework.

In the mechanisms of land appropriation, the lands cultivated during the first years of the Vice-Royalty of Peru were much smaller than the agricultural frontiers of the Inca Empire. Because of the disorder and dislocation caused by the conquest, many of the new lands that had been cleared in the forest and which were in the hands of the councils established by the Incas were abandoned. The natives tended to return to their land of origin, abandoning the new lands, which were resettled many years later.

In addition to this phenomenon, the Spanish invasion abruptly terminated the vertical development of the Andean economies, and destroyed the self-sufficiency that they had enjoyed. The change in structure and the reduction of the population prevented the communities from occupying more distant areas.

In view of the reasons outlined, and since there were no agricultural markets, at first there was no greed to acquire lands, so that later it was easy to seize the abandoned lands.

The conquest and the expansion during the colonial era were privately financed by the war enterprise and this was combined with rewards, concessions, attributions, and privileges granted to the conquerors. The Indian Company (a private enterprise of the conquest) was regulated by the State and was based on the rapid recovery of invested capital. The methods used to reward the services provided were varied and complex. Three deserve special mention, since they derive from the technical and social links between agriculture and the Latin American latifundio (large landed estates) and therefore from the use of resources: grants, direct donations and Indian villages and inhabitants, which were granted to the Spanish colonists by Royal Decree. The direct donations were concessions granted for different reasons, chiefly as rewards for war services. It is logical then that they were not circumscribed to the defined limits but since so little control was exercised over them, they expanded, thereby creating the latifundio system.

The gifts of land with a royal title were usually granted with full enjoyment of the profits, the only requirement being that they should be arable, although at the beginning they were not considered as private property. Indian villages with their inhabitants were granted to the Spanish colonists, who thus received a group of natives who had settled on a certain area of land and who were to be taught the catechism.

Agriculture was organized around the development of urban centers and Spain's demand for products. The dominant agricultural method was Iberian, and it incorporated some pre-Columbian forms. Although a considerable amount of European technology and machinery was introduced, the store of native technical knowledge was only partially lost since many areas were still cultivated in the traditional pre-Columbian manner.

During the seventeenth century, the self-reliant farmers produced corn and beans as the main crops. But this self-reliance clashed with the expansion of export crops that were usually produced by the latifundistas (large landowners).

At first, the Spanish Antilles exported their cattle products to Mexico. During the eighteenth century, however, there was considerable expansion, with the introduction and the boom production of tobacco and sugar, especially in Cuba. Production was no longer sent to Mexico but to the Iberian Peninsula.

What occurred in the Spanish Antilles was repeated in Hispanic America: one way was to orient the production in each colony to the Spanish market, a system that led to the fragmentation into economic zones where a given crop or item predominated. This economic specialization was partly influenced by the marked demographic decline of the seventeenth century which brought about considerable changes in the agricultural sector. Certain labor-intensive crops were abandoned for more extensive ones. This occurred throughout Latin America and it was the first step towards restructuring the society: the partial replacement of the native community by the large colonialist's ranch which was the farming unit controlled by the Spaniards.

Economic growth was slower in Central America than in Mexico and the Antilles. In Guatemala, the large haciendas and the highly feudal native communities merged. In Honduras and Nicaragua, the only noteworthy feature was extensive cattle rearing, in Costa Rica, because of the contribution of the Galician settlers, self-sufficient agriculture began to develop in the middle of the eighteenth century. International trade was centered on a single export item: indigo.

In the intricate organization of New Granada, economic and geopolitical functions were intermingled mainly because it was the center of Spanish military power. Its main export was gold. Its agriculture was structured to be self-sufficient and the main crop was wheat.

In the east, at this time, Venezuela showed greater dynamism and integration. Its most important products were cocoa, indigo, coffee, and cotton. The coast lands were mainly under cocoa cultivation. Livestock rearing was developed in the mountainous periphery and on the plains and this made it possible to export mules and cattle to the Antilles and animals for food to the coast.

The Vice-Royalty of Lima based its economy on mining, especially gold and silver. But the large mining centers in upper Peru, combined with the administrative separations decrees (Vice-Royalty of La Plata), created a subsistence economy based on agricultural activity.[5] Here it is noteworthy to highlight the sharp difference between the coast lands and the mountains. The old irrigation systems continued to be used on the coast and agricultural production was geared to the Spanish-American market: brandy, wine, cotton, sugar, and rice. The mountain range had a dual function: mining production and subsistence agriculture. This agriculture was based on corn and potatoes and a very special kind of livestock was reared: sheep and Camelidae. The hillsides were abandoned and terracing gradually disappeared. Subsequently, when they were resettled this definitely conservationist practice was not observed.

Quito and the General Captaincy of Chile were an adjunct of the Vice-Royalty of Peru. Cocoa was cultivated on the Quito coast and it was marketed overseas through Guayaquil. The mountain range whose population was predominantly native continued to be oriented towards self-sufficient production and cotton was cultivated in certain micro-climates and wheat in the colder areas.[6]

In Chile, in the eighteenth century, there was both mining and agricultural activity. Vast areas were under wheat cultivation, which was sold on the Lima market but wheat was cultivated only in the central region since the south of the country was in the hands of the Araucans or Mapuches. The resistance lasted for three centuries at considerable cost to the Crown for investment in material and human resources. The Araucan territory was incorporated long after independence.

In the River Plate, Buenos Aires was the channel for the growing activity of the coast, the lands in the surrounding areas and Sante Fe. Livestock expansion in these areas was curtailed by the activities of the natives and only a part of the wet pampas was cultivated under the hacienda system.

In the eighteenth century, growth east of the Paraná was impressive. The work of the Jesuits made it possible to have a peaceful Guaraní labor force. The missions produced mostly maté and cotton. After the Jesuits were driven out, the missions decayed considerably and their former boom was transferred to Paraguay, and the livestock expansion of the Brazilian southwest arrived there.

In the southern lands that could be easily adopted there was significant economic activity. Cuyo specialized in the production of wine and Tucumán and the adjacent areas combined the production of tropical crops with livestock rearing, which served both upper Peru and the coast as well as Buenos Aires.

Although agriculture was of great importance in providing food for local consumption and for export, the cultivated areas of Latin America were very small compared with those of today. The demand for self-sufficiency was limited to the needs of a very small population and exports were conditioned by transport and the limitations of international demand. It should be remembered that the vast expansion of the agricultural frontier did not take place during these periods but during the second half of the last century and especially during this century.

III. HOW AGRICULTURAL DEVELOPMENT HELPED TO CONSOLIDATE THE LATIN AMERICAN NATIONS

The new States attempted to consolidate themselves around the exploitation of the natural resources. The hegemonic groups built power on the basis of land and mine ownership.

In Peru, Bolivia, and Mexico, State power derived predominantly from mining. Peru and Bolivia gradually integrated mining power with land power. Mexico was the only exception and this was manifested in the deep-seated political instability of the last century. In mixed economies such as Chile's, the bourgeois also followed the path of integrating mining and agriculture. In Chile, the 1833 Constitution had placed all State power in the hands of the latifundista faction, but because of the surplus it generated the mining sector gradually invested in agriculture mainly through the purchase of haciendas.

Consequently, except in the case of Mexico, power was structured around agriculture or in terms of agreements or the integration of agricultural and mining groups. This fact was most important in land settlement and in the way the ecosystems intervened because the latifundista groups dealt with resources from a political and economic perspective. But the force of State power was neither homogeneous nor centralized. In the internal structure of each country, there were pressure and fighting to establish control over one region or another. Two basic factors impacted on this. On the one hand, there was the economic importance of one region in relation to the others and on the other, the physical location of a region as a channel or depot of the products of others. The attainment of greater economic importance depended, therefore, on the export potential. Thus, Peru had no serious problem but Colombia was riven by civil strife. The location of the port of Buenos Aires was fundamental in establishing the control over the coastal area.

The restructuring of power met with a series of obstacles because it was difficult to reorganize a system of production to cope with the new inroads into the international market. Here, England's presence which had been of great importance during the period when the new republics were breaking away and becoming independent took the form of a penetration of that country's interests and helped forge the first ties of dependency. These ties became so close that production systems were created to dovetail with the development of English industrial development.

The history of political changes in Latin America is closely linked with the rise of specific basic products which produce an economic surplus.[7] Thus, in Venezuela , the conservative hegemony was toppled or crumbled because of the coffee crisis. In Guatemala, a new political style emerged when coffee growing was developed. Honduras and Nicaragua depended on the power that came to them mainly through their livestock-rearing activity, El Salvador from indigo growing, Mexico after its liberation and despite its vast mining resources could not overcome the cotton crisis and the country was torn by long internal struggles. In Costa Rica, political stability was organized around a middle class of coffee producers who resisted any attempt at military intervention and established the basis of a stable democracy.

Countries such as Ecuador, Colombia, Brazil, a part of Mexico and Venezuela and the countries of Central America and the Caribbean behaved differently depending on the vagaries of tropical products. To the sugar and tobacco of previous centuries was added coffee and cocoa expansion. The tropical crops served to establish the economic occupation of the territories, but the place they occupied in the economies of the countries varied considerably. In Brazil, tropical products played an important role in development, the social structure depended on how coffee was organized and the trade and services sectors were organized around agricultural activity. However, in other countries, especially in Central America, the organization of export crops combined the systems of inclusion in the economic structure with those of enclaves. These systems did not foster dynamic development and surpluses were sent to the countries at the center.

The expansion of the agricultural frontier of all these countries was small, especially during the first half of the last century. Tropical crops were cultivated on a small part of the agricultural lands, usually around the shipping ports. The subtropical and temperate areas were generally organized into haciendas and livestock rearing was the main activity. There was almost no penetration into the tropical regions and the forest only served as an energy source.

Since the middle of the last century, Peru has depended on a new source of wealth: fertilizer. Side by side with this single product exploitation, the country was beginning a process of liquidating land communities, and the haciendas on the coast continued to produce sugar and cotton surpluses. When fertilizer declined, nitrate increased spectacularly in the south.

The Pacific war consolidated Chile's position and created serious problems for the Peruvian economy. The Chilean mining sectors, almost without conflict, initiated a new society which until then had been dominated by the latifundistas in the central area. With the rise of nitrate, more exports to the international market and the absence of any basic contradictions between the dominant economic groups, Chile became a country with sustained growth and political stability which was only interrupted, in 1891, during a short civil war, that culminated in the suicide of President Balmaceda.[8] Power had been structured around the latifundia and they were always relatively more important than they may have been through mere control of the economy.

Cuba continued to be a colony of Spain and its economy was based increasingly on sugar. Its fine woodlands were gradually cut down to produce the energy needed for its sugar mills. Since the labor force was composed of slaves, the country was able to withstand the vicissitudes of the market.

In the rest of the Caribbean, export agriculture continued to be centered around sugar and to a lesser extent other tropical crops.

As coffee production soared, it impacted significantly on the good coffee-growing areas such as those in Brazil, Mexico, Colombia, Venezuela, El Salvador, and Guatemala. The coffee cycle was linked to the problem of demand and also to the speculations of the intermediary and financing sector. In 1906, Brazil established a system to prevent overproduction, which although it prevented total bankruptcy, stabilized the price at a low level. The experiences of the coffee institute and the host of misfortunes experienced laid the foundation for the latifundista sector to begin establishing a market organization for its products. It was therefore possible to consolidate a power structure with the union of the landowners as the base.

In Argentina and Uruguay, the increase in cattle was tied to wheat and maize, which were the basic crops in the grain expansion.

The spectacular extension of the railroads brought Santa Fe and the south of Córdoba under cultivation of these crops. In 1870, there had been only 732 km of railroad, in 1890 just 20 years later there were 9254 km.[9] The grain trade was controlled by a few exporting firms. The trading interests, combined with the financial groups in Buenos Aires, dominated this expansion. The livestock rearing groups, particularly from the province of Buenos Aires, continued to wield their influence in the political structure.

These groups of landowners had such a large surplus that they were able to invest in technological innovations. The exceptional ecological conditions of the wet pampas and the structure of an ownership system, the large livestock landed estate (which by definition underutilized resources) prevented the deterioration which had taken place in other commodities, such as coffee, from occurring. However, it must be pointed out that, in the peripheral areas of the pampas, sheep rearing in the last century had already had some deteriorating effects.

Crops which were important for the social, economic, and political structure emerged early in the century. In the low-lying and wet areas of Central America, in Honduras, Nicaragua, Guatemala, Costa Rica, and Panama, and in South America, in Ecuador, Colombia and Venezuela, banana growing expanded so significantly that bananas became the main export product of several Central American countries. Rubber was exploited for a short time and as will be seen later on, was also briefly included as an export product. Another important item was tannin extracted from the quebracho bark, mainly in El Chaco.

Agricultural development in Latin America during this period was chiefly based on a structural and technological system that had been imposed by the Europeans. The most outstanding feature was that agriculture was pursued primarily in the temperate areas. In the tropical region, agriculture was limited to areas of influence on the coast and to enclaves of cotton, coffee, cocoa, and sugar plantations.

IV. PREDOMINANT AGRICULTURAL SYSTEM FROM THE DECADE OF THE 1950S: CAPITALIST MODERNIZATION AND TECHNOLOGICAL DEVELOPMENT

The Latin American countries have tried during the postwar period to modernize their agriculture by implementing several development strategies. The major objective of these strategies was to modernize agriculture by promoting greater investment of surpluses generated within the sector itself and by fostering capital investment from other sectors or from abroad. Internally, the dominant group in the agricultural sector has made the largest investments in infrastructure, thus contributing to a global concentration which dovetails with the prevailing system.

In areas of comparative advantages, where production in many cases has been oriented towards exports, there has been a marked intensity of the phenomena and processes already mentioned. Here, agricultural modernization has received strong State support, irrigation works have been implemented where investment is unprofitable, agricultural activity has required energy and the demand for inputs has given rise to distribution centers. Increased production has led to the construction of warehousing and packaging centers or the establishment of processing agro-industries. Agricultural machinery has created the need for maintenance and repair workshops. All of these activities which require skilled labor have influenced the establishment of training centers, colleges, health services, business services, etc. In other words, concentration has generated more concentration, and also linkages of the agricultural industry with other sectors of the economy.

One of the chronic problems of the Latin American farms has been their profitability. The prevailing system has accentuated the differences in this aspect as well: farms with comparative advantages, which cultivate export crops that can generate surpluses on the one hand, and on the other, farms which produce for national consumption with problems of profitability and virtually no surpluses, cheek-by-jowl with a host of small subsistence farmsteads. Moreover, there are medium-sized holdings, some of which have been influenced by the method and garner surpluses, which build up capital and usually form associations, national or transnational consortia, while others, on the other hand, fail to accumulate capital, split up, are underexploited, and become part and parcel of a subsistence economy.

By manipulating public investments, the influential groups which are normally connected to the high-income landowners have benefited from the investment flows and have enjoyed huge subsidies. Other State policies covering prices, loans, and inputs have also given preferential treatment to these sectors. Furthermore, the model of technology generation has been adapted almost automatically to items in these areas because of the demand for them, their value as foreign exchange earners and the influence of the transnational corporations, which control the technologies and supply the technological inputs.

The model for the creation, adoption, and dissemination of technology has tended therefore to reproduce a type of modernized agriculture, to make it more dependent on the use of technological inputs and to make it more specialized to meet the requirements of the international market and of the new consumption patterns of domestic markets. In the postwar period, this modernization exploded with such force that it brought in its train a significant qualitative change in production methods and in labor relations in Latin American agriculture. Although the traditional forms and systems have been maintained, the orientation of the process has changed with the modernizing forms clearly dominating and adding to the pressure for technological change in agriculture.

The changes in agriculture have been the result of the overall changes in Latin American societies.

The "modernization" of the countryside has been identified with capitalist development. Although large strides have been made in agrarian reform, there have been no substantive changes in the degree of land concentration. When the capitalist style of production expanded, it was able to dominate the factors which promoted it and has been adapting its interests to the behavior of the other sectors, as for example, the peasant economies. The breakdown of the small farmstead is a proces which normally accompanies capitalist development.

The capitalist forms of production have also made a number of adjustments and the changes have been made primarily on the basis of consortia which are more efficient than the traditional agricultural capitals. There has been a penetration of national interest of other sectors, the most representative of these being trade and agro-industries. Thus, it has been possible to establish vertical structures, which have integrated the productive processes right through to exports of manufactured products.

Together with the national interests, transnational interests have also been imposed. In Central America and the Caribbean, where foreign interests have always been present, integration, in many cases, has started at the base, that is from the level of land ownership. Although present in Latin America, this phenomenon has not been as common, since the transnational presence has tended to avoid conflicts and has been directed not to land ownership but to marketing and industrialization processes and on many occasions vertically integrated structures have been developed, frequently in monopsonic relations. This situation has left producers in a clearly passive position in terms of the control of land tenure and subject furthermore to possible social conflicts with the wage earning sector or to the climatic vicissitudes of agriculture.

The capitalist development process has monetarized the peasant economy further, and this has led to a greater use of the land and reduced the forms of precarious land ownership and leasing.

Technological progress has been the main factor responsible for growth in Latin American agriculture but generally speaking, on this continent, progress has been slow compared with the achievements of the countries of the center under similar ecological conditions.

Various explanations, usually simplistic ones, have asserted that the reasons for the slow technological progress are, on the one hand, the economic and structural conditions of Latin American agriculture (especially the problems relating to the profitability of investments) and on the other, to the low cultural level of the producers. While not discounting the importance of these factors, the reason should be sought rather, in whether there is linkage or delinkage of the overall process of the creation and implementation of new know-how.

Possible technological innovations have been linked to the influence of hegemonic groups, which are identified with the technological process itself. In agriculture, the hegemonic groups have been born of social groups linked to the appropriation of the surplus produced directly on the land, of groups which have access to the surplus through vertical processes originating from the marketing of these products and, finally of groups having access to the surplus created by the sale of technological innovations and the relevant inputs. Consequently, it is feasible to infer that these three groups have moved far beyond the agricultural sector and also beyond the borders of each country.

The pressure of these dominant groups has influenced the supply of technologies by the State. The State has played a preponderant role in this area, because the agricultural holdings are small and numerous and also because of the difficulties inherent in biological research, especially in view of the tremendous influence of the climatic differences. In Latin America, State supply has depended to a greater or lesser extent on demand and especially on the form of pressure from the production sectors. There is no doubt that the ratio between supply and demand has depended on the relationships between the dominant groups and the nature of the State. Thus, in the majority of Central American countries, the predominant interests of the medium- and large-landholding sectors has created a private and highly specified system of technology generation and transfer.

In the policies relating to the technological process of Latin America, the public agencies have institutionalized the power relations which already existed.[10] The tendency to transfer institutional models of generation and dissemination is related to the images with which suitable agricultural technology is defined and these images in turn have depended on the agricultural development models of the countries at the center. Consequently, there has been a tendency to use an institutional model; similar to the one which creates the technology that is considered "suitable" but which tends to ignore important alternative technologies.[11]

The models of the developed countries are geared to a specific agricultural and general situation, where the production factors have different costs and where the structure of ownership allows for a certain degree of similarity among interest groups. The situation in Latin America shows notable dislocations in its models, seeing that in many countries the research institutions rather than being clearly directed to the overall needs of the sector are oriented to reproducing foreign research which is tailored only to a specific group of producers. In the majority of these countries, experimental stations have shut themselves off and their only contact with the environment is through one stratum of farmers. The researchers, many of them with limited resources, frequently pursue studies with no priority for the environment or with very limited scope for application to it.

The penetration of the new model of development in agriculture has assumed very special features, because of the sector's peculiarities. In more open sectors, with neither the tradition nor the cumbersome structure of agriculture that model has been imposed more quickly. The manufacturing industry, for example, accelerated the modernization process when faced with the possible obsolescence of products, a situation which does not arise in agriculture. In agriculture, what happens more often is that there is a shift from one item to the other but the different responses to the model are based on systems and forms which are more suited to accepting modernization and others which are more resistant to it.

Unquestionably, the problem of dissemination has had to take on various aspects such as the cost of new inputs, guarantee in marketing, the attitude towards risk, etc. The technologies, which are usually promoted, present obstacles to their application in small holdings as has been observed in many cases of research.[12]

Technological change, which is oriented towards the modern sectors which are virtually monopolized by the large farmers, has caused the development of technologies based on empirical knowledge to be relegated and postponed in order to accommodate imported technologies. Almost all the traditional knowledge is to be found in the peasant's knowledge. The large capitalist holdings have to a large extent determined what technologies have been chosen, since the concept of economic efficiency is usually associated only with large productive units.[13]

The analysis of the model for generating, adopting, and disseminating technology may be interpreted as a criticism of efforts by certain individuals and institutions to make scientific and technological progress. There is no doubt that all the countries have an abundant supply of technological knowledge as a result of the research done by institutions or individuals. All the South American countries have a central agricultural research institution and so the problem is not whether they exist but whether they are adapted to the technological process both in their links and relationships and in their dimensions and areas of operation. The main objection to what these institutions are doing could be that they are run by technocrats and that they are out-of-step with the necessary considerations of the socio-political framework.[14] The levels of delinkage in this relationship depend on a number of determining factors. Very often, progress has been made as a result of the outlook of the director of an experimental station or of a research worker or of a specific group, but generally the systems are to a greater or lesser extent disconnected.

In the period before World War II, agricultural innovation did not affect the rate of increase of productivity in the same way as in recent decades. The main innovations achieved in those early years were soil preparation in both dry and irrigated land, the introduction of livestock breeds, and a number of farming methods such as pruning, crop rotation, etc. Although other innovations did exist they were only applied to a limited extent.

The pace of technological change in Latin America has changed considerably since the green revolution. Systematic genetic research was undertaken to obtain better varieties of products, and the selection was outstanding because of its yield, early maturing, climatic adaptation, and resistance to pests and diseases. Many pesticides began to be used, new irrigation systems were adopted, and new techniques for soil preparation and development were designed and fine-tuned. Agricultural machinery revolutionized all agricultural work and changed the productivity indicators of the labor force. In addition to providing traction and power, it brought all kinds of sowers, harvesters, and packers. Physical and chemical analyses of soils, nutrient determination, water quality, moisture, etc. were improved. The breakthroughs in meteorology contributed to a better understanding of climatic phenomena and helped to determine the qualities of various locations more accurately. In respect of fruit growing, apart from the vast number of new varieties, the new technology improved grafting and pruning methods, the use of hormones and plant regulators, and leaf analysis methods to detect nutritional deficiencies and determine the best periods for harvesting. Techniques for better and more efficient use of water have been added and research has been done on all types of herbicides for weed control.

The animal genetics experts made great progress in the livestock-production technologies. There are now new breeds and new hybrids with better productive characteristics such as weight increase, early maturity, food conversion, robustness, resistance to disease, greater productivity and better quality of non-meat products such as milk, wool, etc.

In the forestry sector, there has been improvement in the species and varieties, in the quality of wood-bearing species, in growth, thickness, etc. The increased productivity is largely the result of the many technical qualities of the new forestry machinery.

In the region, other new technologies have constantly been incorporated, such as the use of plastic, radioactive isotopes, climatic control systems, such as greenhouses, combating hail storms with rockets and nets, frost prevention, etc. There has been ongoing improvement in pest and disease prevention and control in plants and there now exists a host of new fungicides, eel-worm killers, insecticides, and mite killers.

The important place that the green revolution has been given in the changes in agriculture in Latin America requires a special analysis of this process and the facts outlined in the previous chapters explain its real importance.

All the countries in Latin America introduced many technological innovations. Brazil and Paraguay increased their production far beyond expectations. From the beginning of this decade, the increase in yields had been declining in almost all the countries and a series of problems limiting growth began to appear. These problems were termed "second generation problems" and were, according to many researchers, more political than technical. The solutions to them therefore depended largely on the politicians and to a lesser extent on the large landholders.[15]

The optimistic projections for the green revolution failed to materialize because of their technocratic simplicity. They did not take into account its link with the new method and furthermore, no consideration was given to the fact that they had to be implemented in areas where social problems were a limiting factor. Those projections considered only the genetic origin, without taking into account the complex evolution that had taken place. There is no question that the revolution originated in the significant improvements and progress in genetics (especially in quantitative genetics) during this century, therefore, the link between the progress made in this science and the development of cybernetics cannot be overlooked. The systems of improvement and in particular, those of plant selection, have been based on a better understanding of genotypes, particularly in the mechanisms of reproduction and heredity. The development of statistics has made it possible to devise varied experiments where, essentially, the phenotypic quota of the environmental variation can be measured. In this way, plant improvement was not reduced to just the traditional techniques of observation and natural and artificial selection but new methods and systems were introduced, testing the different combinations which the technical knowledge provided.

But to the question of whether there were no high productivity species and varieties before this development took place, the answer is that there were genetic banks which did have such potential as the species and varieties of the "green revolution". The reasons why these species and varieties were not disseminated have to do with the views held by each country and with the role of their respective agricultural systems in the world as a whole.

Since the "genetic birth", the green revolution has been linked to two basic factors: water and energy. Its evolution within the structural limits of each country was closely linked with irrigation works and the series of technological means ("packages") which included energy contributions.[16]

The new species and improved varieties made such a strong first impact that they displaced the old ones. This displacement began on the more fertile soils since the entire genetic potential could be developed on them. As a result, the picture of impact of the green revolution was multiplied, since the new varieties found an almost unlimited environment. The strong support given to irrigation works, as a rural development strategy, contributed to the firm establishment of significant changes. The increase in production was due therefore not only to greater productivity but also to the fact that many dry soils were now being irrigated. Subsequently, when areas of poorer quality were farmed, importance of the whole series of technological packages not immediately evident at the start was revealed.

The technological change triggered by the green revolution became a weapon of social polarization.[17] The assertion that the green revolution was oriented to technologies adapted to the dominant class appears to be without foundation;[18] the explanation for the technological orientation should be sought in the hegemony of the developed countries and in the appropriation of the surpluses that flowed from the technological change and the use of inputs.

An evaluation of the consequences of the green revolution confirms the argument that they contributed to this polarization. Tuchman bases this assertion on the Mexican case.[19] Usually, the right technical and cultural conditions did not exist so that the new technologies could reach the lower income sector. In other words, the implementation of the genetic improvements is a part of a technological package to which the peasant has no access. The extreme artificialization on the one hand, and the cultural specialization on the other, are two factors which reduced the peasants chances of survival.

Furthermore, the high expectations of the genetic improvements led to "the indiscriminate inclusion of new technologies in regions where the soil was not suitable and in many cases, caused even more erosion, the subsequent decline in yields, desertification, etc."[20]

The green revolution should therefore be regarded as an important technological change but which has been implemented to suit interest groups or enterprises. Its effect on development and especially on the physical environment has not confirmed the expectations of the experts.

V. CONCLUSIONS

There is no gainsaying that agricultural development in Latin America has reached a stage where its direction will be determined by the role that the countries assign to this sector.

The creation of foreign exchange, at any cost, to face the external debt problem cannot be the basis of this sector's strategy. The worsening of some food consumption problems, the deterioration of the environment, its effect on the availability of resources, and the increasingly unfavorable conditions of the sector in relation to the terms of trade are influencing more elaborate strategies which go beyond regarding agriculture as an interest sector to "earn foreign exchange".

The method of development which currently prevails and which has been gaining ground since the beginning of this decade not only does not solve rural development problems and peasant poverty but it makes them worse. Agricultural growth was never further away from rural development. What is more, in the last 30 years, the Latin American countries lost a significant degree of their food security because of the productive specialization demanded by the international market.

Deforestation which has been taking place at a truly impressive pace, the erosion of a significant percentage of the soil and its depletion, the silting of the river beds, the salinization of vast irrigated areas, the micro- and macroclimatic changes caused by the environmental changes, the loss of numerous species of flora and fauna, some of them unique and irreplaceable and finally, the serious desertification taking place in many areas of the region demand a rethinking of agrarian strategies and policies and a more scientific and political view of what is meant by "development" and "modernization" in agriculture.

REFERENCES

1. **Borah, W.**, The historical demography of aboriginal and colonial Latin America: an attempt at persepective, Actas del XVII Congreso Internacional de Americanistas, 1966.
2. **Borah, W.**, The historical demography of aboriginal..., *op.cit.*, in *Population and Economics*, University of Manitoba Press, Winnipeg, Canada, 1970.
3. **Denevan, W.**, The aboriginal population of tropical America, in *Population and Economics*, P. Duprez (Ed.), Winnipeg, University of Manitoba Press, 1966.
4. **Furtado, C.**, *La economia latinoamericana desde la conquista Ibérica hasta la Revolución Cubana*, Estudios Internacionales, Editorial Universitaria, Santiago, Chile, 1970, p.21.
5. For more details on the productive structure see, **Santamaría, D. J.**, La estructura agraria del Alto Perú a fines del siglo XVIII, *Desarrollo Económico*, Vol. 18, No. 72, Buenos Aires, January—March 1979, pp. 579—596.
6. **Miller, E. V. A.**, Agricultural Ecuador, *Geographic Review*, New York, 1959, pp. 183—207.
7. For more details, see, **Di Filippo, A.**, *Raíces históricas de las estructuras distributivas de América Latina*, Cuadernos de la CEPAL, No. 18, Santiago, Chile, 1977.
8. For further details see, **Cariola, C. and Sunkel O.**, *The Growth of the Nitrate Industry and Socioeconomic Change in Chile: 1880—1930*, CLACSO and Joint Committee on Latin American Studies, Santiago, Chile.
9. **Cortés Conde, Roberto**, Patrones de asentamiento y explotación agropecuaria en los nuevos territorios argentinos (1890—1910), *Tierras Nuevas*, El Colegio de México, 1973, pp. 105—120.
10. See, 1) **Sagasti, F. and C. Guerrero**, *El desarrollo científico y tecnologico de America Latina*, IDB/INTAL, Buenos Aires, 1974, p.200.
11. **Urquidi, V. and A. Nadel**, Algunas observaciones acerca de la teoría económica y el cambio técnico, El Trimestre Ecomónico, México, Vol. XLVI (2) No. 183, pp. 211—234.
12. **Ferrer, A.**, *Tecnología y política económica en América Latina*, Editorial Paidos, Buenos Aires, 1974.
13. **Herrera, A.**, Tecnologías científicas y tradicionales en los países en desarrollo, *Comercio Exterior*, Vol. 28, No. 12, Mexico, December 1978, pp. 1462—1476.
14. **Perrin, R. and D. Winkelmann**, Impediments to technical progress on small versus large farms, *American Journal of Agricultural Economics*, Vol. 58, No. 5, December 1976, pp. 888—894.
15. **Bolvitnick, J.**, Estrategias de desarrollo rural, economía campesina e innovación technológica en México, *Comercio Exterior*, Vol. 26, No. 7, pp. 813—826.
16. **Gastal, E.**, Los sistemas de producción y la planificación de la investigación agrícola, *Desarrollo rural en las Américas*, Vol. VII, No. 1, January-April 1975, 55.
17. **Street, J. H. and D. J. Dilmus**, América Latina y la brecha tecnológica, *Comercio Exterior*, Vol. 28, No. 12, Mexico City, 1978, pp. 1505—1515.
18. **Brown, L. R.**, *Seed of change: The green revolution and development in the 1970's*, Pall Mall Press, London, 1970.
19. **John, C.**, A review of governmental policies and techniques for keeping farmers farming, *Natural Resources Journal*, Vol. 19, No. 1, January 1979, Albuquerque, New Mexico, pp. 119—144.
20. **De Janvry, A.**, The political economy of rural development in Latin America: An interpretation, *American Journal of Agricultural Economics*, Vol. 57, No. 3, August 1975, pp. 490—499.
21. *Ibid.*, p. 498.
22. **Tuchman, B. H.**, The green revolution and the distribution of agricultural income in Mexico, *World Development*, Vol. 4, No. 1, Washington D.C., 1976, pp. 17—24.
23. *Revista de Comercio Exterior*, Editorial: Alimentación, crisis agrícola y economía campesina, Vol. 28, No. 6, Mexico, June 1978.

10 THE INDUSTRIAL MODEL AND ITS IMPACTS ON SMALL FARMERS: THE GREEN REVOLUTION AS A CASE

Kenneth A. Dahlberg

I. INTRODUCTION

It is only in the past several decades that the pervasive impacts of industrial society upon both social systems and natural environments has led to widespread re-examination of its assumptions and prospects.[1-3] Unfortunately, there has been little linking of this larger debate with the vigorous discussions and debates that are going on regarding the nature, assumptions, and prospects of industrial agriculture — discussions that are themselves fragmented between analyses of what is happening at home and those focusing on the Third World. Such fragmentation is one of the characteristics of industrial society — where both the intellectual and the practical worlds are divided up into progressively more specialized compartments.

II. THE INDUSTRIAL MODEL

This essay will seek to overcome some of this fragmentation by providing an overview of the broad characteristics of the industrial model, its historical application to society, its specific impacts on agriculture and farmers, and its spread to the Third World through what is termed the Green Revolution. To do this, we need to examine some of the major cultural values and assumptions involved in the evolution of industrial society.

A. Underlying Cultural Values and Assumptions

These can best be sorted out by their level and historic compass.[4] At the broadest and deepest level, there is a set of Western cultural assumptions and values that are fundamental to our worldview. Combining Judeo-Christian and Greco-Roman roots, this Western worldview includes basic beliefs regarding man's separation from and dominion over nature, the hierarchical nature of society, and the role of both reason and law.

The Renaissance and Reformation laid the foundations for the Industrial Revolution by encouraging a shift away from the religious values and institutions of the Medieval period towards the secular state, individualism, rationality, science, and exploration.[5] This process of secularization also involved a transformation of society's visions of the future. Rather than a transcendental heaven (or hell), the possibility of humankind building its own "utopia" gradually emerged. From Thomas More's original vision onwards, this new future was seen as primarily urban and was based upon reason and engineering — as illustrated in the concluding section of Goethe's *Faust*.

This early vision of a rational, civilized, urban, and more humane society was given both direction and specific content as the Industrial Revolution gradually developed both the technologies and institutions to embody this new vision. Ironically, however, these developments depended upon and were made possible by a range of important improvements in agriculture — termed by some the "agricultural revolution."[6,7] Once the building blocks were available, the embodiment of this vision was facilitated by a key concept, functional specialization, which involves a particular way of conceptualizing and organizing work whereby the production process is defined as the predominant factor to which both humans and the environment are subordinated, rather than vice versa.[8]

The power of functional specialization to shape institutions and technologies derives from the increasing importance of the value placed upon efficiency — something that is assumed to result from its application. Today, functional specialization has become self-reinforcing, not only through the vested interests of current institutions, but through its intellectual progeny — a range of disciplines, especially economics — which make its claims to efficiency seem plausible through neglect of a range of externalities.[9]

B. Historical Application to Society

The above suggests that a slow historical process was

involved whereby a particular set of institutions and technologies were built upon the values emerging from the Reformation and Renaissance, which themselves grew out of the larger Western world view. Many earlier values of the Medieval period remain with us either as values that are consistent with industrialism or as recessive values which may be drawn upon should we enter another period of broad social transformation. In this regard, the rapid growth of academic interest and programs in Medieval studies over the past decade suggests both a critical questioning of industrialism and a search for alternatives.

The growth and progressive interlinking of the institutions and technologies forming the industrial model — the secular state, representative democracy, separate and formalized economic systems, state supported scientific, technological, and educational systems — all have become dependent upon continued industrial growth and expansion. That is to say, they have a vested interest against seeking alternative institutional and technological approaches (and values) even as the various resource and social foundations upon which industrialism has been built are increasingly threatened.

At the most fundamental level, these vested interests involve an almost unconscious focus on certain aspects of society to the neglect of others. Industrial society has four clear foci: the industrial, the urban, formal systems, and technological systems. The corresponding areas of neglect are the agricultural, the rural, informal systems, and natural systems.[4,10] It can also be noted that formal and technological systems have been historically and institutionally associated with men and male role models, while informal and natural systems have been more associated with women and female role models.

All of these neglected areas and systems — which, as suggested above, provided the natural, human, and capital resources upon which industrialism was built — have been progressively incorporated into the formal systems of industrialism. This has often involved the enclosure of land, forests, fisheries, or other common property resources. This process typically involves extending the scope and level of formal organizational power — whether of the state or corporation — and placing lower level groups and environments under the jurisdiction and control of higher level institutions.

One effort to balance larger community or organizational needs with local or individual needs has involved the creation of federalism and representative democracy. This system, which allowed for adaptive change, has over time been progressively captured both by the power of large vested interests and by the requirements of operating large-scale industrial systems. Along with this, the process of incorporation continues even today, and in two important ways: one is the incorporation of the traditional resource regimes and common property resources of the Third World into those states, and thereby into the international industrial system; the other, occurring within the industrial countries, involves the extension of the concept of "ownership" (whether state or private) to plant and animal species.[11]

III. THE INDUSTRIAL MODEL AND AGRICULTURE

A. Contextual vs. Universal Approaches

One of the greatest areas of difference between agriculture and the industrial model lies in the great contextual variability of agriculture — where climate, weather, soil, topographic, resource, cultural, social, and institutional variations all profoundly affect its viability — as contrasted to industry, where the universalism and rationalism of modern science and technology promote blueprint approaches whereby local variation is fitted, often procrusteanly, to the design.

Tension and conflict between universal and contextual approaches is found in other areas as well. Legal, historical, humanistic, and artistic fields traditionally reflected the variability and contextuality of their subject matter. Today, however, academics in these fields are under increasing pressure to apply strictly rational and universalistic models and approaches.

An added difficulty, particularly as it applies to agriculture, is that when modern researchers and analysts view it through their universalistic lenses, contextual variations are filtered out, not to mention evolutionary adaptations. What this means is that we have a hard time understanding both the broad interactions between agriculture and the larger society as well as the changing nature of agriculture itself — whether in the past or the present.

I have argued elsewhere that we need to do at least three things in a contextual approach: (1) view the broad evolutionary changes which have been involved in the changes from hunting and gathering societies to agricultural societies to industrial societies; (2) view how the various "development plays" (each involving a century or more) take place within the larger evolutionary theater; (3) examine how shorter term policy matters fit into these larger (and longer-term) contexts.[12]

Because most historical studies do not employ such a broad contextual approach, however, they do exhibit considerable confusion in trying to understand the larger historical development and transformation of modern agriculture — whether in Europe or in its exportation to other parts of the world. In Europe, the debate has centered on identifying the period and nature of the "agricultural revolution" referred to above. In a lucid discussion, Grigg has argued that in different analyses, the "revolutionary" period is defined according to what is perceived to be the source of the revolution.[13] Analysts of Marxist bent have emphasized radical changes in land tenure — such as those culminating in the enclosure acts — and the resulting changes in farm size and labor supply. A second approach emphasizes the development and adoption of new farming methods and technologies — which correspond to a somewhat different historical period. A third school seeks to apply contemporary measures of output, rates of growth in productivity, etc. to determine when the most radical changes occurred. Using this latter approach as his measure, Grigg concludes that "it seems that European agricultural change was evolutionary rather than revolutionary until this century. It is the period since the 1930s that merits the title."[13]

Each of these approaches has its defenders and critics — the debate is often being fueled either by the absence of key data which might strengthen or weaken a particular position or by the complexity of the variations found between countries, regions, crops, technologies, farm size, labor systems, etc. Beyond that, there is the problem that most approaches do not explicitly develop their theory of social change. Marxist interpreters do, but their theory — like that of other 19th century social theorists — is often too universalistic and monocausal. It does, however, have the virtue of suggesting the need to examine structural issues, the larger shifts between agricultural and industrial sectors, rural and urban regions, as well as between classes.

Whatever the contextual variations involved, one can see in looking back that several broader patterns have evolved. One is the gradual replacement of labor with capital (mechanization) and energy (fertilizers). The former was emphasized in 19th century U.S. agriculture because of the need to farm large acreages with few laborers. The latter was emphasized in 19th century Europe where land was scarcer and labor was somewhat more plentiful. Since World War II both of these substitutes for labor have been promoted in both regions, along with a variety of other scientific and technological inputs. Linked with this has been the decline in the numbers of farmers per hundred in the population — even though employment in the total food system has been much more stable.

Important as it is to examine past gradual shifts in the relative factor proportions — and even more so the dramatic shifts projected for the coming decades[2,14] — this does not enable us to get a clear long term picture of the interactions between natural systems, social systems, and technological systems. To do this, we need to describe and analyze the successive overlays of institutions and infrastructures that have progressively incorporated agriculture into an ever growing and more centralized industrial society.

B. The Incorporation and Transformation of Agriculture

In the industrial world, these processes can be seen to involve five major systems: (1) security systems; (2) political/legal/administrative systems; (3) economic systems; (4) infrastructural systems; and (5) cultural systems. The general process has involved four major trends: (1) a progressive formalization of each of these systems; (2) an increase in the size of each of these formal systems through their penetration and incorporation of previously informal and natural systems; (3) an increase in the power of each of these formal systems through their gradual interlinking and mutual reinforcement in a progressively more centralized and integrated state; and (4) the increasing role of military-industrial complexes in each major industrial country.

Without trying to describe the complex interactions between these systems and trends, it may be useful to note how each has affected agriculture. The security system would seem to have little importance historically. However, one of the major shifts between medieval and modern society involved the shift in the locus of power and security from the local lord to some higher ruler. Peasants and farmers have always been afraid of local wars and their impacts upon their farms and families, not to mention the impacts of foreign wars upon them in terms of taxes and conscription. Equally, local lords and higher rulers worried about peasant revolts and uprisings. The creation of both formal police and military systems reduced many of these threats. Today, the concerns and language are very different. We speak of "national food security" and the possible use of food as a "weapon". What this indicates is that increasingly agriculture is becoming fully incorporated into national security and economic planning — something that means a corresponding loss of autonomy for both farmers and agriculture as an economic sector.

The expanded power and scope of political, legal, and administrative systems has had different impacts at different times. The expansion of democracy when societies were largely rural meant that these systems served rural constituencies — or at least their rural elites. As industrial societies have become urbanized, political power has shifted and one of the basic dilemmas for farmers has been the political pressures generated for "cheap food policies" for the urban masses. Beyond this, farmers have increasingly lost political power vis-a-vis large banking, agribusiness, food processing, distributing, and marketing interests. This has meant that increasingly they have had to fit into the structures and policies dictated by these other groups. Finally, farmers have been increasingly subjected to a host of administrative rules and regulations — from the local to the national level — involving health and safety, land use, environmental impacts, etc. This is not to say that these rules and regulations are not necessary given the structure and impact of modern industrial agriculture; however, their universal and "blueprint" character may make local adaptations more difficult in the future.

The scope and implications of the formalization and transformation of the economies of industrial countries has been detailed by Polanyi.[15] It is here that perhaps the greatest changes have occurred for farmers — who have gone from largely independent yeomen to dependent cogs in a much larger production system.[16] This dependence reflects not only the declining political power of farmers, but their increasing incorporation into larger market and infrastructural systems over which they have little influence. What political power they have is often exercised by and for the benefit of the large and corporate farmers — something that makes structural reforms in the direction of sustainable agriculture particularly difficult.

The expansion and transformation of infrastructural systems has also been dramatic—particularly the shift from reliance on renewable to nonrenewable resources already mentioned. While providing ways to deliver large, concentrated amounts of energy for mass production systems, these "hard path" systems extract increasing economic, health, environmental, resource, and social costs.[17,18] It is not only energy systems that are "hard path" in structure; production systems, transportation systems, irrigation systems, and communications systems all share similar characteristics of being centralized, energy and capital intensive, and complicated to manage and repair. Thus, the greater their effective scope, the greater energy they require, the greater power and control they exercise, and the greater entropy they create elsewhere. Perhaps the most threatening long-term example of this is the loss of genetic and biological diversity created by the expansion and spread of industrial systems.[11,19]

Finally, the expansion and transformation of cultural systems has also affected farmers and rural culture. In Europe, formal education was always urban oriented and elitist. In the U.S., the needs of an expanding rural population were met through the one room school house and the creation of the Land Grant system — both of which were practical and rural in orientation and available to the masses. However, a complex set of changes in the 20th century — deriving from the introduction of automobiles, rural electrification, the spread of radio (and later TV), school consolidation and state teacher certification — undermined rural culture and made both urban and consumer ways of life appear more attractive than rural ways.[20,21] Also, even though they are still sometimes seen as a repository of rural conservatism and virtue, farmers — like the rest of society — have lost many of the religious values which encouraged frugality and saving for the future.

IV. COLONIALISM AND PRECURSORS TO THE GREEN REVOLUTION

Many of the processes, trends, and practices involved in the incorporation and transformation of agriculture in the industrial world are especially visible in the historical export and imposition of Western ideas, institutions, technologies, and infrastructures upon the non-Western world — first through the conquest of the New World, and then colonialism. Throughout, the assumption and rationale was that Western culture, civilization, economic and political institutions, and religion were superior, thus justifying the use of the West's genuinely superior military technology to impose or encourage them. Many of these attitudes are still found in post-independence approaches to "development" and "technology transfer".

This is seen most clearly in the treatment of the indigenous peoples of Africa, America, and Oceana who were either physically expelled from desirable lands or forced to become subservient. Once some sort of legal rationale for control was established — often through ingenuous treaties — these self-sufficient people were forced into both labor and economic markets through head taxes, labor levies, and trade monopolies. Attempts were made not only to convert them to Christianity, but to discourage or prohibit traditional religious practices. Formal education — with its implicit rejection or denigration of oral cultures — was also promoted.

The most dramatic, but little known impact of this collision of cultures was the massive loss of life — conservatively estimated to be some 28,000,000 people.[22] These indigenous peoples are still among the most misunderstood and mistreated in the world whether they reside in the Amazon jungle or on a North American reservation. They are mentioned here because they provide the most extreme example of the conflicts between Western culture and industrial society and non-Western and non-industrial societies and show most clearly how Western societies have sought to impose their beliefs and institutions. The approaches pursued with the various civilizations of North Africa, the Middle East, and Asia were more subtle and often tempered because Westerners were faced with larger numbers and military power, as well as the presence of formal governing and economic systems.

The great trading companies of the 17th and 18th centuries were supplanted with formal colonies as nationalism and great power rivalry in Europe made them militarily strategic. The infrastructural patterns laid down — whether in the form of roads, railroads, canals, irrigation systems, or telegraph lines — were in a funnel pattern designed to extract minerals or cash crops and ship them to the metropolitan country. The administrative structures were designed to maintain law and order and to facilitate the exploitation of the colony's resources while also opening markets for processed goods. As higher education was gradually introduced, the urban, elite, and academic model prevalent in Europe was used—something Third World countries still suffer from today.

In agriculture, the Green Revolution had many precursors — although in two rather different forms. Tropical crops such as sugar cane, indigo, coffee, tea, groundnuts, and rubber were developed along industrial lines. That is, large-scale production facilities were created—called plantations rather than factories — often displacing small-scale local subsistence producers through land expropriation. Labor was also handled in an industrial manner—with the resulting creation of a dependent, if not a slave work force. The scale, structure, and management of these plantations all reflected Western notions of industry and economic rationality. Their direct linkage with the factories and markets of the metropolitan country was a major factor in creating the dual economy of most colonies.

In the white settler colonies, precursors to the Green Revolution could be seen in the practice of importing temperate zone crops, animals, and cultivation techniques, not to mention Western agricultural research institutions.[12,23,24] Settlers and colonial adminstrators not only believed in their superiority over local crops and practices, but also shared the still common Western cultural belief in the neutrality of technologies — that is, that technologies are neutral in and of themselves and can thus be *transfered* to other locations with the expectation that they should work as well there as in their place of origin.

Important as these precursors were as both precedents and indicators of deep-seated assumptions about agriculture, it must be recognized that their impact on the vast majority of rural subsistence peasants, nomads, and villagers was very small. The foundations for the later rapid expansion of the Green Revolution lay more in the creation of centralized infrastructures and centralized administrative control systems which enabled relatively small Westernized enclaves — symbolized by Western-style cities — to penetrate the colonies and extract their resources. After independence, these centralized systems were progressively expanded through various "development" programs to encompass and control the rural countryside and small and subsistence farmers. The Green Revolution has been an important element in this process.

V. THE GREEN REVOLUTION

A broad overview of the industrial model and its development has been presented to clarify the forces at play in its superimposition on the colonies. It was also shown to have had a major role in transforming and incorporating agriculture at home, while at the same time setting the stage for a similar process in the Third World — a process facilitated by the export of industrial institutions and infrastructures as well as certain agricultural precursors to the Green Revolution. While the following review of the process of the Green Revolution's development, its impacts, and the different ways in which it has been evaluated is brief, the broader context in which it has been set will hopefully bring out a number of important considerations little dealt with in the extensive literature on the topic.[25,26]

A. The Process

Two key components involved in the development of the Green Revolution directly reflect the industrial model: (1) the priority given to increasing production; and (2) the belief in the neutrality of technology. The first was clearly the underlying assumption of both the plant breeders sent to Mexico by the Rockefeller Foundation in the 1940s and the Foundation officials themselves. Both felt that increasing production was the way to meet the rising spectre of hunger — seen to be caused by rising populations — while also improving agriculture in Mexico and other Third World countries. By this time both U.S. and European agriculture were increasingly dominated by the "production paradigm", and it was this paradigm that was embodied in, and exported through the Green Revolution.

The success of this paradigm in the U.S. resulted in part from the failure of various attempts to include matters of rural welfare and reform in USDA programs. These were squashed in Congress by Southern agriculture committee chairmen who feared that they might upset the rural power balance in the South and encourage desegregation.[16] Besides these political disincentives, U.S. experiment stations were (and still are) dominated by disciplinary specialists — largely from the natural sciences — which has meant that measurable increases in crop production have been favored over more complex and difficult to measure improvements in rural nutrition, soil conservation and improvement, general rural development, etc. In addition, increases in production through new hybrids or chemical imputs tended to favor the first adopters — generally the larger and better-off farmers, whose farm organizations in turn supported further work in this direction in the state legislatures.

As mentioned, the dominant and unexamined assumption then — and still largely today — was that increases in production are the best way to solve the problems of hunger. This convenient, but dubious assumption permitted the U.S. plant breeders in Mexico to focus on doing what they knew best — developing new varieties — while at the same time building a whole new set of international institutions and careers. To have questioned this assumption and to have seriously asked the question "how can we best improve the rural welfare of the masses of poor peasants in the Third World?" would have led to an entirely different scenario. Since most poor peasants in Mexico farm dryland, they would have had to be the focus. Production studies done later showed that traditional improved varieties do almost as well as high yielding varieties under dryland conditions and that it is more important to focus on the monitoring and managing of soil moisture as well as the creation of credit systems.[27,28]

Moreover, a welfare focus would have gone beyond a farm-level analysis to bring in a concern for everything from nutrition to marketing arrangements to land and labor patterns to soil erosion problems, not to mention the adequacy of regional and national development plans.[29] Plant breeders would have been only one component of larger interdisciplinary teams and efforts that would have included anthropologists, rural sociologists, nutritionists, ecologists, small animal breeders, and many others — all of whom are still only occasionally represented in the various international agricultural research centers.

The second key component in the Green Revolution involved a still dominant Western belief in the neutrality of technologies. The plant breeders sent to Mexico by the Rockefeller Foundation did not realize that the high yielding varieties they developed were neither economically, socially, nor politically neutral. Economically, the seeds were the product of a U.S. agricultural heritage where the relative factor proportions (little labor, lots of land and energy resources, and lots of capital) led logically to a definition of efficiency in terms of labor efficiency (achieved by substituting energy and capital for labor), an approach which also allows for high production as long as energy is cheap and easily available.

In the Third World, most countries have different relative

factor proportions (lots of labor, little arable land, few energy or capital resources) that traditionally and logically have defined efficiency in terms of energy and capital efficiency. Also, security of production was more important than high levels of production. A lack of awareness of these basic differences still exists today, and most development aid suffers from the failure to recognize that these basically different concepts of productivity and efficiency require different technologies, institutions, and economic approaches than those found in the energy, resource, and capital rich countries.

Socially and politically, the seeds tend to favor those farmers and regions that are already better off. Even though these tendencies were entirely predictable, and even vigorously argued against by early critics such as Sauer,[30] the developers of the "high yielding varieties" (HYVs) followed a universalistic "blueprint" approach, with adaptations to local conditions seen as necessary only for such physical differences as photoperiod and the length of the growing season. While there have been efforts in recent years to adapt the HYV package to semi-arid and low-input settings, the international agricultural research centers are still dominated by the perspectives, assumptions, and priorities of plant breeders.[31]

B. The Impacts

Given that technologies in general (1) reflect the natural and institutional environments where they were developed; (2) have inherent design characteristics and operating requirements; and (3) will have different impacts in different receiving environments, it is not surprising that the widespread introduction of the HYV package — the technological core of the Green Revolution — has led to very different impacts. The receiving environments have varied in almost every aspect: from natural conditions (soil conditions, water, weather, and climate regimes, pests and pathogens) to social conditions (size of farms, land tenure systems, openness to innovation) to economic conditions (availability of credit, labor, and markets).

Not only have there been varied impacts, but varied interpretations of those impacts. As Chambers has pointed out, there has been a polarization of interpretations from the beginning: between the "positive optimists," mainly the biological scientists and the leaders of the various organizations involved in the development and promotion of the Green Revolution, and the "negative pessimists," largely social scientists concerned about its political economy — that is, who gained and who lost.[32] Chambers argues that the optimists made two main errors in their assessments: (1) they assumed that the early gains made in favorable locations could be rapidly spread elsewhere; and (2) that hunger was a problem of production, not poverty. The pessimists, for their part: (1) were much too narrow and selective in the areas they examined; and (2) tended to attribute most changes to the HYV package rather than other possible causes such as population pressures, land subdivision, inequal water distributions systems, etc.[32]

The above suggests several important points for evaluating both shorter- and longer-term impacts of the Green Revolution. First, that much more "profiling" of the various receiving environments needs to be done so that one can evaluate how the original HYV package affected both production and the political economy in any given *type* of receiving environment. Second, more attention needs to be paid to "generational" effects; that is, to the various iterations and adaptations that occur over multiple growing seasons — in the production systems, in the technological inputs, and in the political economy of both individual farmers and laborers, and their village/region.

In addition, there is a need to evaluate a range of costs and benefits that are not currently included in most analyses: those that relate to the interacting energy, environmental, and resource systems upon which agriculture depends. This will require institutions which can encourage greater interdisciplinary knowledge and teamwork by bringing together people working on ecology, energetics, ethnobotany, conservation biology, social forestry, common property systems, agroecology, water systems, resource regimes, etc. Some of the common elements found in these fields are (1) the use of systems approaches; (2) the use of longer time horizons; and (3) their inclusion of informal as well as formal systems in their work and analyses. Even so, they often are not fully aware of the non-neutrality of technologies, nor of the longer-term evolution of industrial society and its structures.

VI. CONCLUSIONS

What then about the small peasant farmer in the Third World? Given all the contextual variation described above, can we still identify any general trends or impacts? Before trying to answer this question, it is perhaps useful to review the impacts of the industrial model on small farmers in the industrial world. The general impacts of the industrial model on agriculture — leading to its incorporation and transformation — have already been discussed. What was shown was that a complex set of interactions and overlays of different systems over several centuries led to a post-World War II situation where rapid growth and extension of "hard path" industrial approaches to agriculture transformed both its resource base and its structure.

This has had two major consequences for small farmers. One is the increasingly rapid decline of small farmers in the traditional sense — of those family farmers who chose farming both as their livelihood and as a way of life.[33] These are being replaced today with a new type of small farmer — the part-time farmer who earns a good portion of his or her livelihood off the farm. Technological changes, crop specialization, cheap food policies, unstable markets, international grain traders and speculators, the decline of rural culture, and rising land values have also all contributed to pressures to increase the scale of production — the "get big or get out" mentality. The non-neutrality of policy and price support instruments — which enable larger and corporate farmers to take greater advantage of them — have exacerbated the situation.

It is of course ironic that these pressures toward greater size and centralization are accelerating precisely at a time when the high economic, social, health, environmental, and resource costs of larger units are becoming clearer.[34] Thus, the battles between those advocating conventional approaches and those promoting alternative, smaller-scale, and more sustainable and regenerative approaches. All of this suggests a twofold reevaluation of the continued export of the industrial model through Green Revolution approaches, particularly its newer and high technology elaborations — biotechnology and genetic engineering. First, there is the question of whether we should be exporting and promoting what may well be obsolescent domestic approaches and models. Second, there is a need to examine traditional and more sustainable agroecosystems in the Third World — which are much more contextually adapted for — in terms of their potential to be upgraded.

Other chapters in this volume do that. What needs to be pointed out here is that any significant movement towards more locally adapted, small farmer oriented, and sustainable agricultural systems will be going against both the infrastructural and institutional grain that has been built up during both the colonial and post-independence periods. As Diwan and Livingston have argued, "alternative development strategies" will have to replace "conventional development strategies" if there is to be any longer term change in direction.[35] And beyond national level planning and strategies, the many interacting and interlinked trade, resource, market, and monetary regimes that increasingly lock all countries into a world industrial system call for a critical examination and the development of global strategies for decentralization.[36]

The intellectual and creative task here is great — for it calls for both a vision of an ecologically viable future based on a combination of current and emerging values and goals as well as the imagination to chart traversable paths from current obsolescent institutions to new ones compatible with the new vision. Crucial to this latter task is the basic challenge of how to transform the larger systems and regimes of the industrial model given the power of the vested interests that control and benefit from them. In a sense, the task may be comparable to that of decolonialization — where creative minorities both inside and outside the metropolitan governments were able to encourage the dismantling of an obsolescent and unjust system.

Another part of the task — and equally crucial — involves the main thrust of this volume: the search for locally adapted, ecologically sustainable, and socially just agricultural and food systems. As yet, there is more focus on agricultural than food systems, but the search for locally adapted and regenerative agricultural systems will encourage a gradual shift towards a food systems approach, which in turn will reinforce efforts at a rural development that is not just the handmaiden of urban/industrial interests and pressures.

Researchers in this area need to be cautious that they do not make the same types of assumptions that Green Revolution researchers made. For example, just as a focus on production does not necessarily move one towards reducing hunger, a focus on small farmers does not necessarily move one towards regenerative and socially just systems. The interactions between ends and means need to be constantly monitored, particularly because of our strong cultural belief in the neutrality of technologies. Beyond this, researchers need to see the creative dimensions of their work and seek to stretch their imaginations. As Bayliss-Smith concluded after reviewing the energy efficiencies of rice systems in Asia: "The analysis of energy costs and energy returns shows that a truly appropriate technology for the poorest regions of the tropical world has yet to be revealed."[37]

REFERENCES

1. **Meadows, D. H., et al.,** *The Limits to Growth*, Universe, New York, 1972.
2. *The Global 2000 Report to the President*, Council on Environmental Quality, Washington, D.C., 1980.
3. **Valaskakis, K., Sindell, P. S., Smith, J. G., and Fitzpatrick-Martin, I.,** *The Conserver Society: A Workable Alternative for the Future*, Harper Colophon, New York, 1979.
4. **Robertson, J.,** *The Sane Alternative: A Choice of Futures*, River Basin, St. Paul, MN, 1979, chap. 1.
5. **Tawney, R. H.,** *Religion and the Rise of Capitalism*, Penguin, London, 1938.
6. **Blum, J., Ed.,** *Our Forgotten Past: Seven Centuries of Life on the Land*, Thames and Hudson, London, 1982.
7. **Chambers, J. D. and Mingay, G. E.,** *The Agricultural Revolution 1750—1880*, Batsford, London, 1966.
8. **Dahlberg, K. A.,** The technological ethic and the spirit of international relations, *International Studies Quarterly*, 17 (1), 55, 1973.
9. **Madden, P.,** Beyond conventional economics — an examination of the values implicit in the neoclassical economic paradigm as applied to the evaluation of agricultural research, in *New Directions for Agriculture and Agricultural Research*, Dahlberg, K. A., Ed., Roman & Allanheld, Totowa, NJ, 1986, chap. 11.
10. **Chambers, R.,** *Rural Development: Putting the Last First*, Longman, London, 1983.
11. **Dahlberg, K. A.,** Redefining development priorities: genetic diversity and agroecodevelopment, *Conserv. Biol.*, 1(4), 311, 1987.
12. **Dahlberg, K. A.,** *Beyond the Green Revolution: The Ecology and Politics of Global Agricultural Development*, Plenum Press, New York, 1979.
13. **Grigg, D. B.,** The agricultural revolution in Western Europe, in *Understanding Green Revolutions*, Bayliss-Smith, T. P. and Wanmali, S., Eds., Cambridge University Press, Cambridge, 1984, chap. 1.
14. **Gever, J., Kaufmann, R., Skole, D., and Vorosmarty, C.,** *Beyond Oil: The Threat to Food and Fuel in the Coming Decades*, Ballinger, Cambridge, MA, 1986.
15. **Polanyi, K.,** *The Great Transformation*, Beacon Press, Boston, 1957.

16. **Heffernan, W. D.,** Review and evaluation of social externalities, in *New Directions for Agriculture and Agricultural Research,* Dahlberg, K. A., Ed., Roman & Allanheld, Totowa, NJ, 1986, chap. 10.
17. **Lovins, A. B.,** *Soft Energy Paths: Towards a Durable Peace,* Ballinger, Cambridge, MA, 1979.
18. **Lovins, A. B. and Lovins, L. H.,** *Brittle Power: Energy Strategy for National Security,* Brick House, Andover, MA, 1982.
19. **U. S. Congress, Office of Technology Assessment,** *Technologies to Maintain Biological Diversity,* OTA-F-330, Washington, D.C., U. S. Government Printing Office, March 1987.
20. **Berry, W.,** *The Unsettling of America: Culture and Agriculture,* Sierra Book Club, San Francisco, 1977.
21. **Gilk, P.,** *Nature's Unruly Mob: Farming and the Crisis in Rural Culture,* North Country Anvil, Millville, MN, 1986.
22. **Bodley, J. H.,** *Victims of Progress,* Mayfield, Palo Alto, CA, 1982.
23. **Crosby, A. W.,** *Ecological Imperialism: The Biological Expansion of Europe, 900—1900,* Cambridge University Press, New York, 1986.
24. **Weiskel, T. C.,** Agents of empire: steps toward an ecology of imperialism, paper presented at the Forest History Society meeting, Durham, NC, May 1, 1987.
25. **Karim, M. B.,** *The Green Revolution: An International Bibliography,* Greenwood Press, New York, 1986.
26. **Anderson, J. R.,** International agricultural research centers: achievements and potential, a study prepared for the Consultative Group on International Agricultural Research, draft of August 31, 1985, World Bank, Washington, D.C., 1985.
27. **Cummings, R. W., Jr.,** Review of Plan Puebla, (mimeographed), Rockefeller Foundation, New York, 1974.
28. **Redclift, M.,** Production programs for small farmers: Plan Puebla as myth and reality, *Economic Development and Cultural Change,* 31, 551, 1983.
29. **Feder, E.,** *Perverse Development,* Foundation for Nationalist Studies, Quezon City, Philippines, 1983.
30. **Oasa, E. K. and Jennings, B. H.,** Science and authority in international agricultural research, *Bulletin of Concerned Asian Scholars,* 14, 32, 1982.
31. **Busch, L. and Lacy, W. B.,** Sorghum research and human values, *Agricultural Administration,* 15, 205, 1984
32. **Chambers, R.,** Beyond the Green Revolution: a selective essay, in *Understanding Green Revolutions,* Bayliss-Smith, T. P. and Wanmali, S., Eds., Cambridge University Press, Cambridge, 1984, chap. 18.
33. **U. S. Department of Agriculture,** *A Time to Choose: Summary Report on the Structure of Agriculture,* Department of Agriculture, Washington, D.C., 1981.
34. **Dahlberg, K. A., Ed.,** *New Directions for Agriculture and Agricultural Research: Neglected Dimensions and Emerging Alternatives,* Roman & Allanheld, Totowa, New Jersey, 1986, chapters 1, 9—12, and 17.
35. **Diwan, R. K. and Livingston, D.,** *Alternative Development Strategies and Appropriate Technology: Science Policy for an Equitable World Order,* Pergamon, New York, 1979.
36. **International Union for the Conservation of Nature and Natural Resources (IUCN),** *World Conservation Strategy,* IUCN, Gland, Switzerland, 1980.
37. **Bayliss-Smith, T. P.,** Energy flows and agrarian change in Karnataka: the Green Revolution at micro-scale, in *Understanding Green Revolutions,* Bayliss-Smith, T. P. and Wanmali, S., Eds., Cambridge University Press, Cambridge, 1984, 170.

11 THE FARMING SYSTEMS RESEARCH AND EXTENSION APPROACH TO SMALL FARMER DEVELOPMENT

D. C. Baker and D. W. Norman

I. INTRODUCTION

For about two decades, there has been an increasing focus on small farmer development strategies throughout the world. Farming systems research (FSR) has been a useful tool assisting in this process. As initially conceived, the farming systems (FS) approach was viewed as a means of improving the efficiency of agricultural research resources. However, in the early 1970s, it became evident that the integrated rural development (IRD) and basic needs[1] approaches lacked an engine which could sustain development in the long run.[2,3] In the face of lagging agricultural productivity growth, rapid population growth, and modest growth in formal sector employment opportunities, many countries initiated farming systems projects during the 1970s and early 1980s.

The distinguishing feature of the FS approach — as a development strategy — is an emphasis on the creation of rural income opportunities by providing limited resource farmers with relevant improved production technologies. The approach represents a "bottom-up" development strategy in the sense that a diagnosis of farmers' circumstances is used to establish research and development priorities. Non-agricultural activities and macro-economic budgetary, monetary and trade issues generally are excluded from the approach.

Interest in the FS approach peaked in the mid-1980s, at which time there were more than 250 farming systems projects worldwide and many countries had started to institutionalize farming systems programs. Much of the impetus for the approach came from donor agency funding, particularly USAID. Leadership in developing methodological guidelines was provided by the international agricultural research centers (IARCs). After a brief period, during which there were unrealistically high expectations based on successes in areas where there was an equable environment for agricultural production, a retrenchment and reassessment led to a phasing out of many FS projects beginning around 1985.

After more than 10 years in the forefront of small farmer development strategies, the FS approach is likely to have a lasting impact on development thinking as well as on agricultural research methodology. This chapter gives a brief review of the approach and an assessment of its contribution. The first section discusses the rationale for the approach. This is followed by an overview of the research and extension methodology. Some examples of FS projects and programs are then presented before turning to an assessment of contributions and factors which have prevented the approach from living up to expectations. The final section projects future directions for the approach.

II. RATIONALE FOR THE APPROACH

In most countries, the main objective in introducing the FS approach has been to improve agricultural sector performance by increasing the productivity of resources used in agricultural production. The approach has three key features which distinguish it from earlier strategies for improving agricultural sector performance. First, it is couched in systems terminology and attempts to follow the principles of systems analysis. Second, it recognizes the key role small farmers must play in rural and national development. Third, it rejects a historically dominant "top-down" approach to technology research in favor of a "bottom-up" perspective of the research and development process. This section clarifies the rationale for the FS approach with reference to these key features.

A. Systems Analysis

Systems analysis is holistic, not reductionist. While reductionism does not deny that things are part of larger wholes, its limitations,[4] Dillon[5] concluded, have often led researchers away from real world problems. Systems analysis, on the other hand, starts with a logical conceptualization of the system to be improved.[6] Once the system is identified, factors limiting the performance of the system are evaluated, a target for improving the system is established, and alternative actions for achieving the target are assessed.

The search for solutions using systems analysis cuts across the boundaries of the traditional disciplines and has a pragmatic orientation. A distinguishing feature of pragmatism is the belief that research should lead to "workable solutions" to identified problems.[7]

B. The Importance of Small Farmer Development

The meteoric spread of the FS approach was remarkable not so much because of the acceptance of systems analytical procedures, which were adopted in many areas of applied research in the 1970s, but because of an acknowledgment of the importance of small farmer development. Agriculture in the 1940s and 1950s was widely viewed as best contributing to supporting the development of the urban/industrial sector. Therefore, the prescription for economic growth was to promote the transfer of labor from agriculture to formal employment in industry.[8-10]

However, by the late 1960s — thanks to the work of several individuals such as Johnston, Mellor and Kilby[11-13] in pointing out the contributions of agriculture to the supply of foreign exchange and food, and the value of agricultural income in creating domestic demand for industrial goods — the importance of agriculture in the development process was established.

Nevertheless, there continued to be an under investment in agricultural research, particularly research on small farmers. Contributing to this was the diffusion model[14,15] focusing on extension efforts designed to help farmers better allocate their resources and on the transfer of technologies from Europe and North America. This complemented the "community development" approach, which emphasized the provision of outside assistance in planning local development projects.[16]

The limitations of these approaches led to the high payoff input model,[14] in which development proceeds through research leading to the introduction of appropriate, improved production technologies. The immediate rationale for the model was derived from the dramatic successes of IRRI and CIMMYT in developing high yielding varieties — and their associated "Green Revolution" inputs. Also, a series of comparative international studies demonstrated that the annual rates of return to resources invested in agricultural research consistently exceeded 30 to 60%.[17-19]

Schultz[20] played a seminal role in the shift to a small farmer orientation when he posited the "efficient but poor hypothesis." Schultz's hypothesis was confirmed in a number of empirical studies during the following decade.[21] During this same period, a wide array of micro-economic studies were being carried out on all aspects of the rural economy, reflecting the "growth with equity"[16] era of development thinking. Three consistent findings which consolidated an interest in small farmer development were:

1. Small scale operations — whether small farms or small industrial firms — generally were often found to have as high output-capital ratios as did large firms and nearly always had greater labor-output ratios.[11,22] Therefore, it was possible to have efficient growth and increased employment generation. Also, in Africa government efforts to initiate large scale agriculture through settlement schemes, plantation and state farms have generally ended in failure[21] — further improving the attractiveness of small scale operations.
2. In both Latin America and Asia, there was evidence that agricultural income inequality was associated with the size of farm operations.[23] Technological change — even the supposedly scale-neutral Green Revolution technologies — was shown to have a tendency to increase inequality.[24-26]
3. Rapid rates of population growth and urbanization, both of which reduce the capacity of governments to provide adequate services and infrastructure, were shown to be functions of the pattern of the pace and pattern of rural income growth.[27-29] Since small farmers continue to dominate the rural economies of most low income countries, it was logical to give increased attention to means of increasing small farm income.

C. Shortcomings of Conventional Technology Research

The growing interest in small farmer development stimulated a reassessment of the relevance of traditional agricultural research procedures for developing technologies for small farmers. Agricultural research historically has started with predetermined commodity foci reflecting national (or colonial) policy goals.[30] Research stations were set up to generate improved production technologies, or to screen technologies transferred from elsewhere. The features of the technologies produced generally took into account national factor endowments,[14] the supply of technologies from other countries, and the mandates of institutions supporting the research, including donor agencies. Since research was carried out in isolation from farmers — and generally from extension agents — the resulting technological recommendations rarely addressed the diversity of circumstances affecting small farmers.[31]

During the early 1970s, it was apparent that few top-down recommendations were being accepted by small farmers and, when accepted, even fewer lived up to expectations when used on-farm. For example, research conducted by IRRI scientists made it clear that results from on-station research did not represent on-farm results due to nontransferable environmental differences (yield gap I between on-station and potential on-farm results) and due to a variety of biological and socioeconomic constraints (yield gap II between actual and potential on-farm yields).[32] In relation to yield gap II, several diffusion studies demonstrated that yield potential often was not the main factor affecting the spread of new technologies.[33]

At the same time conventional research procedures were being challenged, there was a growing appreciation of existing small farm production systems. Many indigenous practices such as intercropping[34] were found to be more appropriate than recommendations based on experiment station findings. Also, several anthropological studies revealed that the indigenous linguistic and knowledge categories of farmers express an understanding of the constraints and potential of farming systems gained through practical experience.[35] Collinson[36] pointed out that most of the apparently sub-optimal practices of small farmers can be explained by compromises made to increase overall farm systems performance. Johnson[37] further argued that small farmers continually are experimenting and adapting their practices to changing circumstances, even if they do have a limited capacity to make large changes over a short period of time.

Based on the above observations and findings, several principles for improving the efficiency of research resources were developed. One set of principles related to the goal of increasing the supply of technologies. The main principle was that technologies needed to be adapted to on-farm technical environments through on-farm experimentation. It was also recognized that the supply of technologies could be increased by building upon the good points of existing farming systems (in order to minimize the time spent "rediscovering the wheel") and by taking into account the indigenous technical knowledge of farmers.

Several additional principles addressed the goal of reducing the gap between the demand for technologies and the supply of technologies. Among the most important were the following:[38]

1. Farmers should be viewed as the clients of agricultural research and development programs.
2. Increased priority should be given to improving communication and collaboration among the various participants in the research process.
3. When assessing the appropriateness of technologies, multiple criteria should be used, relating to technical feasibility, economic viability and reliability, and social acceptability.
4. Proposed changes generally should be evolutionary rather than revolutionary in nature.

During the 1970s, the above principles (to varying degrees) guided the development of farming systems research (FSR) methods at several IARCs, and in a few national agricultural research programs. The methodological guidelines developed for FSR became the foundation for a fuller elaboration of the FS approach in the 1980s.[39]

During this time, however, parallel movements such as appropriate technology and women in development, were taking place. Although there were obviously many common concerns about smallholder development, between these movements and FSR, FSR has in many cases had a better record, having as it usually does better links with science, existing infrastructure, etc.

III. IMPLEMENTATION METHODOLOGY

The cornerstone of the FS approach is farming systems research (FSR) methodology. In FSR, interdisciplinary teams produce recommendations which are suited to particular local situations. This involves on-farm experimentation and working directly with farmers. Longer term technical problems identified through on-farm research are fed up to experiment station researchers. Ideally, policy and support systems problems are also fed up to relevant planning and development agencies.

Many guidelines are available on FSR methods.[36,40-47] As might be expected, there is a great deal of diversity in the terminology used and the specific methods prescribed. Nevertheless, essentially all the FSR guidelines propose a similar sequence involving four processes (or stages of research):[48]

1. Describe and diagnose existing farming systems in order to determine the constraints preventing improved farm system performance.
2. Design an experimental program for investigating technologies which might increase farm systems productivity by breaking or avoiding resource constraints.
3. Test, evaluate and adapt the most promising technologies in representative farming systems circumstances (technical environment and resource levels).
4. Disseminate information about technologies (or institutional changes) which have a high probability of improving farming systems performance.

In this section, the stages of research are characterized. Then, the various measures of success used in FSR are discussed.

A. Stages of Research

The usual starting point in systems description and diagnosis is to generate information on the technical environment, farming activities, production technologies, resource endowments, resource use patterns, farmers' goals and perceptions, and the performance of agricultural support systems. Information is generated through a number of sources, including reviews of secondary sources, rapid appraisal surveys, formal surveys, case studies, participant observation, and direct measurement of technical variables. Since diagnosis serves as the first step in an iterative problem solving sequence — aimed at improving the welfare of farmers — priority generally is given to low-cost and quick methods of information generation, and to developing a qualitative understanding of farming systems.

A notable feature of FSR diagnosis is the identification of particular areas and/or groups of farms to serve as the targets of technology development research. The target areas or groups of farms typically are called recommendation domains[40] or research domains[43] (RDs). Farms within RDs ideally have similar circumstances and, therefore, similar paths to improved productivity. RDs usually are defined on the basis of the research mandate and a rapid appraisal survey, but revisions often are made following additional diagnosis and testing.

The design process which is of critical importance links diagnosis and the testing process. After determining what causes the problems identified in systems diagnosis, the researchers rank the problems with reference to their impact on system performance, and several alternative technologies for addressing each problem are identified. The alternatives generally are based on experiment station research results, farmers indigenous technical knowledge, and the heterogeneity in existing practices. *Ex ante* assessments then are made on: (a) the contribution of each technology to the solution of each problem, (b) the number of farmers who are likely to adopt each technology, and (c) the research and extension resource requirements necessary to test and disseminate each technology. The *ex ante* assessments are used, in conjunction with the original research mandate and knowledge about available resources, to develop a series of testable hypotheses about the specific technical and economic benefits from new biological inputs, component husbandry practices, systems of practices, and/or practices by environment effects. Appropriate on-farm experiments for testing the hypotheses are then designed.

There are many approaches to on-farm experiments. The FSR literature focuses primarily on a conceptual continuum based on trials management to characterize experimental formats.[38] The continuum is from researcher managed and implemented trials (RMRI), to researcher managed but farmer implemented (RMFI), to farmer managed and implemented (FMFI). Which format is used is a function of the trial objectives, resources and skills required, and the risk involved. Except when the emphasis is on technical assessments, there is a presumption in favor of maximizing farmer involvement in managing and implementing trials. In practice, some aspects of most trials are managed and implemented by both researchers and farmers. Therefore, as an alternative, trials also have been classified on the basis of their function, ranging from exploratory, to refinement, to verification testing.[44]

FSR guidelines generally include dissemination as part of the sequence, but specific procedures for disseminating findings remain underdeveloped. For years, it was implicitly assumed that the involvement of farmers and extension agents in the technology development process would ensure that relevant technologies were produced and, in turn, that appropriate technologies would be self-spreading. By the mid-1980s, evidence had accumulated that high profitability and acceptability of an innovation — when evaluated by farmer cooperators — did not necessarily translate into high adoption rates. As a result, in recent years more attention has been given to institutional innovations in extension, communication procedures, and modes of farmer participation.

B. Measures of Success

There are three "boundary issues" which preclude reliable quantification of the contributions of FSR to small farmer development. First, the relative contributions of conventional research and FSR are not separable since they are complementary activities. Second, the adoption of technologies depends on a wide range of circumstances, including the performance of support systems, which are not under the control of FS teams. Third, because the FS approach encompasses technology development and institutional change, significant results often are realized only after 10 to 25 years.[39] Over such a time horizon, any assumptions about what would have taken place in the absence of FSR would be quite arbitrary.

Due to the problems encountered in measuring the impact of FSR on small farmer development, FSR generally has been assessed on the basis of success in identifying and screening improved production technologies. The various measures used to gauge success can be divided into four categories: technical performance, profitability, risk, and farmer adoption rates. The first three measures relate only to the features of the technologies developed, and are often measured on a small sample (or case study) basis. The last measure encompasses success in dissemination as well as technology development research and, correspondingly, has been less frequently used.

Historically, yield increase has been the main technical performance measure used in agricultural research. Some farming systems practitioners feel that often too much emphasis has been, and continues to be placed on, maximum yields as a measure of success, which detracts seriously from issues such as yield stability and tasks such as technology delivery.[49] Yield increase has continued to be a crucial criterion in FSR but has been supplemented with additional measures reflecting: (a) sources of utility besides quantity and, (b) success in utilizing scarce resources. As examples, one can cite reductions in the time required for tillage or planting operations, increases in percent seed field emergence, or land use ratios. A cropping intensity index has been commonly used in cropping systems research.[46] While these measures are not unique to FSR, their use has become common due to the availability of information from systems description and diagnosis.

The popularization of FSR has stimulated greater use of economic measures of success, which in the past have often received scant attention on the part of agricultural researchers. Based on farm management budgeting techniques, a key measure of success is the change in return per unit of scarce resource (generally land or labor, but also cash). Enterprise budgeting usually has been used when comparing systems involving multiple changes, while partial budgeting (which focuses only on the changes in costs and returns) has been used when assessing changes in specific component practices.

Most FSR reference manuals also propose various measures of the riskiness of a technology. One simple approach has been to use a minimum return ("risk premium") to the additional cash investment.[50] At a more sophisticated level, analyses which take into account the distributions of outcomes, such as stochastic dominance analysis, have been used. Another commonly used measure is the stability of yields or net returns across environments which can help in deciding whether different recommendations should be made according to different environments (including management) with different yield potentials.[51]

The most important measures of success in FSR are documented changes in farming practices and farm income. For example, many of the Asian cropping systems projects have reported the percent of farmers adopting recommended cropping patterns. Elsewhere, FSR projects have cited adoption rates for animal traction or new tillage practices, and changes in fertilizer rates, seeding rates, insecticide application rates, etc. When citing adoption rates, there is an implicit presumption that farmers have actually benefited from the changes.

Despite the prodding of sponsoring institutions (particularly donor agencies), there are few examples which document changes in farmer income or consumption levels (over a large area) due to FSR — in part because of the boundary problems cited above. As an alternative, mathematical programming has on occasion been used to project the whole farm income effects of proposed changes in practices.[52]

In conclusion, although some perhaps would agree that the case for FSR is still to be proven, the fact is in many countries support for conventional research is weakening, as shown by cutbacks in research budgets. Given support and time, FSR should be able to reverse this trend. Also FSR, in recognizing the diversity of farmers, has been instrumental in developing non-blanketed, targeted and conditional recommendations. Unfortunately, however, conventional extension programs — and even the Training and Visit approach[53] — are not yet usually geared to accepting and delivering such recommendations.

IV. OVERVIEW OF FS PROGRAMS

The FS approach did not emerge fully developed at a single time and location. During the 1970s there were a few landmark programs which demonstrated the effectiveness of FSR methods. In the late-1970s and early 1980s, the initial experiences with FSR were documented through a few key conferences[48] and "state-of-the-art" reports.[41,42,45] This led to an expansion phase, between 1980 and 1985, during which a majority of the low-income countries launched FSR projects. Although the record of FSR projects was mixed, many countries, IARCs and universities had made commitments to institutionalize FS programs by the mid-1980s.

A. Early Landmarks

In Latin America, FSR techniques were first introduced through cropping systems research projects set up in 1973 and 1974 in Colombia,[54] El Salvador,[55] and Costa Rica.[56]

The first, and best known, national Latin American program was established in the Guatemalan Institute of Agricultural Science and Technology (ICTA). With leadership provided by Hildebrand, ICTA social scientists used rapid appraisal survey techniques, referred to as "sondeos", to identify bean seeds and planting labor as the key constraints. A new cropping pattern was developed which led to an increase in the area cultivated by 40% and resulted in 75% more maize, 40% more sorghum, no reduction in the amount of beans, and a 33% increase in income. Reviewing the impact of ICTA after 10 years, Fumagalli, Ortiz and Castello[57] reported that because of the success of ICTA, Guatemala had registered significant increases in basic grains production that enabled it to shift from being an importer to an exporter of grain.

In Africa, the anglophone tradition of FSR[58] often is traced to multidisciplinary on-farm research carried out through Ahmadu Bello University, Nigeria, in the 1960s and early 1970s. Through on-farm trials, improved packages for cotton, maize and cowpeas were screened for their technical feasibility and economic profitability. The research showed which aspects of cotton, cowpea and maize recommendations best fit into farmers' circumstances.[59] Later, together with inputs from other countries, the procedures developed and used at Ahmadu Bello University served as the basis for formulating the "stages typology" of FSR methods introduced at the Rural Economy Institute in Mali in 1976.[48]

The best known precursor of a francophone approach to FSR in Africa was the Experimental Units program in Senegal, which lasted between 1968 and 1980. By the mid-1970s, improved farming systems had been developed involving the introduction of animal traction, improved seeds, fertilization and pesticides.[60] Eventually, improved production systems were established on nearly half of the farms located in the Experimental Units. Reviewing the contributions of the Experimental Units, Bingen and Faye[61] concluded that the program made a significant contribution to agricultural development in Senegal. Bingen and Faye also noted that the Experimental Units program contributed to the development of an effective extension model involving farmer groups and that both researchers and farmers developed a better understanding of the organization of farming families.

Among the IARCs, the best known and most influential FSR during the 1970s was the rice based cropping systems research led by the International Rice Research Institute (IRRI). Using IRRI cropping systems methodology, intensified cropping systems were developed and extended to farmers in several Asian countries.[41] In India, the International Crops Research Institute for the Semi-Arid Tropics (ICRISAT), generally eschewing minor system changes, developed a new farming system for the regions dominated by vertisol soils. Ryan and Sarin[62] present data showing that in preliminary testing, the return to the increased cost was 258%. Later, in farmer managed trials, the return to the farmers' added expenditure was 244%.[63,64]

B. Expansion Phase

During the FSR expansion phase in the early 1980s, the IARCs played a leading role in promoting FSR, with much of the funding for FSR projects coming from the international donor agencies. In addition to IRRI and ICRISAT, The International Maize and Wheat Improvement Center (CIMMYT) developed and actively promoted its own approach to on-farm research with a farming systems perspective. CIMMYT's approach and training materials[40,50] were used in several Latin American countries and were influential in Africa, in large part because of an active demonstration program led by Collinson.[36] The International Potato Center (CIP) in Peru developed a farmer-first approach to FSR and used the approach to introduce a diffused light potato storage technology in more than 25 countries.[65]

As more FSR programs developed in national settings, the rate of expansion exceeded the development of methodology and, before long, stories of failure outweighed success stories. Nevertheless, there were projects in many countries which identified modified production practices that outperformed existing practices. Some examples which have been cited in the FSR literature include the following:

1. Sudan — In research in the Nuba Mountains, there was found to be a high benefit-cost ratio for the introduction of improved legumes. It was also shown that animals would benefit from dry season feed supplementation with legume forages.[66,67]
2. Nigeria — In various trials in Bida, the addition of a short season cowpea variety had benefit-cost ratios ranging from 7 to 12. Benefits were also achieved through the introduction of a serrated sickle and increased rice hill density. On properly selected sites, a change from ridge to flat planting of rice increased yields by 23% and returns per day of labor by 30%. Also on appropriate sites, a change in planting densities increased yields by 14%.[68]
3. Burkina Faso — A Purdue University farming systems team showed that tied ridges and phosphate fertilizer were of significant agronomic and economic benefits.[69] With donkey traction, tied ridges and fertilizer increased returns per hour by 39% and returns per hectare by 113%.[70] Nagy et al.[52] later reported that in a farmer implemented trial, less than 8% of the farmers would have lost cash by shifting to the recommended system.
4. Nepal — At several cropping systems sites, recommended cropping patterns gave returns over variable costs which were generally above 30%.[71] At some locations there was dramatic evidence of adoption. For example, between 1981 and 1984 in Sukchina, the proportion of farmers using improved varieties increased 20 to 98%; there was a 45% increase in fertilizer application rates; and the area left fallow after a single season rice dropped from 45 to 2%.[72]
5. Panama — INIAP research showed that an increase in plant density had a marginal rate of return of 687%. If a new herbicide was used at the recommended higher population density, there was an additional return of 716%. As a result, by 1982 there had been a high degree of acceptance of the recommendations relating to weed control (61% of farmers), density and row planting (71%), and zero or minimum tillage methods (44%).[73] Also in Panama, FSR methods were used to develop a dual purpose cattle production system — involving mineral salts, a health package and paddocks for calves — which was readily accepted by farmers.[74]
6. Botswana — The Agricultural Technology Improvement Project (ATIP) showed that yields could be increased by 77%, on average over five drought seasons, through the addition of an early season plowing operation. The net gain per additional hour invested was 50% higher than the urban minimum wage. In addition, the return to a fungicide treatment for peanut seed was above 400%. The return to hand planting labor for gap filling was two to three times the minimum wage. Several additional promising interventions were screened and promoted including an improved donkey harness (from Kenya) and a hand operated rotary injection planter (from IITA).

C. Institutionalization

Although the contributions of FSR to small farmer development could not live up to the early claims of its proponents, the logic of the systems perspective and on-farm research stimulated many countries, universities and IARCs to begin institutionalizing FSR programs. By the mid-1980s, for example, the research services in nine African countries had been reorganized to accommodate FSR. Ten universities in the United States had established FSR programs focused on American agriculture. In Asia, six major universities formed the Southeast Asian Universities Agroecosystems Network (SUAN), which is an organizational framework for promoting and supporting FSR-type activities.

Although support for FSR was seemingly increasing among national planners in some low-income countries in the mid-1980s, USAID decided to shift the emphasis of its funding to traditional commodity research and biotechnological research.[39] As part of its shift in emphasis, USAID ended funding for the Farming Systems Support Project (based at the University of Florida) which had played a major FSR networking and training role. The declining support from USAID, and the International Development Research Center (IDRC) in Canada, dramatically reduced the pace of FSR institutionalization after 1985.

After 1985, the institutionalization of the FS approach proceeded primarily in three contexts: at the IARCs, through regional networks, and in countries which had already institutionalized national programs before the ending of donor subsidies. Most of the IARCs continued to have active FSR programs, even if not independently labeled as such. The International Institute for Tropical Agriculture (IITA), for example, devotes 40% of its budget to FSR.

Regional networks were particularly important in supporting FSR in the mid-1980s. In Asia, the cropping systems network (ACSN) began a shift to a FS approach with the crop-livestock workshop at Los Baños in April, 1983. Three years later the shift was formalized at a follow-up workshop held in Khon Kaen, Thailand (July, 1986) when the ACSN was renamed the Asian Farming Systems Network. In Africa, the West African Farming Systems Network (WAFSN) held its inaugural meeting in 1986. Also, many of the projects supported by the Southern African Development Coordinating Council (SADCC) had explicitly adopted elements of the FS approach. In the Caribbean, CARDI was sponsoring FSR in eight countries. Also, many of the Collaborative Research Support Programs (CRSPs) were supporting FSR activities.

V. ASSESSMENT

There is no simple way to assess the FS approach. As a small farmer development strategy, the approach involves modified research and development processes, evolutionary changes in farmer (and researcher) practices and behavior, and a restructuring of agricultural institutions. Moreover, the approach has been applied in different ways at different times in different settings.

A. Contributions

When initially conceived, the primary contribution of the FS approach was to identify interventions to improve the farming systems productivity. As the FS approach has become more fully elaborated, there have been four additional contributions: (a) development of better agricultural research methods, (b) improvements in institutional performance, (c) documentation of factors affecting existing farming systems, and (d) promotion of farmers' perspectives.

1. Interventions Identified for Increasing Productivity

In several locations, FSR teams have shown that farm profitability can be increased through modified tillage practices, new varieties, adjustments in plant populations, fertilizer or herbicide applications, increased cropping intensity, etc. In addition, substantial progress has been made in determining the technical factors affecting the success of competing agronomic practices such as ridge vs. flat planting, animal traction vs. hand hoeing, and intercropping vs. sole cropping. Even where specific improved practices have not been identified, the technical and social science diagnosis associated with on-farm trials has helped establish a foundation for technology development research programs for the next 10 to 20 years.

In general, FSR has been more successful in identifying interventions for equable environments. Perhaps the most dramatic examples pertain to Asian cropping systems research, where large increases in farm profitability have been demonstrated through the addition of short cycle crops.[75] In harsh production environments, fewer interventions have been identified since it often is necessary to develop substantially modified farming systems, rather than introduce minor changes based on exploiting flexibility in existing systems.

2. Improved Agricultural Research Methods

The development of agricultural research methods through FSR has been nothing short of revolutionary in nature. Through FSR, technical research has in most countries shifted from a single-location to a multi-location orientation.[76] In addition, the range of technical factors considered when evaluating technologies has increased. For example, it is now usual practice to consider the threshing characteristics of new varieties, complementary products such as cowpea leaves, and the value of intermediate products — such as manure in producing crops.

One of the most commonly acknowledged contributions of the FS approach has been a move away from a technical perspective of agricultural research. With the inclusion of agricultural economists, greater attention has been given to the returns to labor or capital, and to risk factors. Later, as anthropologists started to play a greater role, the social acceptability and impact of innovations were acknowledged to be important criteria in agricultural research. On the other side, FSR has provided an opportunity for social scientists to begin making more tangible contributions to the development process.

FSR has stimulated a number of innovations in research procedures for rapid appraisal surveys and on-farm trials. Progress has also been made on formal and informal techniques for eliciting farmers' assessments of problems and practices, including the use of farmer groups.[77]

Closely related to the spread of FSR has been the introduction and use of micro-computers. The increased capacity to do on-site data analysis has been coupled with a commitment to collecting less data. Taken together, a bias has been established in favor of pragmatic research, minimizing the time required to complete analyses, and making sure analyses are understandable. This bias is now impacting on on-station research programs and on macro-policy analysts — reflecting a spread of the FS perspective beyond the bounds of formal FSR.

3. Improved Institutional Performance

The FS approach has contributed to improved institutional performance through its bottom-up development perspective. In many countries, results from FSR have been used to target commodity research programs, and to make sure on-station research is relevant to farmers' needs. The emphasis on adjusting to farmer circumstances has led to more relevant policies as well as technologies. For example, in Botswana, research findings from one FSR team were used to help design a major farmer assistance program targeted to resource poor farmers. In Mexico and Panama, insights from FSR led to revisions in national credit programs.[30]

The FS approach has further contributed to improved institutional performance through the development of linkages between on-station research, farmers, and extension agents. As a result of improved linkages, commodity researchers now more commonly visit farmers' fields compared to 10 years ago and, at least in some countries, extension agents are actively involved in technology assessment.

4. Data on Farming Systems

A major contribution of the FS approach is the information which has been generated about existing farming systems. Until the 1970s, most research had a single-discipline perspective and covered a narrow range of factors affecting farm systems performance. Particularly in Africa, there was an inadequate base for development planning.[21] Through FSR, much information has been generated on farm-household relationships, the roles of women, inter-household linkages, livestock and crop-livestock linkages, agro-ecological issues, and farm-level resource constraints. The insights about traditional farming systems gained through FSR may enable the elaboration of a comparative theory of agricultural development.

5. Appreciation of Farmers' Perspectives and Farmer Participation

The contributions of farmers to the development process has been documented at several conferences — see, for example, Matlon et al.[78] and papers given at a workshop held at the University of Sussex in July 1987. Ashby,[79] Lightfoot[80,81] and many others have demonstrated the value of farmers actively participating in on-farm trial work. As a result, most donor agency project documents and national agricultural plans now explicitly recognize farmers' goals and give emphasis to who gains — rather than just growth perspectives. There is also greater awareness of the heterogeneity among farmers, farmers' indigenous knowledge, and the range of factors affecting farmer behavior.

B. Factors Limiting Success

Before considering these in detail, the issue of funding should be mentioned. As earlier emphasized, donor agencies have played a major role in providing funds for FSR. FSR is expensive in terms of recurrent costs vis-a-vis station based research, where fixed and sunk costs dominate. The practical problems of keeping FSR activities going after the withdrawal of donor funding limits the potential for success and highlights the need for an approach to FSR that minimizes the demand for resources. This of course is a major determinant for the institutionalization of the FS approach.

Despite the substantial long-run contributions of the FS approach, there is a consensus among donor agencies and development specialists that the FS approach has not lived up to short-run expectations. Several assessments point up different reasons why, but disagreements continue over the validity and relative importance of the different reasons.

1. Unit of Analysis

One key issue is whether the normal unit of analysis used in FSR — individuals, farms or groups of farms with similar practices and circumstances — is appropriate. Both more aggregated and disaggregated units have been proposed as being more viable. A growing literature on intra-household dynamics suggests that specific individuals, or classes of individuals, should serve as the units of analysis.[82,83] On the other hand, FSR has also been criticized for giving too little attention to policy issues and the macro-economic implications of agricultural development programs.[84] As a result, there have been calls to reject farming systems as the unit of analysis in favor of agro-ecological systems,[85] agro-forestry systems,[86] food systems,[87] etc.

Among FSR practitioners, an unresolved "unit of analysis" issue is proper use of the recommendation domain (RD) concept. Implicit in the RD concept is an assumption that research can focus on groups of farms and still sufficiently take into account the circumstances determining adoption behavior. However, since there remains little evidence on adoption response, the level of appropriate aggregation is by default based on researchers' intuition. Perhaps individual farm planning is necessary or, in contrast, perhaps it is sufficient to develop general sets of alternatives and let farmers decide whether proposals make sense.

Even if one accepts the RD concept, there remains the problem of when RDs should be defined. Important dynamics might be missed if RDs are defined before researchers understand the social organization of production.[88,89] Waiting too long to focus research on particular problems of specific RDs, however, can greatly increase research resource requirements.

2. Scope of Research

In spite of the broad conceptual framework of the FS approach, FSR projects and programs have tended to be narrowly focused. Most FSR teams have focused on improved crop production technologies. In some cases the emphasis on crop production resulted from a diagnosis that one or a few main crops could serve as a "leverage" point through which the most substantial gain could be achieved.[36] In other cases, FSR teams have been assigned a "predetermined focus" to do research on one enterprise or one facet of an enterprise.

An active debate among FSR practitioners is whether the scope of research has been too much predetermined by institutional mandates. Norman and Collinson[38] have argued that most FSR success stories have derived from programs with specific, and relatively narrow, mandates to focus on particular commodities. However, teams operating with predetermined mandates often have not addressed the most important issues from farmers' perspectives. Fresco and Poats[90] label this the problem of "real vs. perceived holism."

On balance, it does appear that too many important phenomena were left out of early FSR analyses. As Johnson[91] argued, the scientists involved in FSR at the IARCs during the 1970s did not really take a holistic approach, encompassing human and institutional change as well as technological change. Moreover, due to institutional mandates, livestock often were ignored and, in general, a submissive approach[92] was taken to support systems improvement.

The potential shortcomings of predetermined FSR became more evident as FS work developed in national programs.[93] Consequently, during mid-1980s, there was a trend toward abandoning the emphasis on particular enterprises and quick turnaround time in favor of a broader systems perspective, encompassing livestock,[94] food consumption,[95] nutrition,[96,97] gender issues,[83] and the requirements for national development planning.[98] Despite adjustments, Eicher,[84] among others, has recently argued that the scope of FSR continues to be too narrow to have a significant impact.

3. *Project Format*

Most FSR has been carried out on a project basis, independent of on-station research programs and isolated from agricultural policy debates.[84] This has in many cases led to a fragmented set of initiatives, an inability to undertake long-term research — due to the vagaries of donor agency funding — and failure to develop credibility among planners, on-station researchers and extension agents.

Institutionalized FSR programs clearly are preferable to donor funded projects, but there has been little agreement on how best to institutionalize the FS approach. In some countries, a separate institutional structure has been created for FSR, but this now appears to be only a marginal improvement over the FSR project format. Consequently, Stoop,[99] and others have argued that on-farm research and the FS perspective should be gradually introduced into existing institutions. Nevertheless, the institutional requirements for FSR cannot be underestimated. As Poats et al.[39] point out, success in FSR requires a reorganization to support and coordinate field teams and to integrate national and location-specific development priorities, even if no separate institutional structure is created for FS work.

4. *Lack of Technologies*

One of the main limitations on FSR has been the lack of technologies which are ready for farmers after modest adaptive research. FSR was never intended to substitute for applied commodity and component design research. But many FSR teams have been assigned a poor farmer-harsh climate focus, where there are few if any on-the-shelf technologies ready for location-specific, adaptive research. Consequently, FSR has involved a mixture of technical diagnostic, design and adaptive research to a much greater extent than was initially envisioned. Particularly in harsh environments, there is a long time required for on-farm technical diagnostic and design research — and the results often are not spectacular in a short-run context. The tendency to ask FSR teams to do more than they should, rather than only investing in FSR when the conditions were appropriate, has substantially contributed to the impression that the FS approach has not lived up to expectations.

5. *Lack of Trained Personnel*

There continues to be an insufficient supply of the skills required for multidisciplinary systems analyses. The lack of systems training has resulted in poor problem definition, use of inappropriate methodology, and a tendency for FSR researchers to concentrate on topics relating to their disciplinary expertise — rather than on farmers' problems. While disciplinary orientations can be overcome through the use of multidisciplinary teams, experience has shown that in-depth collaboration among team members has been relatively rare.

The lack of trained personnel applies to the FSR technical assistants brought in by donor agencies,[100] as well as to national scientists. Few universities have paid attention to the type of training or retraining which is required for FSR.[101] As a result, many donor sponsored FSR projects have been staffed with individuals having little or no training or experience in on-farm research methods or team research approaches. When weaknesses have been identified in the procedures of one or more team members, the tradition of academic freedom and individualism has hindered efforts to make those individuals adjust their methods.[102]

Relevant training and experience are critical ingredients in lending credibility to FSR and creating conditions conducive to the institutionalization of the approach.

6. *Role of Farmers*

Through the early 1980s, most FSR focused on evaluating the technical feasiblity and economic viability of modified practices and varieties. Farmers generally had a passive role. Farmers were the clients of research[42] and generally implemented the trials, but the goals of research often were the agronomic and economic analyses of input-output relations and profitability.[31] In many cases, the farmer perspective got lost.

A number of social scientists have recently been promoting a farmer first approach to FSR.[103,104] Implicit in a farmer first approach is a belief that research methods and priorities will be more relevant and, therefore, adoption rates will be increased — even if the returns per farmer adopting a technology are lower. The main argument against a farmer first perspective is that farmers do not have sufficient knowledge of technological options or national planning goals to be given a controlling role in FSR. The issue remains unresolved and, in practice, most FSR teams try to achieve a balance between researchers' and farmers' perspectives when determining the research agenda.

TABLE 1
Evolution of the Farming Systems Approach

Characteristic	1970s to Early 1980s	Adjustment by mid-1980s	Further adjustment
Increasing Trends to Continue			
Support system perspective			
Extension linkage	Less	More	More
Area development mandate			
Work through channels			
Increasing Trends Subject to Reversal			
Feed-up research priorities			
Emphasis on farmer first	Less	More	Less
Sub-RDs within ecological zones			
Baseline diagnosis			
Decreasing Trends to Continue			
Implemented by isolated teams			
Rely on donor funding	More	Less	Less
Emphasis on quick turnaround time			
Dominated by expatriates			
Focus on private profitability			
Decreasing Trends Subject to Reversal			
Focus on equable areas			
Pre-determined commodity focus	More	Less	More
Feed down from technical research			
Researcher managed trials			
Trends Yet to Begin			
Focus on adoption rates			
Focus on social profitability (sustainability)	Little	Little	More
Inclusion of macro-analysis			

VI. EXPECTED TRENDS

There was a substantial evolution in the FS approach between the late-1970s and the mid-1980s, and there is every reason to believe that the approach will continue to evolve. A summary of recent adjustments and expected further adjustments is presented in Table 1.

In general, the narrowly focused FSR developed at the IARCs has given way to a more comprehensive, longer-horizon systems approach. Farmers are playing an increasing role. At the same time, the domination of FSR by donor agencies and expatriate technical assistants has started to give away to localized programs. While there continues to be a substantial investment in the diagnosis of problems, there is declining interest in describing farming systems and increasing interest in pushing technologies through the testing stage into actual extension.

Two major forces have impacted on the evolution of the FS approach. One has been the expanding number of academic professionals, representing several disciplines, involved in FSR. During most of the 1980s, academic supporters of the FS approach have shown a bias in favor of expansionism and farmer first perspectives. The second major force has been the bureaucratic interests of donor agencies and national agricultural ministries. Bureaucrats have increasingly expressed reservations about the likelihood that FSR can produce short-term, tangible payoffs.

Both the academics and bureaucrats agree that certain developments in the FS approach have been appropriate and should continue. For example, there appears to be a consensus that more attention is needed to livestock, support systems performance, extension linkages, and working through existing channels. Similarly, there is widespread agreement that the trend toward less reliance on donor funding should continue, that factors beside private profitability need to be considered, and that FSR should not be dominated by expatriates.

Most areas of contention about the future of the FS approach stem from disagreements about the viability and affordability of a decentralized, bottom-up approach to development. The dominant academic perspective seems to be that a broad systems focus is needed and that farmers' views should determine the agenda. The bureaucratic perspective emphasizes instead the importance of commodity program priorities as determined by national planning goals, and the value of concentrating efforts in areas with equable environments — so the payoffs will be higher.

The battle is joined over the future of the FS approach, but the outcome is uncertain. Based on past experiences, however, bureaucratic imperatives will probably dominate academic "optimality" perspectives. In such a case, it can be expected that commodity programs and macro-policy issues will receive greater attention during the early 1990s than during the prior 10 years, and that farmers' perspectives may play a declining role in determining national priorities. In any event, multidisciplinary on-farm research is likely to continue in some format, and systems analysis is likely to have a growing impact on national research and development planning processes.

VII. CONCLUSIONS

The FS approach has been an important tool in developing strategies for small farmers for more than a decade, starting in the mid-1970s. By 1986, donor agencies, national agricultural research directors and agricultural researchers had started to revise their high expectations for FSR. Funding for FSR projects and programs has been greatly reduced. In part, the retrenchment can be seen as a "technical" adjustment in the resource allocation process, reflecting the cyclical thinking of donor agencies and development specialists. The reduced support for the FS approach, however, does not reflect what appears to be a continuing — if not increasing — interest in the FS approach and FSR methods in many low income countries.

Regardless of the future of FSR as a distinct entity, the FS approach has made several lasting contributions. The approach has provided a problem solving methodology for developing and evaluating relevant, improved production technologies. The development of production technologies has generally been the short-run objective of FS projects and programs. The approach has also contributed indirectly to improved productivity by inducing the improved performance of agricultural institutions.

To the extent the approach has failed to live up to expectations, it is partially because the expectations were unrealistic. The systems perspective represents an advancement over reductionism. On-farm research is a needed complement to on-station research. Social science perspectives are a needed complement to technical agriculture research. But the FS approach cannot be expected to solve all small farmer development problems. When FSR is properly implemented, it should be possible to identify alternatives for improving small farmer productivity, but there is more to small farmer welfare and to national development than small farmer productivity.

ACKNOWLEDGMENTS

The authors appreciate the constructive suggestions of M. Blackie, D. Byerlee, C. Lightfoot, A. Low, and F. Worman on an earlier draft of this paper. Any mistakes that still remain are, however, the responsibility of the authors. Also, the help of N. Young in drawing up details on the references is gratefully acknowledged. This paper is published with the permission of the Director, Department of Agricultural Research, Ministry of Agriculture, Botswana.

REFERENCES

1. **ILO,** *Growth, Employment and Equity: A Comprehensive Strategy for the Sudan*, ILO, Geneva, 1976, 527.
2. **Lele, U.,** *The Design of Rural Development: Lessons from Africa*, Johns Hopkins University Press, Baltimore, 1975, 246.
3. **Streeten, P. and Burki, S. J.,** Basic needs: some issues, *World Dev.*, 6, 411, 1978.
4. **Rountree, J.,** Systems thinking — some fundamental aspects, *Agric. Syst.*, 2, 247, 1977.
5. **Dillon, J.,** The economics of systems research, *Agric. Syst.*, 1, 5, 1976.
6. **Manetsch, T. and Park, G.,** *Systems Analysis and Simulation with Applications to Economic and Social Systems Part I* (3rd ed.), Michigan State University, East Lansing, 1977, 276.
7. **Johnson, G.,** *Research Methodology for Economists; Philosophy and Practice*, Macmillan, New York, 1986, 252.
8. **Jorgenson, D.,** The development of a dual economy, *Econ. J.*, 71, 309, 1961.
9. **Lewis, W. A.,** Economic development with unlimited supplies of labour, *Manchester School of Economic and Social Studies*, 22, 139, 1954.
10. **Rostow, W.,** *Stages of Economic Growth: A Non-Communist Manifesto*, Cambridge University Press, Cambridge, 1960, 178.
11. **Johnston, B. F. and Kilby, P.,** *Agriculture and Structural Transformation: Economic Strategies in Late-Developing Countries*, Oxford University Press, New York, 1975, 427.
12. **Johnston, B. F. and Mellor, J. W.,** The role of agriculture in economic development, *Am. Econ. Rev.*, 51, 566, 1961.
13. **Mellor, J.,** *The Economics of Agricultural Development*, Cornell University Press, Ithaca, 1966, 403.
14. **Evenson, R. and Kislev, Y.,** *Agricultural Development: An International Perspective*, Johns Hopkins University Press, Baltimore, 1971, 367.
15. **Ruttan, V.,** *Agricultural Research Policy*, University of Minnesota Press, Minneapolis, 1982, 369.
16. **Staatz, J. and Eicher, C.,** Agricultural development ideas in historical perspective, in *Agricultural Development in the Third World*, C. Eicher and J. Staatz, Eds., Johns Hopkins University Press, Baltimore, 1984, 491.
17. **Evenson, R., and Kislev, Y.,** Investment in agricultural research and extension: an international survey, *Economic Development and Cultural Change*, 23, 507, 1974.
18. **Evenson, R. and Kislev, Y.,** *Agricultural Research and Productivity*, Yale University Press, New Haven, CT, 1975, 205.
19. **Griliches, Z.,** The sources of measured productivity growth, U.S. agriculture, 1940—1960, *J. Political Econ.*, 71, 331, 1962.
20. **Schultz, T. W.,** *Transforming Traditional Agriculture*, Yale University Press, New Haven, CT, 1964, 212.
21. **Eicher, C. and Baker, D.,** *Research on Agricultural Development in Sub-Saharan Africa: A Critical Survey*, MSU International Development Paper No. 1, Michigan State University, East Lansing, 1982, 335.
22. **Chuta, E. and Liedholm, C.,** *Rural Non-Farm Employment: A Review of the State of the Art*, MSU Rural Development Paper No. 4, Michigan State University, East Lansing, 1979, 96.
23. **Dorner, P. and Kanel, D.,** The economic case for land reform: employment, income distribution, and productivity, in *Land Reform in Latin America*, Land Economics Monograph No. 3, Dorner, P., Ed., University of Wisconsin, Madison, 1971.
24. **Bieri, J., de Janvry, A., and Schmitz, A.,** Agricultural technology and the distribution of welfare gains, Am. J. Agric. Econ., 54, 801, 1972.
25. **Griffen, K.,** *The Political Economy of Agrarian Change: An Essay on the Green Revolution*, Harvard University Press, Cambridge, MA, 1974, 264.

26. **Pearse, A.,** *Seeds of Plenty, Seeds of Want: Social and Economic Implications of the Green Revolution*, Clarendon Press and Oxford University Press, New York, 1980, 262.
27. **Caldwell, J. C.,** Toward a restatement of demograhic transition theory, *Popul. Dev. Rev.*, 2, 321, 1976.
28. **Easterlin, R., Ed.,** *Population and Economic Change in Developing Countries*, University of Chicago Press, Chicago, 1980, 581.
29. **Harris, J. and Todaro, M.,** Migration, unemployment and development: a two-sector analysis, *Am. Econ. Rev.*, 60, 126, 1970.
30. **Byerlee, D. and Tripp, R.,** Strengthening Linkages in Agricultural Research through a Farming Systems Perspective; The Role of Social Science, *Agricultural Administration*, in press.
31. **Byerlee, D., Harrington, L., and Winkelmann, D.,** Farming systems research: issues in research strategy and technology design, *Am. J. Agric. Econ.*, 64, 897, 1982.
32. **Datta, S., Gomez, K., Herdt, R., and Barker, R.,** *A Handbook on the Methodology for an Integrated Experiment-Survey on Rice Yield Constraints*, IRRI, Los Baños, Phillipines, 1978, 59.
33. **Perrin, R. and Winkelmann, D.,** Impediments to technical progress on small versus large farms, *Am. J. Agric. Econ.*, 58, 888, 1976.
34. **Belshaw, D.,** Taking indigenous technology seriously: the case of intercropping in East Africa, *IDS (Sussex) Bulletin*, 10, 24, 1979.
35. **Brokensha, D., Warren, D., and Werner, O., Eds.,** *Indigenous Knowledge Systems and Development*, University Press of America, Landham, Maryland, 1980, 460.
36. **Collinson, M.,** *Farming Systems Research in Eastern Africa: The Experience of CIMMYT and some National Agricultural Research Services, 1976-81*, MSU International Development Paper No. 3, Michigan State University, East Lansing, 1982, 61.
37. **Johnson, A.,** Individuality and experimentation in traditional agriculture, *Human Ecol.*, 1, 149, 1972.
38. **Norman, D. and Collinson, M.,** Farming systems research in theory and practice, in *Agricultural Systems Research for Developing Countries*, ACIAR Proceedings No. 11, J. Remenyi, Ed., Australian Council for International Agricultural Research, Canberra, 1986, 16.
39. **Poats, S., Galt, D., Walecka, L., Hildebrand, P., Andrew, C., and Reboussin, D.,** *Farming Systems Research and Extension: Status and Potential in Low-Resource Agriculture*, Farming Systems Network Paper No. 4, Farming Systems Support Project, University of Florida, 1986.
40. **Byerlee, D. and Collinson, M.,** *Planning Technologies Appropriate to Farmers: Concepts and Procedures*, CIMMYT, Londres, Mexico, 1980, 71.
41. **Dillon, J., Plucknett, D., and Vallaeys, G.,** *TAC Review of Farming Systems Research at the International Agricultural Research Centres*, FAO, Rome, 1978.
42. **Gilbert, E., Norman, D., and Winch, F.,** *Farming Systems Research: A Critical Appraisal*, MSU Rural Development Paper No. 6, Michigan State University, East Lansing, 1980, 135.
43. **Hildebrand, P. and Poey, F.,** *On-Farm Agronomic Trials in Farming Systems Research and Extension*, Lynne Rienner, Boulder, CO, 1985, 162.
44. **FSSP,** FSR/E *Training Units, Vol. I: Diagnosis in FSR/E, Vol. II: Techniques for Design and Analysis of On-Farm Experimentation*, Farming Systems Support Project, University of Florida, Gainesville, 1986, 367.
45. **Shaner, W., Philipp, P., and Schemehl, W.,** *Farming Systems Research and Development: Guidelines for Developing Countries*, Westview Press, Boulder, CO, 1982, 414.
46. **Zandstra, H., Price, E., Litsinger, and J., Morris, R.,** *A Methodology for On-Farm Cropping Systems Research*, IRRI, Los Baños, Philippines, 1981, 147.
47. **Harwood, R.,** *Small Farm Development: Understanding and Improving Farming Systems in the Tropics*, Westview Press, Boulder, CO, 1979, 160.
48. **IER,** Pour une Programme de Researche sur les Systems de Production Agricole, Institut d'Economie Rurale, Bamako, Mali, 1976.
49. **Blackie, M.,** personal communication, 1987.
50. **CIMMYT,** *From Agronomic Data to Farmer Recommendations: An Economics Training Manual*, completely revised edition, CIMMYT, Londres, Mexico, 1988, 79.
51. **Hildebrand, P.,** Modified stability analysis of farmer managed, on-farm trials, *Agron. J.*, 76, 271, 1984.
52. **Nagy, J., Ames, L., and Ohm, H.,** Technology evaluation, policy change and farmer adoption in Burkina Faso, in *Selected Proceedings of KSU's 1985 Farming Systems Research Symposium, Farming Systems Research and Extension: Management and Methodology*, Farming Systems Research Paper No. 11, Flora, C., and Tomecek, M., Eds., Kansas State University, Manhattan, 1986b, 340.
53. **Benor, D. and Baxter, M.,** *Training and Visit Extension*, The World Bank, Washington, D.C., 1984, 202.
54. **Zandstra, H., Swanberg, K., Zulberti, C., and Nestel, B.,** *Caqueza: Living Rural Development*, IDRC, Ottawa, Canada, 1979, 321.
55. **Hildebrand, P. and French, E.,** Un Sistema Salvadoreno de Multicultivos: su Potencial y sus Problemas, Ministerio de Agricultura y Ganderia, Centro Nacional de Technologia Agropecuria, Santa Tecla, El Salvador.
56. **Moreno, R. and Saunders, J.,** A Farming System Research Approach for Small Farms of Central America, CATIE, Turrialba, Costa Rica, 1978.
57. **Fumagalli, A., Ortiz, R., and Castillo, M.,** A new model for technology transference within FSR/E, in *Selected Proceedings of KSU's 1985 Farming Systems Research Symposium, Farming Systems Research and Extension: Management and Methodology*, Farming Systems Research Paper No. 11, Flora, C., and Tomecek, M., Eds., Kansas State University, Manhattan, 1986b, 111.
58. **Fresco, L.,** *Comparing Anglophone and Francophone Approaches to Farming Systems Research and Extension*, FSSP Network Paper No. 1, Farming Systems Support Project, University of Florida, Gainesville, 1984, 36.
59. **Norman, D., Simmons, E., and Hays. H.,** *Farming Systems in The Nigerian Savanna: Research and Strategies for Development*, Westview Press, Boulder, CO, 1982, 175.
60. **ISRA,** *Recherche et Developpment Agricole: Les Unites Experimentales du Senegal, Compte Rendu du Seminaire Tenu au CNRA de Bambey du 16 au 21 Mai 1977*, ISRA, Dakar, Senegal, 1977, 49.
61. **Bingen, R.J. and Faye J.,** Agricultural research and extension in Francophone West Africa: the Senegal experience, in *Selected Proceedings of KSU's 1985 Farming Systems Research Symposium, Farming Systems Research and Extension: Management and Methodology*, Farming Systems Research Paper No. 11, Flora, C., and Tomecek, M., Eds., Kansas State University, Manhattan, 1986b, 70.
62. **Ryan, J. and Sarin, R.,** Economics of technology options for deep vertisols in the relatively assured rainfall regions of the Indian Semi-Arid Tropics, in *Proceedings of the Seminar on Management of Deep Black Soils for Increased Production of Cereals, Pulses and Oilseeds*, ICRISAT Ed., Ministry of Agriculture and ICRISAT, New Delhi, India, 1981.
63. **Ryan, J. and von Oppen, M.,** Assessment of Impact of Deep Vertisol Technology Options, ICRISAT, Patancheru, India, 1983.
64. **Ryan, J., Virmani, S., and Swindale, L.,** Potentials and Challenges for Increased Income from Deep Black Soils in Relatively Dependable Rainfall Regions of India, Indian Bank Seminar on Innovative Technologies for Integrated Rural Development, New Delhi, India, 1982.
65. **Rhoades, R., Booth, R., and Potts, M.,** Farmer acceptance of improved potato storage in developing countries, *Outlook on Agriculture*, 1, 1983.

66. **Bunderson, W., Woldetatios, T., and Gillard-Byers, T.,** Improved integration of legumes into the agro-pastoral system of the Nuba Mountains, Sudan, in *Selected Proceedings of KSU's 1985 Farming Systems Research Symposium, Farming Systems Research and Extension: Management and Methodology,* Farming Systems Research Paper No. 11, Flora, C., and Tomecek, M., Eds., Kansas State University, Manhattan, 1986b, 486.
67. **Gillard-Byers, T. and Bunderson, W.,** Sequential program planning and adaptive research: an application in the Nuba mountains area of Sudan, in *Selected Proceedings of KSU's 1986 Farming Systems Research Symposium, Farming Systems Research and Extension: Food and Feed,* Farming Systems Research Paper No. 13, Flora, C. and Tomecek, M., Eds., Kansas State University, Manhattan, 1986c, 354.
68. **Ashraf, M.,** A case study of on-farm adaptive research at BiDA Agricultural Development Project, Nigeria, in *Selected Proceedings of KSU's 1984 Farming Systems Research Symposium, Farming Systems Research and Extension: Implementation and Monitoring,* Farming Systems Research Paper No. 9, Flora, C. and Tomecek, M., Eds., Kansas State University, Manhattan, 1986a, 190.
69. **Lang, M., Cantrell, R., and Sanders, J.,** Identifying farm level constraints and evaluating new technology in the Purdue Farming Systems Project in Upper Volta, in *Proceedings of KSU's 1983 Farming Systems Research Symposium, Animals in the Farming System,* Farming Systems Research Paper No. 6, Flora, C. Ed., Kansas State University, Manhattan, 1984, 184.
70. **Roth, M. and Sanders, J.,** An economic evaluation of selected agricultural technologies with implications for development strategies in Burkina-Faso, in *Selected Proceedings of KSU's 1984 Farming Systems Research Symposium, Farming Systems Research and Extension: Implementation and Monitoring,* Farming Systems Research Paper No. 9, Flora, C. and Tomecek, M., Eds., Kansas State University, Manhattan, 1986a, 385.
71. **Timisina, J.,** Some experiences and achievements of cropping systems research in Nepal, in *Selected Proceedings of KSU's 1986 Farming Systems Research Symposium, Farming Systems Research and Extension: Food and Feed,* Farming Systems Research Paper No. 13, Flora, C. and Tomecek, M., Eds., Kansas State University, Manhattan, 1986c, 420.
72. **Singh, B. and Sayre, K.,** Impact of cropping systems program at Sukchaina, Nepal, in *Selected Proceedings of KSU's 1985 Farming Systems Research Symposium, Farming Systems Research and Extension: Management and Methodology,* Farming Systems Research Paper No. 11, Flora, C. and Tomecek, M., Eds., Kansas State University, Manhattan, 1986b, 363.
73. **Caldwell, J. S.,** An overview of farming systems research and development: origins, applications and issues, in *Proceedings of KSU's 1983 Farming Systems Research Symposium, Animals in the Farming System,* Farming Systems Research Paper No. 6, Flora, C., Ed., Kansas State University, Manhattan, 1984, 25.
74. **Sands, M., Mares, V., Sarmiento, M., and Hertentains, L.,** FSR/E experiences in the design and validation of alternatives in a dual purpose cattle production system in Bugaba, Panama, in *Proceedings of KSU's 1983 Farming Systems Research Symposium, Animals in the Farming System,* Farming Systems Research Paper No. 6, Flora, C., Ed., Kansas State University, Manhattan, 1984, 207.
75. **Sangakkara, U.,** Increasing productivity of smallholder rice farming systems in Sri Lanka — a case study, in *Selected Proceedings of KSU's 1985 Farming Systems Research Symposium, Farming Systems Research and Extension: Management and Methodology,* Farming Systems Research Paper No. 11, Flora, C. and Tomecek, M., Eds., Kansas State University, Manhattan, 1986b, 394.
76. **Vincent, W.,** Small farm characteristics, problems, and programs in the Third World, in *Proceedings of KSU's 1981 Farming Systems Research Symposium — Small Farms in a Changing World: Prospects for the Eighties,* Farming Systems Research Paper No. 2, Sheppard, W., Ed., Kansas State University, Manhattan, 1982, 29.
77. **Norman, D., Baker, D., Heinrich, G., and Worman, F.,** Technology development farmer groups: experiences from Botswana, *Exp. Agric.*, 24, 321, 1988.
78. **Matlon, P., Cantrell, R., King, K., and Benoit-Cattin, M.,** *Coming Full Circle: Farmers' Participation in the Development of Technology,* IDRC, Ottawa, Canada, 1984, 176.
79. **Ashby, J.,** The effects of different types of farmer participation on the management of on-farm trials, *Agric. Admin.*, 25, 235, 1987.
80. **Barker, R. and Lightfoot, C.,** Farm experiments on trial, in *Selected Proceedings of KSU's 1985 Farming Systems Research Symposium, Farming Systems Research and Extension: Management and Methodology,* Farming Systems Research Paper No. 11, Flora, C. and Tomecek, M., Eds., Kansas State University, Manhattan, 1986b, 300.
81. **Lightfoot, C.,** Farmer participation in on-farm trials, in *Selected Proceedings of KSU's 1985 Farming Systems Research Symposium, Farming Systems Research and Extension: Management and Methodology,* Farming Systems Research Paper No. 11, Flora, C. and Tomecek, M., Eds., Kansas State University, Manhattan, 1986b, 525.
82. **McKee, C.,** Methodological challenges in analyzing the household in farming systems research: intra-household resource allocation, in *Proceedings of KSU's 1983 Farming Systems Research Symposium, Animals in the Farming System,* Farming Systems Research Paper No. 6, Flora, C., Ed., Kansas State University, Manhattan, 1984, 593.
83. **Moock, J., Ed.,** *Understanding Africa's Rural Households and Farming Systems,* Westview Press, Boulder, CO,1986, 234.
84. **Eicher, C.,** Food Security Research Priorities in Sub-Saharan Africa, OAU/STRC/SAFGRAD International Drought Symposium, Kenyatta International Center, Nairobi, Kenya, 1986.
85. **Gibbs, C.,** Agricultural Systems Research in Asia: A Comparative Discussion of Human Ecology Agroecosystems Research, Farming Systems Research, and Cropping Systems Research, paper presented at Second SUAN-EAPI Regional Research Symposium, Baguio City, Philippines, March, 1985.
86. **Rocheleau, D.,** Criteria for re-appraisal and re-design: intra-household and between-household aspects of FSR/E in three Kenyan agroforestry projects, in *Selected Proceedings of KSU's 1984 Farming Systems Research Symposium, Farming Systems Research and Extension: Implementation and Monitoring,* Farming Systems Research Paper No. 9, Flora, C. and Tomecek, M., Eds., Kansas State University, Manhattan, 1986a, 456.
87. **Riley, J. and Staatz, J.,** *Food Systems Organization and Management Problems in Developing Countries,* ADC Report No. 23, Agricultural Development Council, New York, 1981, 16.
88. **Behnke, R. and Kerven, C.,** FSR and the attempt to understand the goals and motivations of farmers, *Culture Agric.*, 19, 9, 1983.
89. **Cornick, T. and Alberti, A.,** Recommendation domains reconsidered, in *Selected Proceedings of KSU's 1985 Farming Systems Research Symposium, Farming Systems Research and Extension: Management and Methodology,* Farming Systems Research Paper No. 11, Flora, C. and Tomecek, M., Eds., Kansas State University, Manhattan, 1986b, 236.
90. **Fresco, L. and Poats, S.,** Farming systems research and extension: an approach to solving food problems in Africa, in *Food in Sub-Saharan Africa,* Hansen, A. and McMillen, D., Eds., Lynne Rienner, Boulder, CO, 1986, 305.
91. **Johnson, G.,** Small farms in a changing world, in *Proceedings of KSU's 1981 Farming Systems Research Symposium — Small Farms in a Changing World: Prospects for the Eighties,* Farming Systems Research Paper No. 2, Sheppard, W., Ed., Kansas State University, Manhattan, 1982, 7.
92. **Zandstra, H.,** Cropping systems research for the Asian rice farmer, *Agric. Syst.*, 4, 135, 1978.

93. **Norman, D. and Baker, D.,** Components of farming systems research: FSR credibility and experiences in Botswana, in *Understanding Africa's Rural Households and Farming Systems*, Moock, J., Ed., Westview Press, Boulder, CO, 1986, 36.
94. **Flora, C., Ed.,** *Proceedings of KSU's 1983 Farming Systems Research Symposium, Animals in the Farming System*, Farming Systems Research Paper No. 6, Kansas State University, Manhattan, 1984.
95. **Frankenberger, T.,** Adding a food consumption perspective to farming systems research, in *Selected Proceedings of KSU's 1984 Farming Systems Research Symposium, Farming Systems Research and Extension: Implementation and Monitoring*, Farming Systems Research Paper No. 9, Flora, C. and Tomecek, M., Eds., Kansas State University, Manhattan, 1986a, 518.
96. **Smith, M.,** Nutrition and farming systems research and extension, in *Proceedings of KSU's 1983 Farming Systems Research Symposium, Animals in the Farming System*, Farming Systems Research Paper No. 6, Flora, C., Ed., Kansas State University, Manhattan, 1984, 689.
97. **Tripp, R.,** *Including Dietary Concerns in On-Farm Research: An Example from Imbabura, Ecuador*, Economics Program Working Paper No. 82-2, CIMMYT, Londres, Mexico, 1982, 38.
98. **Baker, D.,** Arable Farming Development Priorities in the Central Agricultural Region, Botswana: A Farming Systems Analysis, unpublished Ph.D. dissertation, Michigan State University, East Lansing, 1987.
99. **Stoop, W.,** *Organization and Managerial Implications of a "Farming Systems Approach" for NAR*, ISNAR, The Hague, Netherlands, 129.
100. **Sands, M., Arauz, J., Gaskell, M., Benjarano, W., and Hertentains, L.,** Management experiences from three FSR/E projects in Western Panama, in *Selected Proceedings of KSU's 1985 Farming Systems Research Symposium, Farming Systems Research and Extension: Management and Methodology*, Farming Systems Research Paper No. 11, Flora, C. and Tomecek, M., Eds., Kansas State University, Manhattan, 1986b, 38.
101. **Greene, B.,** Is farming systems research the cart before the horse, in *Selected Proceedings of KSU's 1985 Farming Systems Research Symposium, Farming Systems Research and Extension: Management and Methodology*, Farming Systems Research Paper No. 11, Flora, C. and Tomecek, M., Eds., Kansas State University, Manhattan, 1986b, 177.
102. **Hildebrand, P.,** Role, potential, and problems of farming systems research and extension: developing countries vs. United States, in *Proceedings of KSU's 1981 Farming Systems Research Symposium —Small Farms in a Changing World: Prospects for the Eighties*, Farming Systems Research Paper No. 2, Sheppard, W., Ed., Kansas State University, Manhattan, 1982, 145.
103. **Chambers, R. and Ghildyal B.,** Agricultural research for resource-poor farmers: the farmer-first-and-last model, *Agric. Admin.*, 20, 1, 1985.
104. **Rhoades, R. and Booth, R.,** Farmer-back-to-farmer: a model for generating acceptable agricultural technology, *Agric. Admin.*, 11, 127, 1982.

12 INDIGENOUS APPROACHES TO RURAL DEVELOPMENT: THE AGRARIAN POPULIST TRADITION IN WEST AFRICA

Paul Richards

I. INTRODUCTION

Commitment to self-reliance and a capacity for self-organization are often notable and durable features of agrarian societies. The term "agrarian populism" is sometimes used as a catch-all label for these tendencies. The present paper refines this term somewhat by distinguishing between two (perhaps incompatible) conceptions of agrarian populism: the supply-side populism of urban-based intellectuals and demand-side populism developed by members of rural society on their own account. The paper focuses its main attention on demand-side populism. It proposes that demand-side populism has been an important feature in the political life of a number of agrarian communities in West Africa over the last 200 years. The aim is to open up a debate both on the conditions under which demand-side populism has flourished, and on whether or not this heritage of self-reliance is of continuing relevance to agricultural development in the region today.[1,2]

II. TWO TYPES OF AGRARIAN POPULISM

As a political category, populism, without further qualification, is so broad as to be meaningless. In her excellent book on the subject, however, Canovan[3] has been able to differentiate a number of distinct uses of the term, each of which is internally more or less coherent. Two of these are relevant to the present discussion. The first is the agrarian populism of urban intellectuals (as in Russia in the latter part of the 19th century). It is an example of what I wish to term "supply-side populism". Typically, supply-side populism argues that greater technological self-reliance and political autonomy are virtues that ought to be defended or fostered among rural producers, even perhaps irrespective of local opinion. The second tradition, demand-side populism, is different. Here, practical self-reliance and demands for greater political self-determination are generated from within the farming community itself. Rarely, if ever, is complete autonomy the aim. Typically, demand-side populism reflects a consciousness on the part of rural producers of the overweening power of the state and a desire to seek "a better deal for farmers" by renegotiating both rural-urban terms of trade and the political-administrative links between center and periphery. The best-known examples concern the politics of the independent family farm in the U.S. and Canada (e.g., as represented by the American Peoples' Party) circa 1880—1930.[4] Ukrainian "peasant nationalism" is another important example from roughly the same period.[5]

A. Supply-Side Populism

Broadly speaking, supply-side populism is characterized by a strong identification on the part of scientists and intellectuals with the impoverished masses, especially marginal groups in rural areas. This identification has two regularly recurring aspects. The first is a strong practical sense that poor communities have many hitherto disregarded resources for self-improvement, and that these resources might be brought into action by the catalytic involvement of intellectual volunteers. The second is a belief, often close to religious conviction, that the life of the poor is somehow more honest and authentic than a life of bourgeois sophistication. Many commentators find this the most troublesome aspect of supply-side populism. The Russian populists frequently stand accused (with good evidence) of allowing romantic reaction to cloud their practical and economic judgements. On occasion, even more sinister conclusions are drawn, e.g., that populism represents an unchaining of ideological instincts similar to those involved in the rise of

Fascism. Although Russian agrarian populism had dangerously hyper-nationalist moments, this would be too harsh a judgement.

Some subsequent populists — e.g., the Soviet school of neo-populist agricultural economics centered on the work of Chavanov[6] — have taken pains to deny that there is any necessary connection between the technical and mystical dimensions of supply-side populism. But to others (from Schweitzer to Schumacher to modern Political Ecology) the link between the mystical and technical remains an essential feature. It can be argued, with some justice, that the mystical obsessions of the populists (however mistaken) are rarely as dangerous as the technical obsessions of some of their opponents. This is not the place to rehearse such debates fully.[7] For the present it is sufficient to note that "practical" and "mystical" variants of supply-side populism continue to coexist (and interact). Proponents of "appropriate technologies" and rural development strategies based on strengthening "indigenous technical knowledge"[8] are rightly regarded as standing directly in the line of development of supply-side populism.

B. Demand-Side Populism

Political movements by family farmers in the American west and the Canadian Prairies from c. 1880 to 1920 share a number of contextual features of demand-side agrarian populism — namely, its tendency to flourish in communities dominated by medium-sized family farms, where the rate of technological modernization is rapid and where, in consequence, the level of bank indebtedness is high. A key issue for populists was control by a people's government of commodity marketing and the banking system.

Although sharing some of the darker features and tendencies of supply-side populism in Russia (attacks on finance capital, for example, were often imbued with definite feelings of anti-Semitism) American Populism nevertheless achieved a number of positive results from the perspective of the agricultural community. These included a beneficial measure of client control over the agricultural research and extension system and the creation of a political climate which favored the eventual introduction of government-guaranteed minimum prices for farm produce (supports that some now argue have encouraged an unsound over-capitalization of the family-farm sector[9]).

We should not become too preoccupied with the specific features of U.S. agrarian populism, however. It is only one among a number of distinctive instances of the demand-side phenomenon. The present paper outlines some of the characteristics of two West African cases, and assesses the significance of the political heritage of demand-side populism for African agricultural development initiatives today.

III. DEMAND-SIDE POPULISM IN 19TH CENTURY SIERRA LEONE

During the 1780s, British evangelical philanthropists proposed the establishment of a model colony on the Sierra Leone peninsula as a home for freed slaves. The first settlers arrived in 1787. There were many initial difficulties and several times the venture was on the point of collapse. Strengthened by two new groups of settlers (the Nova Scotians and Maroons) the infant colony survived to be taken over in 1808 by the British Crown as a base for the British West African anti-slavery squadron. Thereafter, the population of the Crown Colony of Sierra Leone expanded rapidly. Slavers caught at sea were brought to Freetown where the legal formalities were fulfilled, the human cargoes discharged, and the ships auctioned. The freed slaves — known as Liberated Africans — were then settled in Sierra Leone, most being assigned to the village settlements of the Peninsula.

Peterson's[10] important history of the colony, to which I am indebted for many of the data and insights in the following account, argues that although Granville Sharp's original model constitution was a failure, the spirit behind the notion of a "Province of Freedom" was given reality by the activities and initiative of the settlers themselves. It would be fair to say, however, that if Nova Scotian and Maroon enterprise was responsible for bringing this about in the capital of Freetown much of the equivalent credit for the rural areas must be assigned to the Liberated African populations.

Although the Liberated Africans hailed from all parts of western Africa from Sierra Leone south to Angola, the greatest number came from those parts of the African coastline where the slave trade was at its most active. In the first half of the 19th century this was southern Nigeria. Yoruba-speaking Liberated Africans were especially numerous because the slave trade had battened on to the civil wars following the break-up of the empire of Old Oyo (Oyo-Ile). Another large contingent came from Igbo-speaking areas along the lower Niger. Thus, whereas the Nova Scotians and Maroons were returning to Africa after a gap of several generations, the Liberated Africans were separated from their life at home only by a matter of weeks or months. Settled in the various peninsula villages, they began to adapt rapidly to a forest-zone environment that was broadly similar to the worlds from which they had been taken.

The Liberated African settlers were nominally under the control of the colonial authority in Freetown. The peninsula villages each generally had a missionary agent or school teacher acting on behalf of Freetown. In the main, however, the less the government tried to interfere the better the Liberated Africans did. The skills and knowledge with which they had arrived worked well in the new environment. On several occasions missionary agents reported that Liberated Africans resisted conversion to Christianity on the grounds that the old gods of Africa were best for dealing with epidemic diseases such as smallpox and other hazards associated with rural life. In other cases (notably among the Yoruba) Liberated Africans were already Muslims. Even those who converted to Christianity generally chose to join dissenting chapels rather than Anglican orthodoxy. Where attempts at government control proved too heavy-handed, Liberated Africans frequently broke away from the established village centers and set up new settlements of their own.

Thus a spirit of practical self-reliance and intellectual independencc thrived in the peninsula villages, perhaps strengthened by the degree of ethnic mixing which was taking place. By mid-century the missionary linguist Koelle estimated that as many as 200 different African languages were being spoken on the peninsula. This represented a substantial pool of African ideas about rural technology, the environment, disease, and social and political organization upon which Liberated African society was able to draw.

The Liberated African legacy encompassed a rich repertoire of ideas about social cooperation. The Yoruba, for example, brought with them experience of rotating credit associations (*esusu*[11]), cooperative trading ventures (*kompin*), and cults such as *egungun* where worship of ancestral spirits went hand-in-hand with the maintenance of strict social discipline. Liberated Africans also absorbed ideas from the neighboring Temne, Sherbro, and Mende people about secret societies and the organizaton of agricultural labor.[12,13] But what was most striking is the way in which many of these ideas were elaborated and developed to suit the needs of Liberated African society. Welfare was of paramount concern. Liberated Africans had in many cases formed themselves into mutual assistance groups even while awaiting assignment to a village for resettlement. These welfare companies covered a wide range of activities, from mutual assistance with farming, house-building and the expenses associated with marriage, to protection against accidents and insurance for widows and orphans. Liberated Africans also organized *esusu* groups, created trading consortia for the purchase of auctioned slaving vessels, and developed new forms of secret society organization (e.g., the Hunters Society, later to be important as an urban association among Freetown civil servants). (This mixing and merging in challenging but not impossible circumstances seems broadly comparable with the situation of recent self-settled refugees in southern Sudan, as described by Harrell-Bond[14]).

In order to make the best of this broad heritage of African thought and technical knowledge, however, it was necessary for Liberated African society to address a number of political problems arising from cultural differences and ethnic rivalries. It was in solving these problems that it seems most appropriate to speak of the emergence of a genuine agrarian populism in rural Sierra Leone in the 19th century.

Inter-village political associations began to emerge among the Liberated African population at an early date. One of the earliest of these was the so-called Secret Society of Abraham Potts (a discharged soldier of the 4th West Indian Regiment). Beginning as a benefit society in Freetown in 1824, it had spread to six colony villages, taking on political and judicial functions by 1827. Governor Campbell took steps to suppress it in July of that year, fearful that it had become an extra-legal government in some of the more remote rural areas.

The Yoruba, in particular, were noted for their secessionist leanings and tendency towards self-government. One commentator (Hamilton) described a "feeling of compatriotism...carried so far, that if a new (man) was apprehended and brought before the magistrate for petty theft, and punished by a small fine, they paid the imposition among themselves, rather than see their countrymen committed to the house of correction for default".[10] In 1826, a group of Yoruba left the colony and settled over the boundary to the east of Waterloo. From there "they threatened the whole eastern district...and presented the central government with the first organized opposition to its policies."[10] Four years later, a further flight of Yoruba from the colony led to armed confrontation with government forces at Cobolo on the Ribbi. In Peterson's judgement "the political dissent of the Yoruba had succeeded by 1833 in allowing them to exercise that element of independence which characterized the life of the resettled Liberated Africans in Sierra Leone".[10]

By mid-century these leanings towards autonomy assumed new and more subtle forms with which the government was finally able to come to terms. The crucial development was the establishment of a political structure known as the Seventeen Nations. The idea emerged in response to a riot between Yoruba and Igbo at Waterloo in 1843, but as Peterson[10] suggests, "its roots lay in the whole experience of the Liberated Africans with the broadened judicial and political functions of their benefit societies." The Seventeen Nations was a council of the various ethnic and national interests represented in each of the colony villages. First proposed by John Macauley, "king" of the Freetown Yoruba, who was sent to mediate after the 1843 riot, the Seventeen Nations provided a forum for settling disputes without need for government intervention. The Waterloo arrangement proved a success and was soon imitated in other colony villages. Each Seventeen Nations council supervised village affairs "in the manner of a government", settling disputes, mending roads, and, in the words of the source cited by Peterson,[10] "preventing social disorder in communities where the spirit of insubordination largely predominates".

Peterson[10] suggests that the name referred to the 17 distinct ethnic groups living in Waterloo when the first council was formed. It is worth noting however that the numbers five, nine and seventeen have special symbolic force in southern Nigeria. Throughout the region the basic unit of the periodic market cycle is 4 days. Local markets may operate every 4 days, but the more important regional and specialized markets tend to work on higher multiples (8-, 16-, or 32-day cycles). In local thought, however, the completeness and functional unity of a market cycle is indicted by counting from market day to market day inclusively. A 4-day interval between markets is described as a "5-5 day" ring (the reduplication indicating that the market system cycles continuously). Important regional market systems operating on 8- or 16-day intervals would be described as "9-9 day" or "17-17 day" rings.

Likewise "nines" and seventeens" sometimes occur in political parlance to indicate both the symbolic completeness but also the equality of elements in a confederal administrative structure. The Ikale of southeastern Yoruba country are unusual among Yoruba groups for having no major precolonial urban centers. Defining their political system, in the 19th century, in opposition to the major centralizing traditions of Benin, Lagos, and Ife they referred to themselves as The Nine Tribes of Ikale. It mattered not that there were in fact 11 constituent elements. The figure nine signified completeness and equality among the units. The Nine Tribes might equally well be translated by a phrase like "Ikale Democratic Federation". (Elsewhere,[15,16] I have suggested that there is much in the recent history of Ikale country with a distinct populist ring to it.)

Did Macauley have some similar notions in mind? Certainly we can see some strikingly populist features in the Seventeen Nations system. Demand-side agrarian populists are not in favor of complete autonomy. The price required to sustain external security may be too great. In the end therefore they must come to an accommodation with central government, while at the same time seeking as much local political autonomy as possible. These were exactly the two objectives Macauley's innovation was designed to secure. Interestingly, the Nine Tribes of Ikale was not an autonomous polity, but a region which maintained a carefully negotiated quasi-autonomy from its more powerful urban neighbors.[15] The desire for autonomy reflects the high values that farming communities place on self-reliance. But it is also a major means of defending local interests against urban-based speculators. It is consistent with this interpretation to find that in the 1880s the Seventeen Nations at Hastings was successful in preventing William Grant, a wealthy Freetown publisher, from acquiring 1000 acres of land near the village, despite the fact that earlier the Governor had granted him permission.

A. The Significance of the Sierra Leone Case

In what ways is this flowering of agrarian populist sentiment in rural areas of the Sierra Leone colony in the 19th century relevant to wider debates about agrarian change in West Africa today? It seems to me that the answer is twofold.

First, the organizational forms characteristic of Liberated African village society were not restricted to the peninsula alone, but can be traced throughout the Upper Guinean coastal region during the later colonial period. Since clearly there was a mutual exchange of ideas and influences between the Liberated Africans and the peoples of the Sierra Leone hinterland it would be foolhardy to claim that the savings clubs, welfare societies and agricultural labor cooperatives important in most villages in modern Sierra Leone and Liberia today are direct copies of Liberated African models. It would perhaps make more sense to think of these as the result of a 19th century synthesis of extra-regional and local influences taking place on both sides of the Colony-Protectorate divide. In a Mende village, Mogbuama, in central Sierra Leone, for example, people are quite clear that some types of labor and savings cooperatives (organized along lines almost identical to those of the Liberated Africa *kompin* as described by Peterson[10]) are modifications of coastal models whereas others derive from up-country influences, e.g., Temne and Mandinka models.[13,17,18] The question of origins, however, is not by itself very important. The more important issue is that the conditions under which such cooperatives flourished in the 19th century remained applicable to much of the Sierra Leone Protectorate during the colonial period — we know, for example, that "company" organization was an important element in the expansion of tidal rice production along the Scarcies, c. 1880—1920.[19] Company organization continues to be important throughout the region today.[20-23] A number of recent development initiatives seek to build upon this undiminished legacy of indigenous cooperation.

The second point is that what happened under perhaps unusually conducive conditions in 19th century Sierra Leone provides a useful guide to the identification of a number of latent populist tendencies in the politics of the regions from which the majority of the Liberated Africans came. The Ikale case has already been cited. Other groups also moved down this path as 19th century warrior leaders ceded power and influence to a younger generation, for whom tree-crop exports were to become a major preoccupation.[24] It was with the massive expansion of tree-crop production during the first four decades of the 20th century that rural politics throughout a region from Asante to Mount Cameroon took a distinct turn towards demand-side agrarian populism.

IV. DEMAND-SIDE POPULISM IN 20TH CENTURY GHANA AND SOUTHERN NIGERIA

The vigorous commercial enterprise of small-scale cocoa planters in Ghana and Western Nigeria is well known from the work of Hill[25] and Berry.[26] Less well known, but equally impressive, is the story of how indigenous enterprise created the adjacent oil-palm export belt in south-central and eastern Nigeria.[27] Recent work has brought out some of the complexity of these transformations, and in particular, the crucial role of innovation by women in the food-crop sector.[28] The actors in these three agrarian dramas are sometimes described as "peasants", though in fact the literature makes it clear that agriculture was often only a relatively small part — a first step — in a complex developmental cycle in which the individual entrepreneurs progressed from planting to commodity trading to transporting, and then in the next generation to urban residence and employment in the professional or governmental sector.

The politics of this transformation has a number of features that mark it out as a clear instance of demand-side agrarian populism (in some respects quite closely resembling the family-farm politics of North America at the same period). In the West African case, alliances of planters, produce buyers and transporters were focal points for several campaigns mounted to secure a "fair deal" for farmers against the depredations of "metropolitan capital". But at the same time as campaigning for fairer prices and taxes, reform of local administration was also prominent on the agenda. Successful direct action was mounted both against the colonial government and major European commercial firms.

The earliest and in some ways the most dramatic of these actions was the "women's war" of eastern Nigeria in 1929.[29,30] This was a series of major incidents in several towns in the palm belt in which rural women confronted the colonial administration over a rumor that they were to be taxed. Igbo women had a tradition of "sitting upon a man" — organizing collective action against men with whom they were displeased (because of abusive or violent behavior, or unwarranted intervention in women's activities). By the 1920s, women in eastern Nigeria had become especially important in the palm kernel trade and in production and processing of foodcrops which were the essential backbone to a greatly expanded rural economy of the palm belt.[28] The rumor of taxation fueled a concern that the "male" world of colonial administration was about to try and extend its authority to cover the commercial activities of women — hitherto regarded as an independent sphere. In order to stave off this threatened colonization of the women's realm, established ideas about sanctioning men were pressed into action in a new context. Women marched upon Aba and other towns in 1929 in effect to "sit upon" the administration.

The subsequent confrontation, during which troops opened fire on the demonstrators, had such serious repercussions that a major review of British administrative policy in the region was required. This made clear the extent to which officials were ignorant of the organization of rural society in eastern Nigeria, and of the ways in which these organizational structures had been changing in response to new economic opportunities. The rapidity and effectiveness with which the initial protest had been coordinated was eloquent testimony to the scope and efficiency of established networks of collaboration among rural women in the area. Clearly, populist capacities for self-organization were as strong among women as among men, and henceforth would have increasingly to be taken into account in the political process.

Let me now note, very briefly, two later examples, to indicate something of the extent and generality of demand-side populism as a feature of economic development in the West African cash tree-crop production zone in the 20th century.

The first concerns concerted action by small-scale cocoa producers in the Gold Coast (Ghana) during the 1930s directed against the monopsonistic activities of the major European cocoa purchasing companies. An effective market boycott in 1937—1938 (the cocoa "hold-up") attracted government attention to the price-fixing activities of the European companies.[31,32] The statutory marketing board system (later extended to all the major cash-crop producing regions of British West Africa) was conceived as an answer to some of these problems, and as an attempt to expand output in a controlled manner by stabilizing prices to producers. In the event, matters never worked out this way. Under wartime emergency conditions followed by high world-market prices in the post-war economic boom boards continued to accumulate surpluses. In time these were drawn upon to fund many urban-oriented developments inherited at Independence by the sons and daughters of the populist agitators of the 1930s.

According to some accounts this is why much of the steam went out of small-farmer demand-side populism as a political force in West Africa from the 1950s onwards. Not all groups, however, were able to gain ready access to urban resources. Accordingly, farmers in the region continued, from time to time, to draw upon the demand-side populist legacy to articulate grievances.[33] The *Agbekoya* movement, active in older cocoa belt settlements around the city of Ibadan in the late 1960s and early 1970s was one such instance.[34] In striking similarity to some of the Yoruba-inspired movements among Liberated Africans in 19th century Sierra Leone *Agbekoya* made its case to the authorities (in this case the state government of Western Nigeria) by demonstrating alternative political organization. The movement's adoption of quasi-military methods and dress appears to have been aimed mainly at undermining the authority of village chiefs by demonstrating a capacity for disciplined self-administration. Perhaps only secondarily was it a strategy for achieving political ends by violent means.

V. THE CONTINUING SIGNIFICANCE OF DEMAND-SIDE POPULISM IN WEST AFRICAN RURAL DEVELOPMENT

This last point about *Agbekoya* is intended to suggest that more needs to be known about the cultural logic of West African demand-side populism, because possibly it is in this area, rather than in the generalities of political economy, that the continuing significance of this tradition is to be found. If the Yoruba who fought the Cobolo War against the British in 19th century Sierra Leone and the Yoruba villagers who participated in the *Agbekoya* uprising in the 1970s are linked, it is perhaps more through their shared cultural inheritance rather than through similarities in economic circumstances. It may be a mistake to seek explanations for the durability of demand-side populism solely or largely in the realm of political economy.

If West African demand-side populism is at heart cultural rather than economic, what gives it cultural durability? Partial answers are beginning to emerge from directions as diverse as agricultural ecology and the study of intellectual culture.[13,35] Take, for instance, the case of seasonal hunger. Seasonal environments in the tropics are inherently risky. These risks are as pronounced in high rainfall regions as in the drier savannas. Village societies recognize these risks as potential leveling mechanisms, against which even wealthy merchants or modern cultivators are not yet invulnerable.

The reality of these hazards reinforces two sets of social tendencies widespread throughout the region: a familiarity with, skill at, and admiration for risk-taking abilities (including the ability to accept the rough with the smooth) and commitment to co-operative arrangement as a means of achieving a measure of social security. It is true that close to the major cities "capitalist penetration" may have corroded values and beliefs on which hazard-coping strategies once rested.[36,37] But in other cases it is apparent that these values are too deeply ingrained in social and individual consciousness to be erased for mere money. Good examples of the way in which survival-oriented values are stubbornly reproduced in defense of a populist moral order are to be found in two recent studies of story-telling performance in farming communities in modern Sierra Leone.[35,38] It is through such means (and with good practical sense) that many rural communities continue to resist "capture" by market forces and the modern state.

It is my contention, therefore, that demand-side agrarian populism in West Africa continues to take much of its impetus from perceptions, attitudes, expectations, and organizational skills formed by the experience of coping with the highly diversified but hazardously challenging environments of the high rainfall zone. The ecological basis for this argument has been developed elsewhere.[16] My present point is to draw attention to the continuing value of this heritage for rural development, especially under conditions of economic or political adversity.[39] In this respect, the ways in which rural Nigerians rehabilitated themselves in the aftermath of the Civil War (1967—1970) is an especially significant story, largely waiting to be told.[40-44]

VI. CONCLUSIONS

Demand-side agrarian populism is a political tradition that stresses and demonstrates the importance of self-reliance and self-organization in rural life. However, it is autonomist only to a point. In most cases the main concern of populists is not to turn away from governments and their supporting chorus of urban-based intellectuals but to secure a fairer hearing. The first requirement is for closer attention to be paid to the ways in which rural life already works.

This is one point on which supply-side and demand-side populisms converge. It is conceivable that the main contribution of supply-siders will be to serve as facilitators for the dialogue around which such mutually improved understanding might be built. It is appropriate that some of the best work along these lines stems from the Nigerian university world where there is a substantial tradition of recognizing and working with demand-side populism as a valid framework for grass-roots development activities.[45-47] This enthusiasm often closely parallels the activities of urban-based "descendents unions" in raising investment resources for the home villages of their members.

Perhaps it is appropriate to conclude with a quotation from a document that is an outcome of one such initiative. The document in question is a development plan prepared by "the central planning committee of the *Egbe omo ibile Awe*" for a small rural town in the savanna zone of western Nigeria. The final chapter of this plan is entitled (significantly) "Seeking Political Leverage" and ends with a statement that effectively encapsulates the authentic spirit of demand-side populism in West Africa today: "(this plan) is an act of faith and a hopeful commitment to the belief that a determined people can by their own effort change the depressing and demeaning circumstances of their life".

REFERENCES

1. **Kofi, T. A.,** Peasants and agrarian economic development: populist lessons for Africa, in *Decolonization and Dependency: Problems of Development of African Societies,* Yansane, A. Y., Ed., Greenwood Press, Westport, CT, 1980.
2. **Williams, G.,** Taking the part of peasants: rural development in Nigeria and Tanzania, in *The Political Economy of Contemporary Africa,* Gutkind, P. W. and Wallerstein, I., Eds., Sage, Beverly Hills, CA, 1976.
3. **Canovan, M.,** *Populism,* Junction Books, London, 1981.
4. **Hofstadter, R.,** North America, in *Populism: Its Meanings and National Characteristics,* Ionescu, G. and Gellner, E., Eds., Weidenfeld & Nicolson, London, 1969.
5. **Conquest, R.,** *The Harvest of Sorrow: Soviet Collectivization and the Terror Famine,* Hutchinson, London, 1986.
6. **Harriss, J., Ed.,** *Rural Development: Theories of Peasant Economy and Agrarian Change,* Hutchinson, London, 1982.
7. **Kitching, G.,** *Rural Development and Underdevelopment,* Methuen, London, 1982.
8. **Chambers, R.,** *Rural Development: Putting the Last First,* Longman, Harlow, 1983.
9. **Solkoff, J.,** *The Politics of Food,* Sierra Club, San Francisco, 1985.
10. **Peterson, J.,** *Province of Freedom: A History of Sierra Leone, 1787—1870,* Faber & Faber, London, 1969.
11. **Bascom, W.,** The *esusu:* a credit institution of the Yoruba, *J. R. Anthropol. Inst.,* 82, 63, 1952.
12. **Rodney, W.,** *A History of The Upper Guinea Coast, 1545—1800,* Clarendon Press, Oxford, 1970.
13. **Richards, P.,** *Coping With Hunger: Hazard and Experiment in an African Rice-Farming System,* Allen & Unwin, London, 1986.
14. **Harrell-Bond, B.,** *Imposing Aid,* Oxford University Press, London, 1986.
15. **Richards, P.,** Core, periphery and urbanity: the politics of settlement in the Okitipupa region, western Nigeria, 1870—1950, Biennial conference, African Studies Association of United Kingdom, Birmingham, September, 1982.
16. **Richards, P.,** *Indigenous Agricultural Revolution: Ecology and Food Production in West Africa,* Hutchinson, London, 1985.
17. **Ames, D. W.,** Wolof co-operative work groups, in *Continuity and Change in African Cultures,* Bascom, W. R. and Herskovits, M. J., Eds., University of Chicago Press, Chicago, 1959, 224.
18. **Linares, O. F.,** From tidal swamp to inland valley: on the social organization of wet rice cultivation among the Diola of Senegal, *Africa,* 51, 557, 1981.
19. **Glanville, R. R.,** Rice production on swamps, Sierra Leone Agricultural Notes no. 7, 1938.
20. **Jackson, M.,** *The Kuranko: Dimensions of Social Reality in a West African Society,* Hurst, London, 1977.

21. **Johnny, M., Karimu, J., and Richards, P.,** Upland and swamp rice farming systems in Sierra Leone: the social context of technological change, *Africa,* 51, 596, 1981.
22. **Karimu, J.,** Strategies for peasant farmer development: an evaluation of a rural development project in northern Sierra Leone, Ph.D. thesis, University of London, 1981.
23. **Seibel, H. D. and Massing, A.,** *Traditional Organizations and Economic Development: Studies of Indigenous Co-operatives in Liberia,* Praeger, New York, 1974.
24. **Berry, S.,** *Cocoa, Custom and Socio-economic Change in Rural Western Nigeria,* Clarendon Press, Oxford, 1975.
25. **Hill, P.,** *The Migrant Cocoa Farmers of Southern Ghana: a Study in Rural Capitalism,* Cambridge University Press, Cambridge, 1963.
26. **Berry, S.,** *Fathers Work For Their Sons,* University of California Press, Berkeley, 1986.
27. **Usoro, E.,** *The Nigerian Oil-Palm Industry,* University of Ibadan Press, Ibadan, 1974.
28. **Martin, S.,** Gender and innovation: farming, cooking and palm processing in the Ngwa region, south-eastern Nigeria, *Journal of African History,* 25, 1984, 411.
29. **Gailey, H.,** *The Road to Aba,* 1969.
30. **Van Allen, J.,** "Aba Riots" or "Women's War"?, in *Women in Africa,* Hafkin, N. and Bay, E., Eds., Stanford University Press, Stanford, 1976, 59.
31. **Howard, R.,** Differential class participation in an African protest movement: the Ghana cocoa boycott of 1937—38, *Canadian Journal of African Studies,* 10, 469, 1976.
32. **Southall, R.,** Farmers, traders and brokers in the Gold Coast cocoa economy, *Can. J. Afric. Stud.,* 12, 185, 1978.
33. **Beer, C.,** *The Politics of Peasant Groups in Western Nigeria,* Ibadan University Press, Ibadan, 1976.
34. **Williams, G.,** Political consciousness among the Ibadan poor, in *Sociology and Development,* de Kadt, E. and Williams, G., Eds., Tavistock Press, London, 1974, 109.
35. **Jackson, M.,** *Allegories of the Wilderness: Ethics and Ambiguity in Kuranko Narratives,* University of Indiana Press, Bloomington, 1982.
36. **Watts, M.,** *Silent Violence: Food, Famine and Peasantry in Northern Nigeria,* University of California Press, Berkeley, 1983.
37. **Shenton, R. and Watts, M.,** Capitalism and hunger in northern Nigeria, *Rev. Afric. Polit. Econ.,* 15, 16, 53, 1979.
38. **Cosentino, D.,** *Defiant Maids and Stubborn Farmers: Tradition and Invention in Mende Story Performance,* Cambridge University Press, Cambridge, 1982.
39. **Dei, G.,** Adaptation and environmental stress in a Ghanaian forest community, Ph.D. thesis, University of Toronto, Toronto, 1986.
40. **Arua, E.,** Yam ceremonies and the values of Ohafia culture, *Africa,* 51, 694, 1981.
41. **Okafor, J. C.,** Edible indigenous woody plants in the rural economy of the Nigerian forest zone, *Forest. Ecol. Manag.,* 3, 45, 1980.
42. **Okafor, J. C.,** Woody plants of nutritional importance in traditional farming systems of the Nigerian humid tropics, Ph.D. thesis, University of Ibadan, Ibadan, Nigeria, 1981.
43. **Uzozie, L. C.,** Tradition and change in Igbo food-crop production systems: a geographical appraisal, Ph.D. thesis, University of London, London, 1979.
44. **Uzozie, E.,** The changing context of land-use decisions: three family farms in the yam-cultivation zone of eastern Nigeria, *Africa,* 51, 678, 1981.
45. **Mabogunje, A. L.,** *The Development Process: A Spatial Perspective,* Hutchinson, London, 1980.
46. **Mabogunje, A. L.,** The dilemma of rural development in Africa, *Region. Dev. Dial.,* 2, 1, 1981.
47. **Igbozurike, U.,** *Agriculture at the Cross-roads: A Comment on Agricultural Ecology,* University of Ife Press, Ile-Ife, 1977.

13 AGROECOLOGY AND RURAL DEVELOPMENT IN LATIN AMERICA

Miguel A. Altieri

I. INTRODUCTION

Perhaps one of the most challenging aspects of Latin America is that any attempt at agricultural development must deal with the tremendous ecological, cultural and socioeconomic diversity that characterizes the agricultural systems of the region. Systems range from small-scale subsistence farms managed by ethnic groups and campesinos using local resources to large-scale commercial farms utilizing modern technology, and from systems located in medium and high altitude Andean mountains to tropical lowland areas. Such agricultural diversity results from local variations in climate, soils, economic relations, social structures and cultural history, making it very difficult for developers to claim that there is only one unique agricultural development strategy able to deal with such complexity.

II. AGRICULTURE AND TECHNOLOGY IN LATIN AMERICA

Attempts at modernizing Latin American agriculture started after World War II when land and labor-saving technologies were incorporated into farming operations and were capable of significantly impacting land productivity. The diffusion of these new technologies was related to agroclimatic and socioeconomic conditions, and was introduced to Latin America in a pattern of sequential cycles[1] (Figure 1). After an extensive initial period in which agronomic practices (i.e., planting dates and densities, etc.) were the most important technology, mechanization became the dominant technology in the early 1970s. Next came the dissemination of improved seeds, especially hybrids. Pesticides and fertilizers represent the last of this technological cycle, which has still to reach its inflection point. In fact, pesticide expenditures for Latin America are expected to triple over the next 10 years, especially in Brazil, Mexico, Argentina and Colombia.[2] Three aspects characterize this process of technological diffusion:

1. Technological change has been concentrated mainly in temperate and subtropical areas where physical and socioeconomic conditions are similar to those in the industrial countries where the technology was generated.
2. Despite some local input production and manufacturing, Latin America has become a net importer of chemical inputs and agricultural machinery, increasing government expenditures and exacerbating technological dependence. (Between 1980 and 1984 Latin America imported about 430 million dollars worth of pesticides, used about 6.5 million tons of fertilizers, and had well over 850,000 tractors.)[2]
3. Technological change has mainly benefitted the production of export and/or commercial crops produced primarily in the large farm sector, marginally impacting productivity of food crops, which are largely grown by the peasant sector.[3]

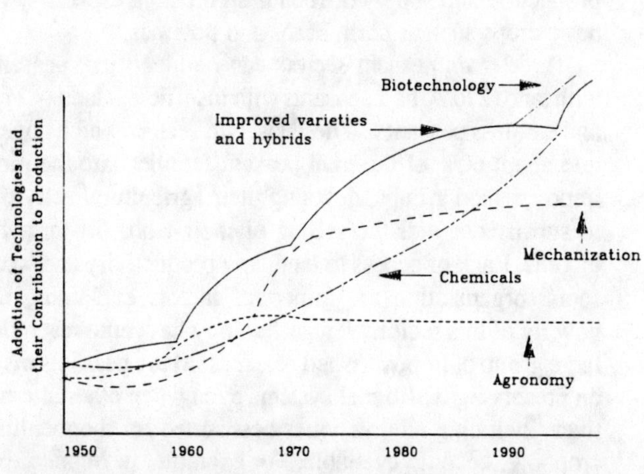

FIGURE 1. Technological cycles in Latin American agriculture. (Adapted from Pineiro, M. and Trigo, E., IICA, San Jose, Costa Rica, 1983.)

A fifth technological diffusion cycle will soon be initiated in Latin America as biotechnology research (mainly promoted by the private sector), starts offering new innovations. Recent advances in biotechnology are starting to challenge countries to reassess their technological development. Since biotechnology research and development is mainly concentrating on specialty and industrial crops, and/or on crops tailored to specific input needs, this introduction will affect agriculture unevenly, worsening economic inequality by directing benefits only to resource-rich farmers who can afford to adopt biotechnology products, thus bypassing resource-poor farmers, especially those confined to rainfed, hillside areas or humid tropical zones.[4] Biotechnology can speed up the technological treadmill, increase production and put downward pressure on prices, which may cause peasants to become increasingly marginal producers without assistance unless the public sector, universities and NGOs undertake a major research effort tailored to meet peasant needs and circumstances. It is vital that the small farm sector maintain its level of contribution to the Latin American market to expand its role as producer, and also presently there is no place else in the economy to absorb these people. To do this, peasant production will require the necessary technological support. Otherwise, due to its decreasing significance, the peasantry and the small farm sector will, in the near future, cease to be a technological challenge and become a social one.

A. The Peasantry in Latin America

In Latin America there are about nine million peasant units with an average farm size of 2.1 ha.[5] These peasant farmers constitute about 80% of the total number of farmers, holding 18% of the total agricultural land and only 7% of the arable land. It is in this sector where about 40 to 50% of the agricultural output for domestic consumption originates (Table 1). Thus, the contribution of the peasant sector to the production and supply of food is significant, especially for basic crops such as corn, beans and potatoes.

Three main peasant sectors can be identified: peasants with land (2 to 20 ha), peasants with insufficient land (<2 ha) and landless peasants.[5] These last two sectors which constitute about 60% of the total peasant families, are the most impoverished groups, devoting their agricultural activities to subsistence and the selling of their labor off-farm for income. Lack of access to land, low productivity and weak social organization are important factors explaining the poverty of this sector. New technology has yet to reach this large group of impoverished peasants. Many peasants insist on preserving traditional systems even when other alternatives, including new varieties best suited for monoculture cropping, become available. For example, in Mexico, between 10 to 25% of the total peasants have adopted improved seed, fertilizers, pesticides and machinery (Table 2). On Colombian hillsides, 15% of the peasants adopted new maize varieties, whereas 65% of the producers adopted them in the lowland valleys. Several factors may explain why some peasants have difficulty in adopting these new techniques:

TABLE 1
Estimated Contributions to Agricultural Production by Commercial and Small Farmers in Latin America

Indicators	Entrepreneurial agriculture (%)	Peasant agriculture (%)
Production for domestic consumption	59.0	41.0
Production for export	68.0	32.0
Production of permanent crops	59.0	41.0
Production of short-cycle crops	47.0	53.0
Production of maize	49.0	51.0
Production of beans	23.0	77.0
Production of potatoes	39.0	61.0
Production of rice	68.0	32.0
Production of coffee	59.0	41.0
Production of sugarcane	79.0	21.0

From Ortega, E., Peasant agriculture in Latin America, ECLAC/FAO, Santiago, Chile, 1986.

1. Modern varieties and global technological recommendations have proved to be seriously unfit to the tremendous ecological and socioeconomic heterogeneity that characterizes peasant farms and households (Figure 2).
2. The physical and socioeconomic conditions of agricultural experiment stations where technologies are generated sharply differ from those of peasants.[6] Therefore the technology generated at these research experiment stations usually does not fit peasant needs nor their ecological and socioeconomic conditions.
3. Many peasants have a resistance to adopting technologies which they perceive as increasing risks by further monetarizing their economies and making them more dependent on the market.[7]

B. Rural Development Programs and the Peasantry

Several institutions are involved in developing and extending agricultural innovations to peasants in Latin America: International Centers (Centro Internacional de Mejoramiento de Maiz y Trigo (CIMMYT), Centro Agronomico Tropical de Investigacion y ENSEÑANZA (CATIE), Centro Internacional de Agricultura Tropical (CIAT) and Centro Internacional de la Papa (CIP)) Ministries of Agriculture with their agricultural experiment station networks and integrated rural development programs (DRIs), agricultural universities and non-government organizations (NGOs). Many of the government institutions are aware of criticisms that claim that their programs lack an appreciation for the ecological and socioeconomic milieu in which peasants operate, exclude peasants as both collaborators and beneficiaries, and promote inappropriate technologies. Certain national and regional institutions have therefore started to

TABLE 2
Level of Modernization and Use of Inputs by Type of Producer in Mexico

Type of producer	Improved seed	Use of fertilizers	Pesticides	Use of tractors	Use of animal traction	Mechanization
Peasant producer	12.0	25.0	11.0	21.0	66.0	14.0
Large-scale entrepreneur	59.0	83.0	77.0	91.0	42.0	90.0

Note: Values are expressed as percent.

From Ortega, E., Peasant agriculture in Latin America, ECLAC/FAO, Santiago, Chile, 1986.

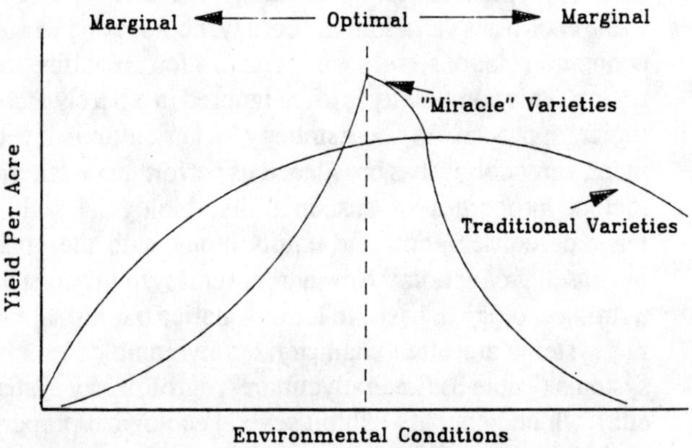

FIGURE 2. Environmental tolerance and yield response of traditional crop varieties managed by peasants, and high-yielding varieties offered by the Green Revolution.

improve their analysis of existing small farming systems, consider limitations imposed by both physical and socioeconomic factors, and design research accordingly.

To date there have been two significant responses to the challenge of tailoring research to peasant needs and circumstances. The first has been Farming Systems Research (FSR),[8] pioneered by the international centers, and which entails an understanding of what farmers are doing, why they have chosen their current practices, and what is required for them to accept a new technology. Although CATIE, CIP, CIMMYT and CIAT emphasize research efforts that encourage peasant participation, focus is still placed on increasing commodity production through technological transfer and innovation, with recommendations centering on high-yielding varieties accompanied by a variable set of chemical pesticides and fertilizers. This has been certainly the approach of national research programs such as those developed by the Instituto Nacional de Investigaciones Agropecuarias (INIAP) in Ecuador, Ministerio de Agricultura (MAG) in Costa Rica and the Instituto de Ciencia y Tecnologia Agricola (ICTA) in Guatemala, in collaboration with International Centers or under direct AID sponsorship.[9-11] Although the participation of social scientists is novel in FSR programs,[12] their efforts seem to be subordinated to the need expressed by plant breeders to understand why farmers are not adopting the new varieties. In other words, the mission of social scientists in agricultural research seems constrained to overcome the social barriers that limit the adoption by peasants of the technologies developed at the centers.

The second response has been Integrated Rural Development (DRI), which combines the economic goal of enhanced agricultural production with the development of fisheries, forestry, handicraft production, off-farm employment, and the provision of health, education and other communal services.[13] So far, results have been modest, given in part to the complexities of the problem and also to the organizational structure and inexperience of the public institutions responsible for implementing them. Although in many DRI districts (i.e., DRI programs in Honduras and Colombia) increased production has led to higher income, this did not result in improved nutrition or welfare mainly because other basic community needs were not addressed.[7,13]

Despite the numerous efforts it is often possible to detect deficiencies in the capacity of International and official Latin American institutions to generate technologies that respond to the needs and characteristics of the peasantry.

C. Novel Rural Development Approaches

In many countries, either due to the economic crisis or government policies, the peasantry has not been given an active role within national development models nor is it perceived economically by the market. This disconnection between the public institutional sector and the peasantry has created a "void", which increasingly has been filled by a number of non-government organizations (NGOs) many of which have emerged under the umbrella of the Catholic church, creating an instance of collaboration and action with the marginalized peasant sector. Many of these NGOs have initiated programs emphasizing research, education, technical assistance and credit provision activities aimed at the development of peasant agriculture. Several of these groups have demonstrated a unique ability to understand the specific and differentiated nature of small farm production and have promoted some successful experiences in generation and transfer of peasant technology.[14]

A large portion of these NGOs are actively immersed in the search of alternative systems of production aimed at:

1. Improving the production of basic foods at the farm level to enhance the family nutritional intake
2. Promoting an efficient use of local resources (i.e., land, labor, agricultural sub-products, etc.)
3. Increasing crop and animal diversity to minimize risks
4. Improving the natural resource base through water, soil and local germplasm conservation
5. Reducing the use of external inputs to reduce dependency, but sustaining yields with appropriate technologies
6. Ensuring that alternative systems have an empowering effect not only on individual families but on the total community.

NGOs have developed several lines of work to reach these objectives:

a. Rescue and re-evaluation of peasant technologies
b. Valuation of traditional food crops (*Amaranthus*, quinoa, lupine, etc.) and conservation of native germplasm,
c. Testing and implementation of organic farming and other low-input techniques
d. Protection and regeneration of the natural resource base through soil conservation, erosion control, water harvesting, reforestation, etc.
e. Development of appropriate farming tools (animal-drawn implements, etc.) and/or energy conserving practices (community forests, biogas, alternative energy sources, lorena stoves, etc.)
f. Food processing and preservation
g. Development of research methodologies to assess peasant knowledge and rationale, as well as rural appraisal techniques to measure the agroecological performance of peasant technologies and farming systems — Considerable research efforts have been devoted by some NGOs to develop guidelines on how to spatially and functionally better organize peasant farming systems to ensure food self-sufficiency as well as ecological stabilization, either at the farm level or at the regional level.

Research efforts conducted by NGOs are complemented by programs of popular education and information dissemination aimed at developing the peasants' own knowledge and technologies in the realm of health, agricultural production and social organization. Many agricultural programs of some NGOs tend to preserve and strengthen "peasant's productive rationale" while supporting peasants in the process of technological adaptation and linkage to markets. Most groups contribute to existing technologies through demonstration farms that incorporate both traditional peasant techniques as well as new viable alternatives.[15]

Although some of the efforts of several NGOs have had considerable impact in promoting grassroots development at the local level, most NGOs understand that rural development strategies will not fully succeed at a more regional level, unless the structural obstacles that underline rural poverty (low productivity and limited access to land) can be overcome.

III. AGROECOLOGY AS A STRATEGY TO AGRICULTURAL TECHNOLOGY DEVELOPMENT

In many areas of Latin America, the urgent need of the most impoverished rural sectors is survival; therefore, improvement of subsistence production is absolutely essential for the welfare of these rural populations. In order to match the agricultural development with the needs and potentials of local peasants, what is required is an agricultural development approach, namely agroecology, that is more sensitive to the enormous variations in ecology, population pressures, economic relations, and social organization prevailing in the region. This complexity is often ignored in a purely technological approach. As a new strategy for agricultural development, agroecology has broadened its performance criteria to include properties of sustainability, biological stability, resource conservation and equity along with the goal of increased production.[16] New agricultural systems must have a strong ecological basis. In Latin America traditional farming systems are often characterized by complex cropping systems (Table 3) (i.e., polycultures, agroforestry systems, etc.), which generally exhibit several ecological properties such as biological diversity, nutrient recycling capabilities, risk aversion, soil and water conservation features, pest suppressive potential, etc.[17] These desirable elements of traditional agriculture should be identified and retained in the course of agricultural modernization and used as starting points in any process of technology development.[18]

A. Latin American Nongovernmental Organizations

Due to its novelty in viewing the question of peasant agricultural development, agroecology has heavily influ-

TABLE 3
Extent of Polyculture Adoption in Selected Latin American Countries

Country	Observations on polycultures
Dominican Republic	40% or more maize grown with other crops
Mexico	20% of maize mixed with other crops
Brazil	6—11% of rice and maize mixed with other crops
Paraguay	33% of beans, 10% of sweet potatoes, 10% of maize mixed with other crops
Venezuela	16% of rice, 33% of maize, 20% of beans, 20% of cassava, 50% of cotton mixed with other crops
Colombia	90% of the beans associated with maize
Guatemala	73% of beans mixed with other crops, especially maize
Latin America	60% maize and beans in crop mixtures

From Francis, C. A., *Multiple Cropping Systems*, Macmillan, New York, 1986, chap. 1.

enced the agricultural research and extension work of many Latin American NGOs. Several characteristics of the agroecological approach to technology development and diffusion make it especially compatible with the discourse of NGOs:

1. Low-input and regenerative agricultural techniques and designs proposed by agroecology are socially activating since they require a high level of popular participation.
2. Agroecological techniques are culturally compatible since they do not question peasants' rationale, but actually build upon traditional farming knowledge, combining it with the elements of modern agricultural science.
3. Techniques are ecologically sound since they do not attempt to radically modify or transform the peasant ecosystem, but rather identify management elements that, once incorporated, lead to optimization of the production unit.
4. Agroecological approaches are economically viable since they minimize costs of production by enhancing the use efficiency of locally available resources, and thus potentially making farmers less dependent on the market and/or industry.

Despite the above characteristics, agroecology has not yet provided the practical and theoretical contributions that one would expect. Worldwide attention on agroecology started after the energy crisis of the 1970s, and in Latin America more specifically after the economic crisis derived from the external debt. So far, what has been achieved during these few years is that the potentials of agroecology have been recognized. There has not been, however, a sum of human, scientific and institutional resources that would foster a more in-depth development of agroecology at the national or regional level.

B. The Contributions and Approaches of NGOs

As explained before, Latin American NGOs emerged as an attempt to fill the vacuum left by the withdrawal of the state (either by policy definition as in Chile or by lack of economic resources due to the debt as in Bolivia, Peru and most Central American countries) from peasant assistance programs. NGOs, however, have not meant to serve as a replacement for the state agencies, but represent a microlevel effort to identify, experiment and demonstrate potentially effective lines of action within the prevailing socioeconomic and political context.[19]

NGOs work either directly with grass-root groups or with other groups (cooperatives, unions, etc.) which are in direct contact with the grassroots. Strategically, NGOs have developed two main lines of work:

1. Assistance programs which include activities directly linked to the peasant systems of production, either in its research and experimentation aspects or in the implementation of alternative technology programs in peasant communities. Research activities are usually conducted in farmers' fields or in experimental fields (of NGOs or collaborating public institutions) which at the same time serve as "demonstration farms".
2. Promotion programs which focus on popular education and social organization. Central to these programs are activities that enhance peasant's awareness about their reality, possibilities and limitations, as well as methodologies that reinforce the value of cooperation and solidarity, so that the community can become the agent of its own change. The selection and formation of promoters or leaders that provide a multiplier effect in each community has been an important strategy followed by NGOs.

C. Alternative Agriculture and Extension

Obviously, not all NGOs conform to these categories; therefore a variety of approaches to alternative agricultural development and extension can be observed. For example:

1. Many NGOs are involved in the extension of appropriate technologies, directly transferred from industrial countries or other LDCs and adapted to the local situations (i.e., lorena stoves, alternative fertilizers, community composts, dry latrines, biogas, alley cropping, windmills, etc.). Although many of these techniques make perfect environmental sense, little acceptance and adoption by local communities can be perceived because of their high cost or because they are culturally alien to the target groups.
2. Other NGOs emphasize the rescue of traditional knowledge and technologies and understanding of peasant rationale, in order to harness self-promoting activities among the peasants. So far few groups have been able to assemble technologies based on local knowledge and which contributed to solving agricultural "bottlenecks" not directly solved by peasants themselves.
3. Another group of NGOs has focused on the problem of rural development as a social process, which means that peasants must become aware of their situation through pedagogical means, in order to develop the capabilities needed to overcome their circumstances. In this regard these NGOs recognize that the poorer the farmer is, the higher the relevance of agroecological approaches, given that farmers have little choice but to resort to their own resources. Professionals involved in these groups not only have appropriate technical skills, but also have skills in popular education.

IV. CONCLUSIONS

A major dilemma facing Latin American agroecologists is to determine how to further increase agricultural productivity in a way that the natural resource base is conserved and that the rural poor significantly benefit. The experience of several NGOs suggests that agroecology can contribute to improve peasants' technical capabilities so that they can enhance their production autonomy in order to deal with their basic needs and to increase their bargaining power with the state and/or other groups.

It has become increasingly clear that improving the access of peasants to land, water, and other natural resources, as well as to equitable credit, markets, appropriate technologies, etc. is crucial to ensure sustainable development.

Securing control and access to resources can only be assured by political reforms or well organized, community-based actions. Agroecology can only hope to provide the ecological basis to manage the resources once they become available to the rural poor. In other words, as an agricultural development approach, agroecology, however, cannot confront the structural and economic factors that underlie rural poverty. This will require a much broader development approach that strongly emphasizes the social organization of the peasantry. It will also require the recognition that poverty is not only an economic state but also a psychological attitude, and that in order to overcome this attitude, suitable pedagogical and methodological tools must be formulated. In this regard, technological issues must assume their corresponding role within a development strategy that incorporates social and economic dimensions.

Given its characteristics (i.e., it generates practical concepts, tools and methods for small farmer development), agroecology can become the focus of a new institutional arrangement capable of effectively dealing with the problem of rural development in Latin America. This new arrangement should be built upon already existing experiences and may require strong working linkages between NGOs, international centers, local universities, community organizations, etc. promoting rural development. The linkages should, however, be based on the shared need for a rural development strategy that adequately incorporates the appropriate technology question. Agroecology can serve as a focus for the training, research and information networking activities which would necessarily be emphasized by the new institutional axis, to further the development of peasant agriculture in Latin America.

REFERENCES

1. **Pineiro, M. and Trigo, E.,** *Cambio tecnologico en el agro latinoamericano*, IICA, San Jose, Costa Rica, 1983.
2. **Burton, D. K. and Philogene, B. J. R.,** *An overview of pesticide usage in Latin America*, Report to the Canadian Wildlife Service Latin American Program, Ottawa, Canada, 1986.
3. **de Janvry, A., Runstem, D., and Sadoulet, E.,** *Technological Innovations in Latin American Agriculture*, IICA Program Paper Series, San Jose, Costa Rica, 1987.
4. **Hobbelink, H.,** *Biotechnology and Third World Agriculture: New Hope or False Promise?* Int. Coalition for Development Action, Brussels, 1987.
5. **Ortega, E.,** *Peasant Agriculture in Latin America and the Caribbean*, Joint ECLAC/FAO, Agriculture Division, Santiago, Chile, 1986.
6. **Chambers, R. and Ghildyal, B. P.,** Agricultural research for resource-poor farmers: the farmer first and last model, *Agric. Admin.*, 20, 1, 1985.
7. **de Janvry, A.,** *The Agrarian Question and Reformism in Latin America*, Johns Hopkins University Press, Baltimore, 1981.
8. **Shaner, W. W., Phillipp, P. F., and Schmehl, W. R.,** *Farming Systems Research and Development: Guidelines for Developing Countries*, Westview Press, Boulder, CO, 1982.
9. **Espinosa, P. and Garrett, P.,** The relevance of gender in farming systems research: experiences in Ecuador, *Agric. Admin. Extension*, 26, 101, 1987.
10. **Pachico, D. and Borbon, E.,** Technical change in traditional small farm agriculture: the case of beans in Costa Rica, *Agric. Admin. Extension*, 26, 65, 1987.
11. **Hildebrand, P.,** Combining disciplines in rapid appraisal: the Sondeo approach, *Agric. Admin. Extension*, 8, 423, 1981.
12. **Rhoades, R. E.,** Breaking New Ground: Agricultural Anthropology, International Potato Center, Lima, Peru, 1984.
13. **Lacroix, R. L. J.,** *Integrated Rural Development in Latin America*, World Bank Staff Working Papers #716, The World Bank, Washington, D. C., 1985.
14. **Altieri, M. A. and Anderson, M. K.,** An ecological basis for the development of alternative agricultural systems for small farmers in the Third World, *Am. J. Altern. Agric.*, 1, 30, 1986.
15. **Page, D.,** Science goes to work for Chile's urban and rural poor, *Grassroots Dev.*, 10, 38, 1986.
16. **Conway, G. R.,** Agroecosystem analysis for research and development, Winrock International Institute, Bangkok, 1986.
17. **Francis, C. A.,** *Multiple Cropping Systems*, Macmillan, New York, 1986.
18. **Altieri, M. A.,** *Agroecology: the Scientific Basis of Alternative Agriculture*, Westview Press, Boulder, CO, 1987.
19. **de Janvry, A.,** Perspectives for Inter American Foundation programs in Chilean agriculture, unpublished material, 1983.

Production Systems

14 AGROFORESTRY: AN APPROACH TO SUSTAINABLE LAND USE IN THE TROPICS

P. K. R. Nair

I. INTRODUCTION

In the recent past there has been a reawakening among scientists and other experts dealing with land use systems to value age-old, integrated land-use systems. With scientific attention focused on multiple cropping and an increased awareness about the inappropriateness of most high-input agricultural technologies for farmers in the tropics, the search for resource-conserving, socially acceptable farming systems had been accelerated. The result is a growing appreciation of the relevance of integrated, sustainable, multiple-input land-use systems and the productive and protective value of trees in farming systems.

On the forestry front, the seriousness of forest destruction has been realized and its alarming consequences analyzed, and it is clear that the major cause of deforestation is the search for more and more areas to produce food and obtain fuelwood. Thus, in the wake of mounting food and fuel shortage pressures and the serious environmental problems associated with deforestation, it is no longer prudent to ignore the conservation benefits and the potential for sustained yields provided by tree-based farming systems.

An increasing awareness about the role and value of trees and other woody perennials in the everyday life of rural communities has been expressed by numerous terms such as village forestry, social forestry, farm forestry, forest village, community forestry, etc. Depending on the priority of objectives, the definitions and interpretations of the terms vary; but in common parlance most of these terms have often been used synonymously, and sometimes even out of context. However, each of the terms has distinct connotations; some are land-use systems (e.g., farm forestry); others are land (forest) management approaches (e.g., community forestry). In a general sense, all these terms denote one or the other form of the integration of trees within land use and rural development systems. Although the trees so used are not necessarily conventional forestry species or managed in a forestry community (natural or planted), they all have, unfortunately, a semantic emphasis on forestry as if they were specific branches of forestry. The term agroforestry has now been established as a generic term for a wide range of integrated land management approaches and systems involving woody perennials. Initially this term caused confusion about its identity vis-a-vis forestry. However, with a better understanding of its role, concepts and principles, agroforestry has now earned a distinct identity of its own. Agroforestry is an approach to land use involving deliberate inclusion of trees and other woody perennials in the crop/animal production fields.[1,2] Efforts are now being made to improve existing indigenous agroforestry systems and to devise appropriate ways to integrate the production of trees and other woody species into the production of agricultural crops and/or livestock in a sustainable manner.

II. AGROFORESTRY SYSTEMS

A. The Concept of Agroforestry

The concepts and principles of agroforestry have been well elucidated in several publications from the International Council for Research in Agroforestry (ICRAF) as well as other organizations. In view of the relative newness of the topic, some of these aspects need to be recapitulated here even at the risk of repetition.

Agroforestry refers to land-use systems and technologies where woody perennials (trees, shrubs, palms, bamboos, etc.) are deliberately used on the same land-management units as agricultural crops and/or animals, in some form of spatial arrangement or temporal sequence. In agroforestry systems there are both ecological and economical interactions between the different components.[1]

This definition implies that agroforestry:

1. Normally involves two or more species of plants (or plants and animals), at least one of which is a woody perennial
2. As a system always has two or more outputs
3. Always has a cycle of more than 1 year
4. Even at its simplist is more complex ecologically (structurally and functionally) and economically than a monocropping system

The most important aspect of agroforestry is that it is an interface between forestry and agriculture, offering a viable combination of production and sustainability. Its major attributes are production of multiple benefits, sustainability and adaptability.[2,3]

B. Examples of Existing Agroforestry Systems

To increase the understanding of agroforestry in general, and the various productive and protective roles it plays in land use systems in developing economies, ICRAF has conducted an elaborate inventory of such systems and practices from different parts of the world.[4] This global project has gathered information on hundreds of agroforestry systems. For example, the Agroforestry System Description Series, being serialized in the publication *Agroforestry Systems*, describes some of the prominent and promising systems. The computerized Agroforestry Systems Register at ICRAF contains entries of about 150 systems and/or practices.

An overview summary of the examples of various agroforestry systems in different parts of the world, derived from this global inventory, is given in Table 1. Table 1 shows that agroforestry is widespread in the developing countries of almost all ecological regions. Direct or indirect contribution to food production is an important function of most of these systems.

C. Types of Agroforestry Systems

An assessment of the roles of agroforestry to food production and to the economy and development of the countries concerned is currently possible mostly in qualitative terms, which is not often very appealing. In order to evaluate the existing systems and develop action plans for their improvement, it is necessary to classify these systems and thus provide a practical and realistic framework. Several criteria can be used. The most common are the system's structure (composition and arrangement of components), functions, socio-economic scale and level of management, and ecological spread, as given in Table 2.[5] Since there are only three basic sets of components that are managed by humans in all agroforestry systems, woody perennials, herbaceous plants and animals, a logical first and simple step is to classify agroforestry systems on their component composition. The three basic types of systems as depicted in Figure 1:

1. Agrosilvicultural (crops, including tree/shrub crops and trees)
2. Silvopastoral (pasture/animals plus trees)
3. Agrosilvopastoral (crops plus pasture/animals plus trees)

Other specialized agroforestry systems such as apiculture with trees, aquaculture in mangrove areas, multipurpose tree lots, and so on can also be specified.

D. Agroforestry Practices and Technologies

A further evaluation of the organization of these various systems shows that an agroforestry system consists of specific examples of practices found in a locality or area, and described according to biological composition and arrangement, level of technical management, or socioeconomic features. An agroforestry practice, on the other hand, denotes a specific arrangement of agroforestry components in space and/or time. Several such practices are involved in system constitution and maintenance of an agroforestry system.

Although several agroforestry systems have been recorded from around the world, the distinctive practices that constitute these systems in various ecoregions and locations are only a few. The most commonly found practices and their essential characteristics are listed in Table 3.

The scale of management and extent of adoption of these different practices in any particular system will vary considerably depending upon a number of factors. Any of these practices can become an agroforestry system when developed or spread to such an extent in a specific local area as to form a distinct land utilization type in that area. One essential point to note here is that an agroforestry practice can be found to exist even in a non-agroforestry land use system. An example is the practice of growing rows of *Sesbania grandiflora* on the bunds of rice paddies in Java, Indonesia. This woody species is harvested for its leaves as green manure, its flowers are eaten as a vegetable, it provides firewood and adds to the fertility of the soil underneath through biological fixation on atmospheric nitrogen. Thus the woody species interacts ecologically and economically with the crop (rice) production system; in other words, it is an *agroforestry practice* in a crop production system.

Another term commonly used in all land use systems including agroforestry is "technology," and it is usually used in combination with a particular production system or land-use system as the prefix; thus, agricultural technology, wood production technology, agroforestry technology, etc. The word or suffix "technology" in this sense indicates an innovation or improvement, usually through scientific intervention, that can be applied with advantage in the management of the system or practice concerned. The inputs that are used for developing such technologies can sometimes be the most significant aspects, and these are called input technologies, e.g., improved varieties of plants, agrochemicals and so on.

TABLE 1
Some Examples of Prominent Agroforestry Systems and Practices in the Developing Countries

Systems		Examples from different geographic regions						
Major systems	**Sub-systems/ practices**	**South Pacific**	**Southeast Asia**	**South Asia**	**Middle East and Mediterranean**	**East and Central Africa**	**West Africa**	**American tropics**
Agrisilvicultural systems	Improved "fallow" (in shifting cultivation areas)		Forest villages of Thailand; various fruit trees & plantation crops used as "fallow species in Indonesia	Improvements to shifting cultivation; several approaches, e.g., in the north-eastern parts of India		Improvements to shifting cultivation, e.g., gum gardens of the Sudan	*Acioa barterli, Anthonontha macrophylla, Giricidia sepium,* etc. tried as "fallow" species	Several forms
	The Taungya system	(e.g., Taro with *Cedrella* and *Anthocephalus* trees)	Widely practised; Forest villages of Thailand is an improved form	Several forms, several names		The "Shamba" system	Several forms	Several forms
	Tree gardens	Involving fruit trees	Dominated by fruit trees	In all ecological regions	The Dehesa system; "Parc arboree"			e.g., 'Paraiso Wood-lot' of Paraguay
	Hedgerow inter-cropping (alley cropping)		Extensive use of *Sesbania grandiflora, Leucaena leucocephala,* and *Calliandra callothyrsus*	Several experimental approaches, e.g., conservation farming in Sri Lanka		The Corridor System of Zaire	Experimental systems on alley cropping with *leucaena* and other woody perennial species	Experimental
	Multipurpose trees and shrubs on farmalnds	Mainly fruit or nut trees (e.g., *Canarium, Pometia, Barringtonia, Pandanus, Artocarpus altitis*)	Dominated by fruit trees; also *Acacia meamsii*-cropping system, Indonesia	Several forms both in lowlands and highlands, e.g., hill farming in Nepal; "Knejri"-based system in the dry parts of India	The Oasis sysstem; crop combinations with the Carob tree; the Dehesa system; irrigated systems; olive trees + cereals	Various forms; the Chagga system of Tanzanian highlands; the Nyabisindu system of Rwanda	*Acacia albida*-based food production systems in dry areas; *Butyrospermum* + *Parida* systems; "Parc arboreé"	Various forms in all ecological regions
	Crop combination with plantation crops	Plantation crops and other multipurpose trees; e.g., *Casuarina* and coffee in the high-lands of PNG; also *Gliricidia* and *Leucaena* with cacao	Plantation crops + fruit trees; small-holder systems of crop combinations with plantation crops; plantation crops with spice trees	Integrated production systems in small-holdings; shade trees in plantations; other crop mixtures including various tree spices	Irrigated systems; olive trees + cereals	Integrated production; shade trees in commercial plantations; mixed systems in the highlands	Plantation crop mixtures; small-holder production systems	Plantation crop mixtures; shade trees in commercial plantatons; mixed systems in small-holdings; spice trees; babassu palm-based systems
	AF fuelwood production	Multipurpose fuelwood trees around settlements	Several examples in different ecological regions	Various forms including some forms of social forestry		Various forms	Common in the dry regions	Several forms in the dry regions

TABLE 1 (continued)
Some Examples of Prominent Agroforestry Systems and Practices in the Developing Countries

Systems		Examples from different geographic regions						
Major systems	Sub-systems/ practices	South Pacific	Southeast Asia	South Asia	Middle East and Mediterranean	East and Central Africa	West Africa	American tropics
Agrisilvicultural systems (continued)	Shelterbelts, windbreaks, soil conservation hedges	*Casuarina oligodon* in the highlands as shelterbelts and soil improver	Terrace stabilization in steep slopes	Use of *Casuarina* spp. as shelterbelts; several windbreaks	Tree spices for erosion control	The Nyabisindu system of Rwanda	Various forms	Live fences, windbreaks especially in highlands
Silvopastoral systems	Protein bank (cut-and-carry fodder production)	Rare	Very common, especially in highlands	Multipurpose fodder trees on or around farmlands (especially in highlands)		Very common	Very common	Very common
	Living fence of fodder trees and hedges	Occasional	*Leucaena, Caliandra,* etc. used extensively	*Sesbania, Euphorbia, Syzigium,* etc. common		Very common in all ecological regions		Very common in the highlands
	Trees and shrubs on pastures	Cattle under coconuts, pines and *Eucalyptus deglupta*	Grazing under coconuts and other plantations	Several tree species being used very widely	Very common in the dry regions; the Dahesa system	The *Acacia*-dominated system in the arid parts of Kenya, Somalia and Ethiopia	Cattle under oil palm; cattle and sheep under coconut	Common in humid as well as dry regions, e.g., grazing under plantation crops in Brazil
Agrosilvopastoral systems	Woody hedges for browse, mulch, green manure, soil conservation, etc.	Various forms; *Casuarina oligodon* widely used to provide mulch and compost	Various forms	Various forms especially in lowlands		Common; variants of the "Shamba" system	Very common	Especially in hilly regions
	Home gardens (involving a large number of herbaceous and woody plants with or without animals)	Several types of home gardens and kitchen gardens	Very common; Java home gardens often quoted as good examples; involving several fruit trees	Common in all ecological regions; usually involving fruit trees	The Oasis system	Various forms (the Chagga home gardens; the Nyabisindu system)	Compount farms of humid lowlands	Very common in the thickly populated areas
Other systems	Agro-silvo-Fishery (aquaforestry)		Silviculture in mangrove areas; trees on bunds of fish-breeding ponds	Occasional				
	Various forms of shifting cultivation	Common	e.g., Swidden farming	Very common; various names		Very common	Very common in the lowlands	Very common in all ecological regions
	Apiculture with trees	Common	Common	Common	Common	Common	Common	

TABLE 2
Major Approaches in Classification of Agroforestry Systems (and Practices)

Categorization of Systems (based on their Structure and Function)			Grouping of Systems (according to their Spread and Management)	
Structure (nature and arrangement of components, especially woody ones)		Function (role and/or output of components, especially woody ones)		
Nature of components	Arrangement of components		Agro-ecological/ environmental adaptability	Scioeconomic and Management Level
Agrisilviculture (crops and trees incl. shrubs/trees and trees) Silvopastoral (pasture/animals and trees) Agrosilvopastoral (crops, pasture/animals, and trees) Others (multipurpose tree lots, apiculture with trees, aquaculture with trees, etc.)	*In space* (Spatial) Mixed dense (e.g.: Home garden) Mixed sparce (e.g.: Most systems of trees in pastures) Strip (width of strip to be more than one tree) Boundary (trees on edges of plots/fields *In time* (Temporal) Coincident Concomitant Overlapping Sequential (separate) Interpolated	*Productive Function* Food Fodder Fuelwood Other woods Other products *Protective Function* Windbreak Shelterbelt Soil conservation Moisture conservation Soil improvement Shade (for crop, animal, and man)	*Systems in/for* Lowland humid tropics Highland humid tropics (above 1,200 m a.s.l.; e.g.: Andes, India, Malaysia) Lowland subhum tropics (e.g.: savanna zone of Africa, Cerrado of South America) Highland subhumid tropics (tropical highlands) (e.g.: in Kenya, Ethiopia)	*Based on level of technology input* Low input (Marginal) Medium input High input *Based on cost/benefit relations* Commercial Intermediate Subsistence

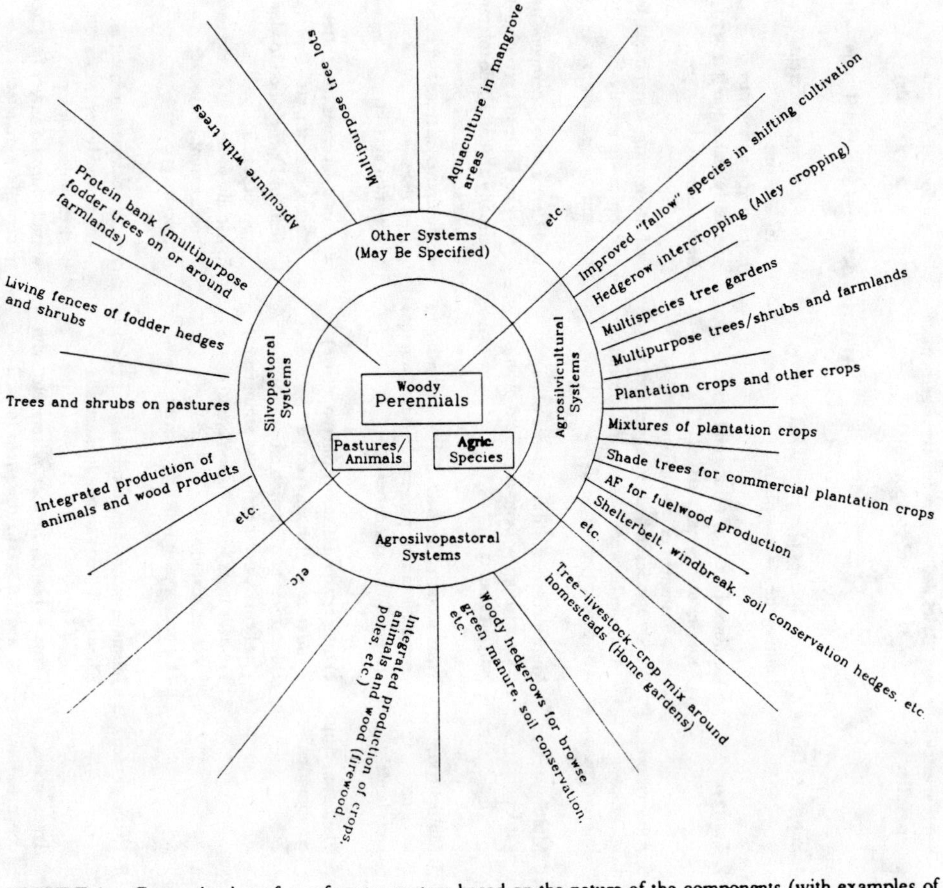

FIGURE 1. Categorization of agroforestry system based on the nature of the components (with examples of common fub-system/practices under each system). (From Nair, P. K. R., *Agroforestry Systems*, 3, 97, 1985. With permission.)

TABLE 3
Major Agroforestry Practices

Agroforestry practice	Brief description (of arrangement of components)	Major groups of components (w = woody h = herbaceous f = fodder for grazing a = animals)	*Main types of component interactions in space(s)/time(t)	Primary role of woody component Prt = protective Prd = productive	Agroecological adaptability
A. Agrisilvicultural Systems (Crops — including shrub/vine/tree crops — and trees)					
1. Improved fallow	Woody species planted and left to grow during the "fallow phase"	w: fast-growing, preferably leguminous h: common agri. crops	t: sequential (in rotational cycles)	Prt: soil fertility improvement Prd: wood products	In shifting cultivation areas
2. Taungya	Combined stand of woody and agric. species during early stages of establishment of plantations	w: usually plantation forestry spp. h: common agric. crops	s: concomitant t: sequential (in rotational cycles)	Prd: addl. income from agric. crops; also management care for plantation species	In all ecol. regions (where taungya is practiced); several improvements possible
3. Alley cropping hedgerow intercropping)	Woody species in hedges; agric. spp. in alleys in between hedges; microzonal or strip arrangement	w: fast-growing, leguminous, that coppice vigorously h: common agric. crops	s: zonal (strip) t: concomitant to intermittent	Prt: soil improvement, and/or soil conservation Prd: addl. products from hedges	Sub-humid to humid areas with high human population pressure and fragile (productive but easily degradable) soils
4. Multilayer tree gardens	Multispecies, multilayer, dense plant associations with no organized planting arrangements	w: different woody components of varying form and growth habits h: usually absent; shade tolerant ones sometimes present	s: mixed dense t: coincident	Prd: various types of outputs Prt: soil conservation and protection; efficient nutrient cycling	Areas with fertile soils, good availability of labor, and high human population pressure
5. Multipurpose trees on crop lands	Trees scattered haphazardly or according to some systematic patterns on bunds terraces or plot/field boundaries	w: multipurpose trees and other fruit trees h: common agri. crops	s: mixed sparse t: interpolated	Prd: prod. of various tree products Prt: fencing, social values, plot demarcation	In all ecol. regions esp. in subsistence farming; also commonly integrated with animals
6. Plantation crop combinations	1) Integrated multistorey (mixed, dense) mixture of plantation crops 2) Mixtures of plantation crops in alternate or other regular arrangement 3) Shade trees for plantation crops; shade trees scattered. 4) Intercropping with agric. crops.	w: plantation crops like coffee, cacao, coconut, etc., fruit trees (esp. in 1); fuelwood/fodder spp. esp. in 3 h: usually present in 4, and to some extent in 1; shade tolerant species	s: mixed dense (1, 2, 4) or mixed sparse (3) t: 1) interpolated 2) coincident or overlapping 3) coincident 4) intermittent or concomitant	Prd: prod. of a large number of produces Prt: soil surface protection; shade and windbreak	In humid lowlands or tropical humid /sub-humid highlands (depending on the adaptability of the plantation crops concerned); usually in small-holder subsistence systems
7. Home gardens	Intimate, multistorey combination of various trees and crops around homesteads	w: fruit trees predominate; also other woody spp; vines, etc. h: shade-tolerant crops	s: mixed dense t: coincident to interpolated	Prd: prodn. of various products Prt: soil surface protection	In all ecol. regions esp. in areas with high population density
8. Trees in soil conservation and reclamation	Trees on bunds, terraces, raisers, etc. with or without grass strips; trees for soil reclamation	w: multipurpose and/or fruit trees h: common agri. species	s: zonal (strip) t: intermittent	Prt: soil conservation Prd: various tree products	In sloping areas, esp. in highlands, reclamation of degraded acid, alkali soils; and sand dune stabilization
9. Shelterbelts and windbreaks, live hedges	Trees around farmlands/plots	w: combination of tall-growing spreading types h: agri. crops of the locality	s: zonal (strip/boundary) t: coincident, interpolated	Prt: shelterbelts/windbreak; fencing Prd: usually firewood	In wind-prone areas
10. Fuelwood production	Interplanting firewood species on or around agric. lands	w: firewood species h: agri. crops of the locality	s: zonal (strip/boundary) t: coincident	Prd: firewood Prt: fencing; shelterbelts; boundary demarcation	In all ecol. regions

TABLE 3 (continued)
Major Agroforestry Practices

11. Trees on rangeland or pastures	Trees scattered irregularly or arranged according to some systematic pattern	w: multipurpose; usually of fodder value h: present a: present	s: mixed sparse t: coincident	Prd: trees as fodder Prt: shade; environmental protection	Extensive grazing areas
12. Protein banks	Production of protein rich tree fodder on farm/range lands for cut-and-carry fodder production	w: leguminous fodder trees h: present f: present	s: zonal t: coincident or interpolated	Prd: fodder prodn. Prt: soil and environ. protection depending on arrangement	Usually in areas with high man:land ratio
13. Plantation crops with pastures and animals	Example: cattle under coconuts in southeast Asia and the South Pacific	w: plantation crops f: present a: present	s: mixed sparse t: intermittent	Prt: shade and protection to fodder and animals Prd: (plantation crop products)	In areas with less human pressure on plantation crop lands
		C. Agrosilvopastoral Systems (trees + crops + pasture/animals)			
14. Home gardens involving animals	Intimate, multistorey combination of various trees and crops, and animals, around homesteads	w: fruit trees predominate; also other woody spp. a: present	s: mixed dense t: interpolated to coincident	Prd: various products Prt: soil surface protection; organic recycling	In all ecol. regions with high density of human population
15. Multipurpose woody hedgerows	Woody hedges for browse, mulch, green manure, soil conservation, etc.	w: fast-growing and coppicing fodder shrubs and trees h: (similar to alley cropping and soil conservation)	s: zonal (strip or boundary) t: intermittent to coincident	Prt: soil conservation Prd: various products, mainly fodder	Humid to subhumid areas with hilly and slopping terrain
		D. Others			
16. Apiculture with trees	Trees for honey production	w: honey-producing trees (other components may be present)	s: boundary, mixed, sparse (or any other) t: (variable)	Prd: honey production Prt: location-specific	Depending on the feasibility of apiculture
17. Aquaforestry	Trees lining fish ponds, tree leaves being used as 'forage' for fish	w: trees and shrubs preferred by fish (other components may be present)	s: boundary planting t: coincident to intermittent	Prd: fish production Prt: bund stabilization around fish ponds	Lowlands
18. Multipurpose woodlots	For various purposes (wood fodder, soil protection, soil reclamation, etc.)	w: multipurpose species; special location-specific spp. (other components may be present	s: zonal or mixed t: separate or overlapping	Prt: various protective Prd: various	Various

TABLE 3 (continued)
Major Agroforestry Practices

*Temporal arrangements of components in agroforestry systems

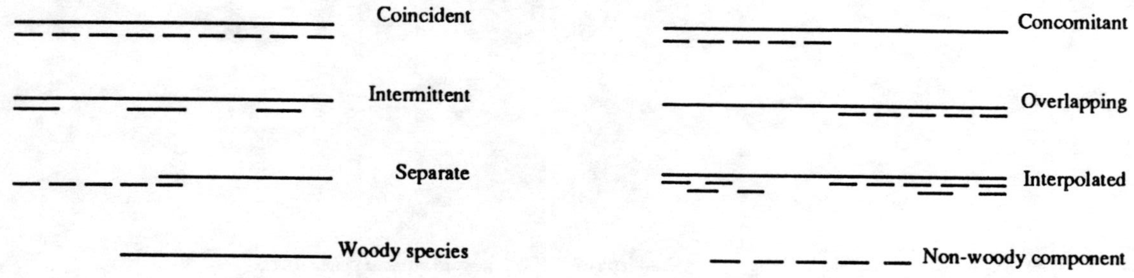

Most of the practices mentioned in Table 3 can be improved scientifically and developed into modern agroforestry technologies. The success potential of an agroforestry practice depends on the extent to which such technologies have been perfected, and made use of in the management of the practice (or the system).

III. PRODUCTIVE AND PROTECTIVE ASPECTS OF AGROFORESTRY

A. Productive Role

Field examples of agroforestry systems and practices presented in Tables 1 and 3 show that they are not only widespread in different ecological regions, but also are important because of the role of woody perennials in producing the basic household needs, protecting the systems and prolonging their sustainability. The main advantages of agroforestry stem from these primary roles of production, protection and sustainability of the woody perennials.

The productive role of woody perennials in agroforestry systems includes production of food, fodder, firewood, resins, fiber and various other wood products. The role of trees in food production is both direct and indirect. The direct role includes various tree products that are used for human consumption: fruits, nuts, flowers, bark, leaves, roots, etc. The indirect role consists of the effects of an array of protective functions of trees such as soil fertility improvement, use of trees for environmental protection (soil conservation, windbreak, shelterbelts, reclamation of degraded lands, etc.), role of trees in silvopastoral systems for animal production, and production of wood for agricultural purposes, implements, food curing and preservation, firewood, etc. However, since most of these benefits and values of trees are not easily measurable, the usual tendency of "experts" is to ignore the significance of this enormous resource.

The most important groups of woody species that have a direct role in food production in indigenous agroforestry systems are fruit trees.[6] Integration of trees in crop production fields is an essential part of traditional farming systems in the dry regions. Two typical examples are the extensive use of *Acacia albida* in the groundnut and millet production areas of sub-Saharan Africa[7,8] and the dominant role of *Prosopis cineraria* in the arid northwestern parts of India.[9,10]

Woody perennials on farmlands also produce fuelwood. Many fast-growing multipurpose species, several of them legumes, suitable for firewood production in different environmental conditions have been identified[11] and there are several agroforestry approaches to integrate them with crops for firewood production.[12]

In the "animal agroforestry" systems, the woody components can be used as a source of fodder to improve livestock productivity or to obtain other commodities such as fuel, fruit or timber. Based on this production objective, silvopastoral systems can be either browse grazing or forest/plantation grazing systems.

Reviewing the role of woody perennials in these systems, Torres[13] summarized the situation as follows:

1. Several species have been identified for different ecological zones, which are potential sources of fodder.
2. Although dry matter productivity from the foliage of "trubs" (trees plus shrubs) seems to be rather small, its reliability in bad years and its presence during the dry seasons makes it particularly valuable; while pod-producing trubs may become a very useful source of energy and protein concentrate.
3. Protein supply appears to be the main nutritive role of trees and shrubs, which could be limited by low intakes.
4. There is a need to undertake studies evaluating the animal production potential of fodder trubs vis-a-vis alternative sources.

B. Protective Role

The protective role of agroforestry stems from the soil improving and soil conserving functions of the woody perennials, as well as the microclimate amelioration and shelterbelt/windbreak effects by the trees. Nair[6] examined the soil productivity aspects of agroforestry and noted that there are various avenues through which the woody perennials, especially legumes, could improve and enrich soil conditions. These include nitrogen fixation, addition of organic matter through litterfall and dead and decaying roots, modification of soil porosity and infiltration rates leading to reduced erodibility of the soil, and improving the efficiency of nutrient cycling within the soil-plant system. Nair[14] summarized these beneficial effects of trees on soils under agroforestry systems, and the mechanisms by which these effects could be accomplished under appropriate conditions; these are presented in Table 4. Discussing this topic, Young[15] identified a few negative aspects of trees on soils such as nutrient depletion by fast-growing, less-competitive species, accelerated erosion under badly managed tree stands, adverse effects of shading and spectral quality changes of light on understorey species, etc.

One of the main protective roles of agroforestry is in soil conservation. The long tradition of planting *Leucaena leucocephala* in contour-hedges for erosion control and soil improvement in Southeast Asia, especially Indonesia and the Philippines, is a typical example. Indirect terraces are formed with time when the washed-off soil is collected behind the hedges. Loppings and prunings from such hedgerow species could also provide mulch to aid in preventing sheet erosion between trees, as demonstrated in Rwanda by Neumann.[16] The presence of more plant cover on the soil, either alive or as mulch, also reduces the impact of raindrops on the soil and thus minimizes splash and sheet erosion. Therefore, as pointed out by Lundgren and Nair,[17] the potential role of agroforestry in soil conservation lies not only in woody perennials acting as a physical barrier against erosive forces, but also in providing mulch and/or fodder and fuelwood at the same time. These aspects of erosion control under agroforestry have been reviewed by Wiersum[18] and Young;[19] the latter also gives examples of agroforestry in erosion control and data on observed erosion rates under these various agroforestry practices.

TABLE 4
Summary of the Potential Beneficial Effects of Trees on Soils

Nature of processes	Processes	Main effect on soil	Scientific evidence
Input processes (augment additions to the soil)	biomass production	addition of carbon and its transfomations	available
	nitrogen fixation	N-enrichment	available
	rainfall	effect on rainfall (quantity and distribution) and therefore nutrient addition through rain	not adequately demonstrated
Output process (reduce losses from the soil)	protection against water and wind erosion	reduce loss of soil as well as nutrients	available
Turn-over processes	nutrient retrieval/cycling/ release	uptake from deeper layers and "deposition" on surface via litter	not adequately demonstrated
		withholding nutrients that can be lost by leaching	not demonstrated
		timing of nutrient release: this can be regulated by management interventions	available
"Catalytic" processes (indirect influences)	physical processes	improvement of physical properties (water-holding capacity, permeability, drainage, etc.) at the microsite as well as at the watershed (macrosite)	available
	root growth and proliferation (enhanced)	addition of (more) root biomass; growth promoting substances; microbial associations	partially demonstrated
	litter quality and dynamics	improvement of litter quality through diversity of plant species; better timing of quantity, and method of application of litter possible	now being increasingly studied in alley cropping and other intercropping experiments
	microclimatic processes	creation of more favorable microclimate; shelterbelt and windbreak effects	available
	(bio)chemical/biological processes (net effects of various processes)	moderating effect on extreme conditions of soil acidity, alkalinity, etc.	partially demonstrated

Source: Nair, 1987c.

Other protective functions of woody perennials in agroforestry include their role as live fences, shelterbelts and windbreaks. Windbreaks are narrow strips of trees, shrubs and/or grasses planted to protect fields, homes, canals and other areas from the wind and blowing sand. Shelterbelts, a form of windbreak, are long rows of trees and shrubs, planted many rows wide perpendicular to the direction of the prevailing wind. Where wind is a major cause of soil erosion and moisture loss, windbreaks can make sustainable production possible. When properly designed and maintained, a windbreak reduces the velocity of the wind, and thus its ability to carry and deposit soil and sand. It can improve the microclimate in the area it protects by decreasing evapotranspiration. In some cases windbreaks have been shown to increase the productivity of the crops they protect. In addition to these, windbreaks can also provide a wide range of useful products, from poles and fuelwood to fruit, fodder, fiber and mulch. One of the well-documented studies on windbreaks is the project in Majjia valley in Niger (average rainfall 425 mm).[20]

These productive and protective benefits (as well as the adverse effects) of agroforestry are either inferred from land-use systems related to agroforestry, or are still only untested hypotheses. The magnitude of benefit or adverse effect that could be experienced will depend upon a number of site-specific factors. Besides, many of these attributes of trees, as compared with annual crops, can be realized only over relatively longer periods of time. One of the main issues to be considered here is the soil fertility improvement in agroforestry through addition of plant biomass vis-a-vis other uses for the biomass (fodder, fuelwood, fuel, etc.). Similarly, the proportion of space occupied by tree components in an agroforestry combination, in order to derive the suggested benefits to any tangible extent, will depend upon several factors such as soil type, plant (tree) species, tree-crop arrangement, etc. All these need to be taken into account while suggesting plant and soil management technologies in agroforestry.

C. Economic Benefits and Costs

The literature about farming systems that incorporate agroforestry practices describes a number of related benefits or opportunities, costs and constraints.[21-26] The principal positive and negative economic features and the underlying hypotheses about agroforestry are summarized in Table 5.

One hypothesis is that many agroforestry practices are

TABLE 5
The Main Benefits and Costs of Agroforestry

Benefits and opportunities	Costs and constraints
Maintains or increases site productivity through nutrient recycling and soil protection, at low capital and labor costs	Reduces output of staple food crops where trees compete for use of arable land and/or depress crop yields through shade, root competition or allelopathic interactions
Increases the value of output on a given area of land through spatial or inter-temporal intercropping of tree and other species	Incompatibility of trees with agricultural practices such as free grazing, burning, common fields, etc., which make it difficult to protect trees
Diversifies the range of outputs from a given area, in order to (a) increase self-sufficiency, or/and (b) reduce the risk to income from adverse climatic, biological or market impacts on particular crops	Trees can impede cultivation of monocrops and introduction of mechanizaton, and so (a) increase labor costs in situations where the latter is appropriate and/or (b) inhibit advances in farming practices
Spreads the needs for labor inputs more evenly seasonally so reducing the effects of sharp peaks and troughs in activity characteristic of tropical agriculture	Where the planting season is very restricted, e.g., in arid and semi-arid conditions, demands on available labor for crop establishment may prevent tree planting
Provides productive applications for underutilized land, labor or capital	The relatively long production period of trees delays returns beyond what may be tanable for poor farmers, and increase the risks to them associated with insecurity of tenure
Creates capital stocks available to meet intermittent costs or unforeseen contingencies	

Source: Arnold, 1987.

characteristic of simple resource-poor situations and are not easily adapted to more intensive agricultural practices. Another, in a sense contradictory to the first, is that the competition between trees and food crops, and the priority that they must give to meeting their basic food needs, will exclude poor farmers, who have very little land, for tree growing. As pointed out by Hoskins,[27] it is not easy to select and describe the crucial socioeconomic variables in a universal way; situations differ depending on the locality, environment and the major traditional production objectives. Issues overlap and will need to be studied case by case.

IV. DESIGN OF AGROFORESTRY SYSTEMS

A. Criteria

A good agroforestry system should fulfill the basic criteria of production (of multiple outputs), sustainability and adaptability. The ways to improve productivity will essentially involve improvement of the biological potential for increased and efficient outputs of tree products and yield of associated crops. Improved production should go hand in hand with achieving conservation goals, and thus satisfy the "sustainability" parameter. This is particularly important in the low-input situations of the small farmer who may not be interested in sustainability/conservation per se. Finally, the new or improved systems have to fit into the socioeconomic mold of the farmer, i.e., they should be adaptable.

B. The D & D Methodology

ICRAF's diagnostic and design (D & D) methodology for agroforestry systems is built upon these three fundamental criteria. Summarizing the 7 years of work in developing the D & D methodology, Raintree, the leader of that project in ICRAF stated[28] that in its most basic form, the logic of D & D embodies a common-sense, problem-solving approach, as illustrated in Table 6. This framework is equally applicable to both research and extension activities. If the technologies envisaged in the design already exist, the design can be used directly as a guide for agroforestry interventions by extension agents and rural development workers. If the envisaged technologies have not yet been developed or are not yet sufficiently well understood to warrant extension to farmers, the design then serves as a basis for identifying the research gaps that need to be filled in order to develop the required technologies.

According to Raintree,[28] there are many ways of applying this basic logic. Which of several variations on the specific procedures is used depends upon the *purposes and resources of the user*, the *geographic scale* and the *time frame* of the application, and *what is already known* about the area. For example, if the user is a national research institution and the purpose is to develop a general plan for agroforestry research within the country as a whole, then a large-scale, broad based, rapid appraisal approach will be needed initially, followed by smaller scale, more focused, in-depth D & D exercises in representative localities. In this same context, if a "macro" D & D has already been carried out, then most of the prediagnostic information for the "micro" D & D will already be known. On the other hand, if the user is an individual or a local organization whose work is already focused within a specific locality, or even on a single farm, then a more complete, intensive and continuous "micro" D & D approach would be appropriate. Thus, there are variable scale methods of application of D & D: "micro" scale is applicable to household management units such as family farms, "macro" scale for a region or country or ecozone, and "meso" scale for a local community or ecosystem such as a small watershed. Understandably there are overlaps between these boundaries and scales.

One of the essential characteristics of the D & D methodology is its iterative nature. As in any other similar procedures used to initiate research or development activities, the initial diagnostic and design process must be followed up by continuous monitoring and evaluation of the situation as it develops over the course of the project. It will thus be possible to refine the original diagnosis based on subsequent, more in-depth information emanating from farmers' fields as well as experiment stations. By adjusting the plan of action to such new information, the D & D process becomes continuous, dynamic and self-corrective.

TABLE 6
The Basic Logic of Agroforestry Diagnosis and Design

Basic questions	Key factors to consider	Mode of inquiry
	Prediagnostic Stage	
Definition and selection of the focal land-use system (which system to focus on?)	Distinctive combination of resources, technology and land user objectives	Differentiating and comparing the different land-use systems
How does the system work? (how is it organized, how does it function to achieve its objectives?)	Production objectives and strategies, subsystems and components	Analyzing and describing the system
	Diagnostic Stage	
How well does the system work? (what are its problems, limiting constraints, problem-generating syndromes & leverage points?)	Problems in meeting system objectives (production short-falls, sustainability problems)	Diagnostic interview: and direct field observations
	Casual factors, constraints and intervention points (all relevant biophysical and socioeconomic factors) biophysical and socioeconomic	Troubleshooting the problem subsystems
	Design and Evaluation Stage	
How to improve the system? (what is needed to improve system performance?)	Specifications for problem-solving or performance-enhancing interventions	Iterative design and evaluation of promising alternatives
	Planning Stage	
What to do to develop and disseminate system improvements?	Research and development needs, extension needs	Research design and project planning
	Implementation Stage	
How to adjust the plan of action to new information?	Feedback from on-station and on-farm research trials, extension trials and special studies	Rediagnosis and redesign (the iterative D & D learning process)

Source: Raintree, 1987.

V. CURRENT RESEARCH ACTIVITIES IN AGROFORESTRY

A. State-of-the-Art of Research in Agroforestry

No matter how elegant and practical an agroforestry research methodology may be, nothing can be accomplished practically unless there is a sufficient number of well-tested and proven technologies suitable and adoptable under different situations. As pointed out by Steppler,[29] the scientific and technical aspects of agroforestry have been beset with much anecdotal material. ICRAF's agroforestry systems inventory[4] showed a bewildering array of agroforestry systems worldwide. It also showed that there are relatively few rigorous experimental data pertaining to performance of agroforestry systems, that the number of trees — multipurpose trees — being used in various systems is in excess of 2000, and that the systems vary from relatively simple with two or three components, to the complex home gardens, which may contain upwards of 50 species plus animals and fish.

The problem of lack of experimental data or even of production data for the various systems is a serious gap in our knowledge. This is partly a reflection of the newness of the science of agroforestry (it was until recently literally considered as a subsistence form of land management) and partly that there are essentially no experimental techniques applicable to agroforestry systems. Most statistical techniques and experimental designs have been developed for monocultures. There is even relatively little experimental work done with annual crop mixtures.

An additional dimension is the question of appropriate input/output measurement. Monoculture and forage mixtures are relatively easy to measure because the output is clearly identified. An agroforestry system has multiple outputs. The agricultural component will probably be easy to measure, yield being the most visible output. The tree component may be much more elusive. There will be visible outputs (direct benefits), for example, fruits, fuelwood, building poles, and others that affect crop production, such as leaf mulch, and fixed nitrogen and yet there are other nearly invisible ones that could have an effect on the entire system, such as recycling of nutrients from subsoil, control of erosion, and increased infiltration rates (indirect benefits).

Yet another difficulty in dealing with agroforestry research is the location-specific nature of the systems and practices. Each system and practice will have its own specific research needs depending on various local circumstances. Furthermore, biological issues cannot be tackled in isolation from the sociocultural and economic aspects and vice-versa.

B. Ongoing Research

Despite the above-mentioned complexities and difficulties of doing research in agroforestry, several research initiatives in various aspects of agroforestry are under way in different places around the world. Some of the prominent among these shall be briefly mentioned here.

1. Alley Cropping

This form of hedgerow intercropping is becoming the subject of research under a wide range of situations. It refers to a cropping system where arable crops are grown in the interspaces (or alleys) between rows of planted trees or woody shrubs, which are pruned periodically during the cropping season to prevent shading to interplanted crops and to provide green manure/mulch for the arable crops. Pioneering research on this technology (some forms of which are old practices in southeast Asia, especially in Java, Indonesia) has been going on in the International Institute of Tropical Agriculture (IITA), Ibadan, Nigeria.[30] It is now being tested in several other places also.

2. Multipurpose Trees and Shrubs

Trees used outside forestry, except a few ornamental and fruit and nut trees, had hitherto been relegated to a status of lower economic significance and importance. But, with the recognition of the importance of agroforestry, such trees and other woody perennials used in traditional mixed systems as well as other such species of agroforestry potential have earned a higher "ranking" and "status" necessitating a distinct identity and definition for them. The term "multipurpose trees and shrubs" (MPTS) thus emerged and acquired wide acceptance in a rather short time to encompass all such woody perennials of agroforestry significance and potential.[31]

One of the greatest potentials of agroforestry lies in using ("exploiting") the vast array of such indigenous MPTS. ICRAF's MPTS inventory and information system has assembled an enormous information base on several hundred such species.[32] There are two major tasks with respect to this multipurpose tree program: first, to rapidly screen the hundreds of species and genera and to classify them based on a limited number of criteria, and secondly, to initiate more detailed studies on selected candidate species arising from the first task. One species of the genus *Leucaena (L. leucocephala)* has been relatively well studied.[33]

3. Nitrogen-Fixing Trees (NFTs)

Most NFTs are MPTS (but not vice versa). By virtue of their ability to fix atmospheric nitrogen and contribute nitrogen to the soil via leaf and fallen litter (litter *sensu stricto)*, or the release of root debris and nodules (root litter), shrubs have a special role in agroforestry.

A comprehensive state-of-the-art reviewer on the topic[34] suggested that improving nitrogen fixation by trees in agroforestry is a leading research priority and a goal that can be achieved not only by selecting the most efficient *Rhizobium* or *Frankia* strains, but also by improving the host plant, one of the quickest methods being through clonal multiplication of elite trees. The Waimanalo (Hawaii, U.S.A.)-based Nitrogen Fixing Tree Association is a good information source on ongoing research on NFTs.

4. Soil Productivity and Protection

In the process of conceptualization and principle-formulation of agroforestry, a number of hypotheses and attributes have been advanced in the recent past. One of the most significant and often repeated among these is the suggested potential of agroforestry as a major practical land management alternative for maintenance of soil fertility and productivity in a variety of situations in the tropics. As stated by Sanchez,[35] the following hypothesis is strongly implied in the agroforestry literature: "appropriate agroforestry systems improve soil physical properties, maintain soil organic matter, and promote nutrient cycling". Most aspects of this hypothesis have not been tested and proven by research. However, past experience and accrued knowledge on the behavior and management of tropical soils suggest that some of the agroforestry options can overcome at least a few of the drawbacks of monocultural (crop) production systems.

An account of soil productivity and management considerations under agroforestry has been given by Nair.[6] Wiersum,[18] Lundgren and Nair[17] and Young[19] discussed the role of agroforestry in soil conservation, taken in its broader sense to mean conservation of fertility as well as prevention of erosion. This was followed by a detailed ICRAF review on the topic, which elaborated the potential of agroforestry as a practical means of sustaining soil fertility, and discussed the research needs in this direction.[36] The role of agroforestry in the context of "land clearing and development in the tropics" was discussed by Nair[37] as a part of IBSRAM (International Board of Soil Research and Management) activities. As mentioned earlier in this chapter, research on soil fertility aspects of alley cropping is being undertaken vigorously at IITA, Nigeria and other research groups.[30] Sanchez[35] reviewed the current trends in research on soil aspects of agroforestry and suggested the directions in which research should proceed.

Thus, there is now substantial scientific activity related to soil fertility and productivity aspects of agroforestry, including field experiments in various phases of implementation at various places. Modeling of soil changes under agroforestry is also in progress at ICRAF.[36]

5. Socioeconomic Aspects

Reviewing the economic considerations in agroforestry, Arnold[38] stated that the established presence of agroforestry practices in many land-use systems today has perhaps been overshadowed by the concerns of particular contemporary problems that underlie much of recent interest in it. As a result, we may have been more concerned with developing new or improved agroforestry practices to cope with such issues as fuelwood shortages than with increasing our understanding of how existing activities function, and what economic contributions they make. Yet some of the most dynamic changes that have occurred in the recent past have been in existing agroforestry systems.

Agroforestry practices contribute to a wide variety of existing farming systems. In a number of these, farmers have been responding to growing pressures on their land and other resources by intensifying or increasing various agroforestry activities. These have proved to be particularly important in enabling them to intensify use of certain types of land as size of land-holding decreases, and to reduce the intensity of management as labor becomes scarcer and more costly. The appropriateness of agroforestry as a response to these changing pressures is often related to the concurrent shift in farmer production objectives towards income generation as farm size falls below the level at which they can meet their basic household food needs. These ongoing changes suggest the need for a closer look at access to land, labor and capital as ways of promoting some agroforestry activities, and at the place of other activities as population densities increase. In particular, the growing importance of income generation as a production objective of farmers points to the need to give greater attention to the markets for agroforestry products, and to the intermediate stages between production and the market. Increased economic research has an important role to play in clarifying and extending understanding of these and other economic considerations in agroforestry.

VI. CONCLUSIONS

A discussion on agroforestry cannot be concluded without a mention of the institutional dilemma facing agroforestry today.[39] Land-use institutions today are inadequate to deal with integrated activities like agroforestry that cut across conventional disciplines of agriculture, forestry, livestock management, etc. As a result of separate institutional developments in these disciplines there are separate laws and policies governing different land-use aspects, separate educational and research institutions and separate administrative departments and ministries. Moreover, the objectives of most of these disciplinary institutions is to optimize the outputs of their mandate commodities per unit of land or input. In integrated approaches, the main objective is to optimize the production of individual components in order to maximize the *total* combined production of different outputs per unit area of land or input.

The expedient way to overcome these limitations and yet work within the available institutional framework is through efficient and pragmatic collaborative and interinstitutional programs. This will undoubtedly need a fundamental rethinking among disciplinary scientists, institutions and decision makers.

Agroforestry offers solutions to many problems. Its promises are extremely encouraging and attractive. However, there are many constraints to be overcome. As a development tool, agroforestry will be helpful only if it can be put effectively into the hands of the farmer, and if the political decision-makers see this as a tool for achieving equity in development.

REFERENCES

1. **Lundgren, B. O. and Raintree, J. B.**, Sustained agroforestry, in *Agricultural Research for Development: Potentials and Challenges in Asia*, B. Nestel, Ed., ISNAR, The Hague, 1983.
2. **Nair, P. K. R.**, Multiple land-use and agroforestry, in *Better crops for food: CIBA Foundation Symposium* 97, 101, London: Pitman Books, 1983.
3. **Nair, P. K. R.**, Agroforestry: a sustainable land-use system for the fragile ecosystems in the tropics, *Malayan Nature J.*, 35, 109, 1982.
4. **Nair, P. K. R.**, Agroforestry Systems inventory, *Agroforest. Syst.*, 5, 301, 1987a.
5. **Nair, P.K.R.**, Classification of agroforestry systems, *Agroforest. Syst.*, 3, 97, 1985.
6. **Nair, P. K. R.**, Soil productivity aspects of agroforestry, ICRAF, Nairobi, 1984.
7. **Felker, P.**, State of the Art: *Acacia albida as a Complementary Intercrop with Annual Crops*, University of California, Riverside (mimeo. 1), prepared under contract for USAID, 1978.
8. **Miehe, S.**, *Acacia albida* and other multipurpose trees on the *Fur* farmlands in the Jebel Marra highlands, Western Darfur, Sudan, *Agroforest. Syst.*, 4, 89, 1986.
9. **Mann, H. S. and Saxena, S. K., Eds.**, *Khejri (Prosopis cineraria)*, Monograph No. 11, Jodhpur, India: Central Arid Zone Res. Inst., 1980.
10. **Shankarnarayan, K. A., Harsh, L. N., and Kathju, S.**, Agroforestry in the arid zone of India, *Agroforest. Syst.*, 5, 69, 1987.
11. **NAS**, *Firewood Crops*, National Academy of Sciences, Washington, D.C., 1980.
12. **Nair, P. K. R.**, Agroforestry and firewood production, in *Biomass*, D. O. Hall and R. P. Overend, Eds., John Wiley & Sons, Chichester, 367, 1987b.
13. **Torres, F.**, The role of woody perennials in animal agroforestry, *Agroforest. Syst.*, 2, 131, 1983.
14. **Nair, P. K. R.**, Soil productivity under agroforestry, in *Agroforestry: realities, potentials and possibilities*, H. Gholz, Ed., Martinus Nijhoff, The Hague, 21, 1987c.
15. **Young, A.**, Effects of trees on soils, in *Amelioration of Soils by Trees: A Review of Current Concepts and Practices*, R. T. Prinsley and M. J. Swift, Eds., Commonwealth Science Council, London, 1986a.
16. **Neumann, I. F.**, Use of trees in smallholder agriculture in tropical highlands, in *Environmentally Sound Agriculture*, W. Lockeretz, Ed., New York, Praeger, 351, 1983.
17. **Lundgren, B. O. and Nair, P. K. R.**, Agroforestry for soil conservation, in *Soil Erosion and Conservation*, El-Swaify, S. A. and Moldenhauer, W. E., Eds., Soil Cons. Soc. N. Amer., Iowa, 1985.
18. **Wiersum, K. F.**, Surface erosion under various tropical Agroforestry Systems, in *Symposium on Effects of Forest Land Use on Erosion and Slope-stability*, C. L. O'Loughlin and A. J. Pearce, Eds., East-West Center, Honolulu, Hawaii, 1984.
19. **Young, A.**, *The Potential of Agroforestry for Soil Conservation*, ICRAF Working Paper No. 42, ICRAF, Nairobi, 1986b.
20. **USAID.** Windbreak and shelterbelt technology for increasing agricultural production, S & T/FNR, USAID, Washington, D.C., 1987.

21. **FAO,** *Forestry for Local Community Development,* Forestry Paper No. 7, FAO, Rome, 1978.
22. **FAO,** *Tree Growing by Rural People,* Forestry Paper No. 64, FAO, Rome, 1985.
23. **Budowski, G.,** Applicability of agroforestry systems, in *Agroforestry in the African Humid Tropics,* L. H. MacDonald, Ed., United Nations University, Tokyo, 1981, 13.
24. **Arnold, J. E. M.,** Economic considerations in agroforestry projects, *Agroforest. Syst.,* 1, 299, 1983.
25. **Wiersum, K. F., Ed.,** *Viewpoints on agroforestry,* Department of Forestry, Agricultural University, The Netherlands, 1981.
26. **Chambers, R. and Longhurst, R.,** Trees, seasons and the poor, in *Seasonality and Poverty,* R. Longhurst, Ed., IDS Bulletin 17, 1986, 44.
27. **Hoskins, M. W.,** Agroforestry and the social milieu, in *Agroforestry: a Decade of Development,* H. A. Steppler and P. K. R. Nair, Eds. ICRAF, Nairobi, 1987, 191.
28. **Raintree, J. B.,** The state of the art of agroforestry diagnosis and design, *Agroforest. Syst.,* 5, 219, 1987.
29. **Steppler, H. A.,** ICRAF and a decade of agroforestry development, in *Agroforestry: A Decade of Development,* H. A. Steppler and P. K. R. Nair, Eds., ICRAF, Nairobi, 1987.
30. **Kang, B. T. and Wilson, G. F.,** The development of alley cropping as a promising agroforestry technology, in *Agroforestry: A Decade of Development,* H. A. Steppler and P. K. R. Nair, Eds., ICRAF, Nairobi, 1987, 227.
31. **Burley, J. and von Carlowitz, P. G., Eds.,** *Multipurpose Tree Germplasm,* ICRAF, Nairobi, 1984.
32. **von Carlowitz, P. G.,** ICRAF's multipurpose tree and shrub information system, *Agroforest. Syst.,* 5, 319, 1987.
33. **Brewbaker, J. L.,** *Leucaena:* a multipurpose tree genus for tropical agroforestry. in *Agroforestry: A Decade of Development,* H. A. Steppler and P. K. R. Nair, Eds., ICRAF, Nairobi, 1987, 289.
34. **Dommergues, Y. R.,** The role of biological nitrogen fixation in agroforestry, in *Agroforestry: A Decade of Development,* H. A. Steppler and P. K. R. Nair, Eds., ICRAF, Nairobi, 1987, 245.
35. **Sanchez, P. A.,** Soil productivity and sustainability in *Agroforestry Systems,* in *Agroforestry: A Decade of Development,* H. A. Steppler and P. K. R. Nair, Eds., ICRAF, Nairobi, 1987.
36. **Young, A.,** Soil productivity, soil conservation and land evaluation, *Agroforest. Syst.,* 5, 277, 1987.
37. **Nair, P. K. R.,** *Agroforestry in the context of land clearing and development in the tropics,* ICRAF Working Paper No. 33, ICRAF, Nairobi, 1985b.
38. **Arnold, J. E. M.,** Economic considerations in agroforestry. in *Agroforestry: A Decade of Development,* H. A. Steppler and P. K. R. Nair, Eds., ICRAF, Nairobi, 1987, 173.
39. **Lundgren, B. O.,** Institutional aspects of agroforestry development, in *Agroforestry: a Decade of Development,* H. A. Steppler and P. K. R. Nair, Eds., ICRAF, Nairobi, 1987.

15 POTENTIAL OF MULTIPLE CROPPING SYSTEMS

Charles A. Francis

I. MULTIPLE CROPPING SYSTEMS AND FOOD PRODUCTION

Deliberate planting of a number of crop species in the same area began with the first conscious efforts to increase and organize food production. Perhaps as a response to observations and knowledge about the natural ecosystems around them, people mimicked these systems of multiple species with their plantings of crops with dissimilar growth habits, maturities, and resource needs.[1] This makes multiple cropping systems the oldest of human food production activities. Multiple species cropping persists in many parts of the world, especially on small farms, for a number of biological, economic, and nutritional reasons.[2] There is an inherent biological stability in more diverse cropping systems, as well as an economic buffering when several crops are produced instead of one or two and offered for sale. The nutritional advantages of growing multiple species are obvious for the small farmer, and coupled with the biological benefits of diversity in each field these help explain the continuing popularity of intensive, multiple cropping systems for farmers with limited resources.

Today's commercial monocultures represent a relatively recent development, and one which has only found success where there are large tracts of land, capital to acquire fossil-fuel based inputs, and management ability to handle sophisticated equipment and technologies. Much of our research planning has been based on the assumption that monocultures would replace traditional multiple cropping systems as newer varieties and modern cultural practices are developed for more agroclimatic areas, and as agricultural modernization progresses.[3] Although there have been changes in some favored areas, the lack of capital for agriculture and problems of infrastructure have prevented this technology from reaching a majority of the world's farmers.

Expensive fossil fuels are one reason. Agricultural economist Vernon Ruttan of the University of Minnesota suggests that, "Agriculture in many countries in the Third World is unlikely to ever effectively enter the fossil fuel era",[4] due to high prices and competition from other sectors in developing countries. If we accept this premise, it is urgent that we examine in more detail the current importance of multiple cropping systems and how they can be improved to meet human needs for food and income.

A. Important Systems in the Americas

Typical of the complex systems which have evolved with farmers in the Americas is the potato (*Solanum* spp.)/maize (*Zea mays*)/climbing bean (*Phaseolus vulgaris*) relay intercrop in Rio Negro in the coffee zone of Colombia. Potatoes are planted during the cool season in January, and maize is planted in rows alongside the potato in April. When the potatoes are harvested in May and June, the soil is used to hill up the adjacent row of growing maize plants. Climbing beans are planted at the base of each maize plant in July before the maize reaches pollination. Beans are always planted on the upwind side of the maize so that tendrils will be blown toward the maize stalks and beans will climb successfully. Both beans and maize are harvested in December and January, and the cycle begins again.

This system has evolved over decades in the region of Rio Negro, and is well adapted to the prevailing temperature regime and the fairly well-distributed rainfall through the year. It is also a system which builds on the available family labor in the region, and meets the goals of producing a range of products for sale and foods for consumption on the farm, all from a limited individual farm area. Appropriate to the soils, climate, economy, and culture of Rio Negro, this complex system has come to prevail in this region. It has persisted in spite of attempts to introduce monoculture technology. The corn/bean/potato system is well adapted to some parts of the coffee zone in Colombia, while other crops and combinations are prevalent at other elevations and in different rainfall regimes in the Americas. Pinchinat et al.[5] describe the principal food crops that are included in these systems: maize, potato, bean, cassava *(Manihot esculenta)*, amaranth (*Amaranthus* spp.), quinoa *(Chenopodium quinoa)*, and squash (*Cucurbita* spp.). All these species are indigenous to the Americas. Such crops are planted together with a number of introduced crops from other continents: cowpea *(Vigna unguiculata)*, sorghum *(Sorghum bicolor)*, peanut *(Arachis hypogaea)*, and pigeon pea *(Cajanus cajan)*. Coffee (*Cofea arabica* and *C. robusta)*, cacao *(Theobroma cacao)*, and banana (*Musa* spp.) are perennial crops often included with the annuals found on small farms.

Although maize is a principal crop in the patterns of much of Central and South America across all altitudes,[6] cropping systems may be dominated by other species more adapted to certain climates and cultures. This includes cassava in the lowlands of Colombia, Brazil, Ecuador, Venezuela, and countries of Central America,[7] sorghum in Brazil's northeast dryland region,[8] and near the coast in El Salvador,[9] or potato in the Andean highlands.[5] The primary crop may dominate the system for climatic, economic, or cultural reasons, but other crops planted in the same field also have important economic and nutritional reasons for being included. The objective is to maximize food production or income from the primary crop, while enhancing and diversifying output with additional crops in the system. These crops often provide the main source of income and food for the family. Francis[10] cited several studies showing the importance of multiple cropping in the total production of food crops in the Americas: beans in Guatemala (73%),[11] beans in Colombia (90%),[11] beans in Brazil (80%),[12] and maize in all of Latin America (60%).[13] Since these crops are produced primarily by small farmers, it is easy to imagine how important this production is to the total well being of the small farm sector as well as to national economies.

B. African Multiple Cropping Systems

Within the wide range of soil and climatic regimes of the African continent are found a near limitless array of multiple species cropping systems. These have been reviewed by Okigbo and Greenland[14] and by Steiner.[15] Different crops such as maize, sorghum, millet *(Pennisetum americanum)*, rice (*Oryza* spp.), plantain (*Musa* spp.), or various root crops may dominate the system, depending on rainfall amount and distribution. Shifting cultivation was once normal in much of Africa, especially in tropical West Africa and the highlands of East Africa. Population pressure has forced shorter and shorter fallow periods until today a permanent pattern of cultivation predominates.[16] There are some areas where semi-permanent fields rotate with bush fallow, with the time of fallow depending on fertility of soils, distance of fields from dwellings, and population pressure.[14]

These crops are important to families for food and income, but also play a vital role in national food production, perhaps more than in any other continent. According to Okigbo and Greenland,[14] the following proportions of these crops in Nigeria are produced in multiple species systems: maize, 76%; millet, 90%; groundnut, 95%; cowpea, 99%; and cocoyam, 86%. In Uganda, Jameson[17] reported high proportions of basic food crops that were produced in multiple cropping systems: maize, 84%; groundnuts, 56%; beans, 81%; cowpeas, 62%; and pigeon peas, 76%.

Systems in tropical Africa, especially in the lowlands, may include more than ten crops grown together to take advantage of different niches in the crop ecosystem.[14] These crops are often concentrated near dwellings, and as dooryard gardens. The forest zone of Nigeria, for example, may have mixtures of cassava, maize, melon, groundnut, pigeon pea, yams of several types, pumpkin, and rice. An illustration of complex systems is shown in Figure 1 from Nigeria.[10] Given the importance of these crops and cropping systems, both to farm families and to national food supply, it is unfortunate that more research emphasis has not been directed toward improving their productivity. Much of the research which has been done on food crops has not addressed the challenges of multiple species in the same field, nor provided relevant recommendations for small farmers.

C. Intensive Asian Cropping Systems

Double cropping has long been a feature of Asian rice systems. Beets[18] summarized a report from Perkens[19] which described early rice used in double cropping in the 11th century in South China. This application of short-cycle varieties allowed two or more crops per year, a feature of many intensive cropping systems today in Asia. The lowland rice-based systems have been surveyed by Beets,[18] Gomez and Gomez,[20] and Harwood and Price.[21] Rice predominates in South and Southeast Asia, both in lowland paddy culture and upland rainfed culture, while drier areas of the Indian subcontinent are dominated by wheat or millet.

In addition to many of the same annual upland food crops which are grown in the Americas and Africa, a number of perennial crops are important components of Asian multiple cropping systems.[20,21] These include papaya *(Carica papaya)*, cacao, coffee, coconut *(Cocos nucifera)*, mango (*Mangifera* spp.), pineapple *(Ananas comosus)*, banana and abaca (*Musa* spp.). These are important sources of food as well as income for small farmers in Asia. However, the family that produces either a food or non-food crop for sale and becomes dependent on the international export market for guaranteed sales is highly vulnerable to changes in the international trade systems. It is desirable both for families and for countries to become as self-sufficient as possible in food production, especially in basic grains and root crops.[4] Perhaps the most research has been done in Asia on multiple cropping systems and how they can be improved. Especially active have been the International Rice Research Institute and the International Center for Research in the Semi-Arid Tropics as well as some national programs. This has benefited the small farmer far more than development of 'miracle' varieties which carry expensive recommendations for high-technology systems of monoculture.

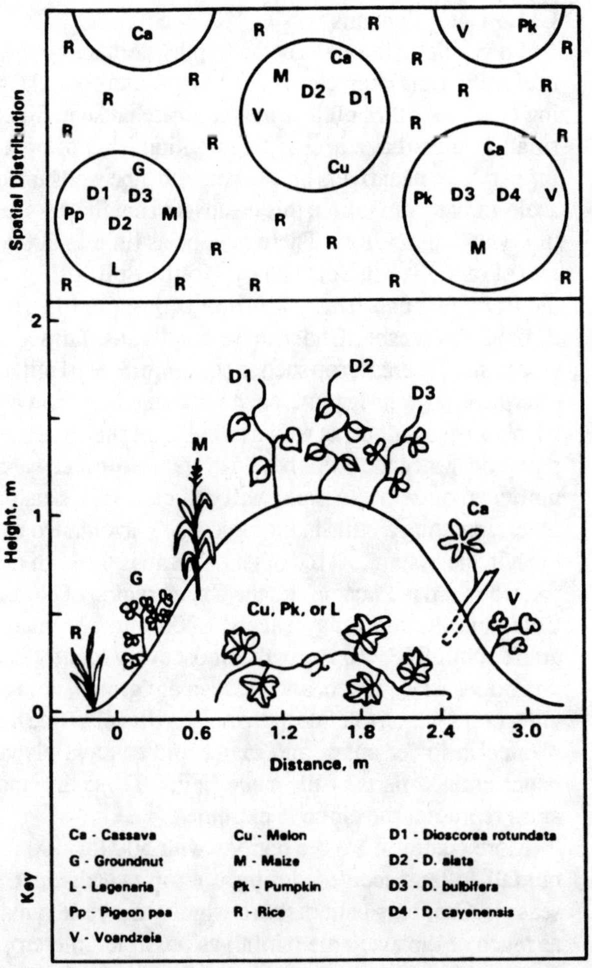

FIGURE 1. Spatial distribution of 12 crop species on and between raised mounds in Abakaliki, East Central State, Nigeria. (From Okigbo, B. N. and Greenland, D. J., in *Multiple Cropping*, Spec. Publ. 27, American Society of Agronomy, Madison, Wisconsin, 1976, 63. With permission.)

D. Overview of Food, Systems, and Stability

Availability of food for people in both rural and urban sectors of each country must have high priority in national planning and implementation of agricultural and economic policy. The preceeding sections by geographic area describe in brief the importance of multiple cropping systems in those regions, and how complex systems contribute to total food supply. Later sections describe in more detail how multiple species systems are biologically and economically well suited to small farm agriculture, and how they make efficient use of scarce resources. Cropping systems which are integrated with small animal production on the farm and which make maximum use of internal resources have an inherent stability which helps to guarantee the family both food and income, in comparison to higher-risk monocultures of one or a small number of crops. A better understanding of both the importance and the future potentials of these systems is critical to national planners and research directors who can dedicate more resources to improving multiple cropping approaches to agriculture. This can benefit both small farmers and the countries which depend on their agricultural production.

II. HOW SYSTEMS RELATE TO CLIMATE AND RESOURCE BASE

Specific component crops and the systems in which they are grown are determined in large part by the rainfall patterns, temperature regimes, soil fertility and topography of each region. That's as far as most agronomists, ecologists, or economists explore in their explanations of why systems appear as they function today. There is a broader range of factors which also influences what crops are grown, how they are grown and by whom. These include economic resource base of farm families; geographic isolation or proximity to roads, sources of inputs, and markets; historical antecedents which have resulted in a certain mix of farm sizes; and political factors which influence the appropriateness and availability of new technology. Thus, the presence of a certain set of crops on a farm is a result of many factors and their complex interactions.[2] Climate and resource base have direct biological and economic influence on cropping patterns.

A. Highland Systems

Hardy[22] classifies lands above 2000 m elevation in tropical America as 'highlands', although this term is somewhat arbitrary. Maize and potato are dominant crops in these areas throughout Central America and the Andean zone. Amaranth, beans of many types and colors, quinoa, wheat *(Triticum aestivum)*, and barley *(Hordeum vulgare)* are among the components of these highland systems. Farm size is often small, and individual operations are characterized by hand or animal traction for basic land preparation and cultivation. In the highlands of Peru and Bolivia, family lands have been divided for generations among the children until some individual fields may consist of little more than a few rows some 20 to 40 m long. One family may farm up to a dozen or more of these fields which have been acquired by inheritance, barter, rental, or purchase. Highland farms are often distant from any means of frequent communication with outside markets or sources of inputs — including information about improved technology. Thus, there has been limited impact of research and development programs on many of these farms.

Similar patterns are found in the highlands of East Africa where crop mixtures are dominated by maize or by sorghum. Where rainfall is adequate for maize, this is the cereal of choice. In areas marginal for maize, sorghum is a less risky crop. The two cereals may be found interplanted near this margin of adaptation for maize, such as in the central highlands of Tanzania. Where there is not enough moisture for sorghum, pearl millet is a preferred crop, again with interplanting of sorghum and millet in the interface areas. Near the equator, bananas grow at relatively high elevations, and potatoes are a common component of these systems.

Multiple cropping systems predominate in the highlands of Asia. In Nepal and Pakistan, for example, wheat and barley as well as maize are grown with legume crops in the same fields, and other short-cycle crops are grown with residual moisture where there is enough to double crop. In most areas of the highlands, crop growth is slow due to limited heat units, and crops have been chosen to take advantage of this long, frost-free growing season if there is ample rain. Because of low temperatures, transpiration rates are low and crops can withstand several weeks of limited rainfall. Farms are often self-sufficient in production inputs because of the isolation from roads and other communications; individual communities are more self-sufficient in food and other needs compared to towns in the lowlands.

B. Medium Elevation — the Coffee Zone

In Colombia, many of the hilly lands from 1000 to 2000 m elevation are cultivated by small landholders who grow maize, beans, banana, coffee, cassava, and a wide range of minor crops. These are often shallow soils, and the region is prone to erosion from intensive cultivation of two or more crops per year. A visitor is struck by the incongruity of intensive cropping of this land which would more logically be used for pastures and grazing, or left in forest, while the level and more fertile lands of nearby valleys have large cattle ranches as well as extensive sugar cane plantations. Today's land use and ownership pattern reflects political decisions made more than 400 years ago, when the valley lands were granted to Spanish military officials and politicians, and small farmers were pushed into the mountains to produce their food crops on marginal land. Agronomic and ecological factors may be less important than political realities which still influence land use today.

Medium elevation farming regions often have better contact with railroad or road systems, and thus farmers may have access to more inputs and to markets when there is excess production. Principle crops include maize, beans, banana, cassava, coffee, and other food and income sources. With the higher available heat units, there is potential for two or more crops per year, even when grown in sequence. Combinations of annual and perennial crops are often found on small farms. If the farmer's goal is to establish a small coffee plantation, bananas of several types and annual crops such as maize, beans, or cassava may be planted among the bananas. Coffee trees are started in this melange in a systematic pattern, and the areas immediately around the trees are kept free of crops and weeds. Thus, there is a source of income and food from short-term crops during the three years the family has to wait for the first substantial harvest of coffee. Older varieties of coffee needed the shade of the banana, although new varieties of quality 'arabica' coffee do not have this requirement. The intermediate elevation is important across continents, with intensive cultivation by small farmers and high population density in many areas.

C. Lowland Systems

Lower elevation areas in the tropics, perhaps below 1000 m elevation, are characterized by year-round potential growing conditions if rainfall or irrigation are present. Since few small farmers have access to irrigation, except on a very limited scale near rivers or lakes or with tube wells for small-scale garden cultivation, this discussion emphasizes rainfed or dryland agriculture. There may be as little as 300 to 500 mm of rainfall as in Botswana or Northeast Brazil, and even then rainfall is unpredictable in time of onset and distribution through the season. Under these conditions, farmers often plant a single cereal crop such as sorghum or pearl millet, and interplant a grain legume such as groundnut, cowpea, or Bambara groundnut, as well as a cucurbit such as squash or pumpkin. Since the crops have different temporal water and nutrient requirements, some will do better each season than others, creating a built-in biological risk-avoidance mechanism in the system. At the other extreme is the high rainfall pattern found in Panama and the Choco region of Northwest Colombia. Rainfall may exceed 8000 mm per year, with ample rainfall for crop growth almost every month. In these conditions, rice, maize, and cassava are grown, along with other crops which tolerate high moisture levels. Rice may be planted in lower areas, and maize and cassava planted in better drained parts of the same fields. These dry and wet areas represent the climatic extremes.

More common are the regions with 500 to 1500 mm of rainfall, often received during one or two distinct rainy seasons. Cropping patterns are designed to take advantage of as much of the available rainfall as possible. Intercropping long-season pigeon pea or pearl millet with short-season sorghum is a common practice in Northern Nigeria,[23] where the drought-tolerant and slow growing legume or millet will take advantage of residual moisture after sorghum is harvested. Sesame *(Sesamum indica)*, soybean *(Glycine max)*, or maize is relay planted at about flowering time with the principal maize crop in the Pacific Coastal region of Guatemala in order to produce a second crop with available moisture. Cassava is a long-season (9 to 15 months) crop which is often interplanted with shorter season annual crops and allowed to continue growth and development into and through the dry season. It can be stored in the ground until needed, and there is little loss of dry matter or food value during the dry season. These strategies all find use by small farmers to gain a food subsistence and some excess for market in the lowland areas of the tropics.

D. Internal vs. External Resources

One characteristic common to small farmers throughout the developing world is lack of capital or equity to purchase production inputs. It is critical that their cropping systems maximize the use of resources or production inputs which are 'internal' to the system.[4] These include sunlight, rainfall, biologically fixed nitrogen, animal manures, and the pest protection afforded by crop rotation and careful choice of crop species and varieties. Multiple cropping appears to

TABLE 1
Agricultural Production Resources that are Derived from Internal and External Sources

Internal resources	External resources
Sun — source of energy for plant photosynthesis	Artificial lights — used in greenhouse food production
Water — rain and/or small, local irrigation schemes	Water — large dams, centralized distribution, deep wells
Nitrogen — fixed from air, recycled in soil organic matter	Nitrogen — primarily from applied synthetic fertilizer
Other nutrients — from soil reserves recycled in cropping system	Other nutrients — mined, processed, and imported
Weed and pest control — biological, cultural, and mechanical	Weed and pest control — chemical herbicides and insecticides
Seed — varieties produced on-farm	Seed — hybrids or certified varieties purchased annually
Machinery — built and maintained on farm or in community	Machinery — purchased and replaced frequently
Labor — most work done by the family living on the farm	Labor — most work done by hired labor
Capital — source is family and community, reinvested locally	Capital — external indebtedness, benefits leave community
Management — information from farmers and local community	Management — from input suppliers, crop consultants

Adapted from References 24 and 25.

enhance this crop protection under a wide range of conditions. Rodale[24] contrasts internal inputs with those which are 'external' to the system and must be purchased or otherwise acquired from outside the farm. These include fossil fuel energy for pumping water, synthetic nitrogen, mined phosphorus or potassium, chemical herbicides and insecticides. All are the result of a manufacturing process which uses nonrenewable fossil fuels, and all represent a cost each season to the farmer who must bring these to the farm. Such external inputs are unavailable to most small farmers because of cost or distance from the source, and technology based on external inputs is often viewed as expensive and unrealistic by the farmer. Table 1 lists a series of resources needed for crop production, and summarizes the contrast between 'internal' and 'external' for each crop need.[25]

The majority of research in the past four decades has focused on how to combine external inputs with desired crops to dominate the natural environment and produce a commercial food or fiber crop for both internal consumption and export abroad. There has been limited attention given to the efficient exploitation of internal resources in systems that build on those cropping patterns that farmers are presently using. There is a great need to research and validate resource-efficient and sustainable cropping patterns that can make maximum use of those resources internal to the farm, and that efficiently use those few purchased inputs that are essential to improve productivity and profitability of the systems.

E. Importance of Input Costs

When the commercial-scale farmer in Nebraska or in the Cauca Valley of Colombia looks at costs of production, the expenditure for seed of hybrid maize or an improved and certified bean variety may be only 4 to 6% of the total production cost. When balanced against the benefits which this high quality seed brings to the cropping system that season, the cost is minimal and few farmers would question the value of the investment. But if the same hybrid or variety were recommended to a small farmer who concentrates on efficient use of internal resources, the seed might represent the only cash cost for producing a crop during the entire production cycle. This is a vastly different perspective from two different farmers considering the same improved technology. It is not difficult to understand why the commercial farmer plants the hybrid maize and the small farmer continues to plant a traditional variety.

An important step in the improvement of productivity is consideration of the resource base of farmers at any point on the economic spectrum, and careful assessment of what is perceived by the farmer as 'useful new technology which will solve a known constraint to production'. We have often been less than successful in moving new hybrids or fertilizer recommendations to small farmers because the technology was not well understood or demonstrated in the field, or because of a prohibitive cost or difficulty in obtaining the input in a timely manner. Those changes in a production system which are easily understood and seen in the field, which require a minimal change from the current system, and which do not have a prohibitive additional cost to the farmer are much preferred by the small farmer. One approach which has been successful in the design and testing of new technology is 'farming systems research and extension', a process which involves the farmer in all steps of the activity from identification of limiting constraints to validation and adoption of new technology.[26] When the farmer is involved in the process of designing and testing technology, we can be assured that costs of inputs will receive the needed attention.

F. Sustainability of Production

Cropping systems which are based on high levels of purchased inputs each season have been shown to increase productivity and profits under some conditions in the developing world. This is relevant only when capital, fertile soils, access to inputs and markets, and educational and management levels are adequate for the new technology. These systems can be sustained, however, only as long as the fossil fuels and production inputs are available at a reasonable cost and the price received for crops more than pays for the investment in inputs. In contrast, many subsistence cropping systems have been shown to be sustainable over a long period of time without such infusion of technology and inputs from outside the farm. Productivity may be moderate to low in some systems, but this often reflects the soil and rainfall resource which predominates at a particular farm site. The real challenge for many of the world's small farmers is to develop alternative systems which make efficient use of internal resources and increase both the productivity and the sustainability of cropping activities. An approach that appears to meet these objectives is the farming systems research/extension method outlined above. If we can combine some of the best and lowest cost components of improved technology, those which do not require radical changes in current systems, with the conventional wisdom of farmers about their environment and the crops which grow well, we can work together to design alternative systems which will work and improve agriculture for small farmers.

III. RESOURCE USE EFFICIENCY IN MULTIPLE CROPPING

Small farmers employ multiple cropping systems in part because of their limited access to resources to sustain monocultures. Although it is difficult to generalize about this resource base, it appears that many small farmers work with limited land areas, relatively poor native soil fertility, and lack of capital to purchase external inputs to apply to crops. Taking advantage of this limited land resource and substituting labor and management for other inputs, the small farmer and family make use of several efficiencies which are inherent in multiple cropping patterns.

A. Cropping in the Time-Space Continuum

Multiple cropping is defined as an intensification of planting in either the time dimension, the space dimension, or both.[27] When crops are planted more intensely than one crop per year in a given field, they take advantage of available light, water, and nutrients through a greater part of the total year's potential growing season. When two or more crops are planted together in the field, either at the same time or in some overlapping fashion, they exploit different parts of the field environment in four dimensions (including time) to more efficiently use available growth resources. This is the nature of the conscious placement of crops in the field for maximum use of scarce resources through as much of the year as possible. Adding the time dimension to systems creates what could be called a time-space continuum.[28]

In addition to the biological dimensions of resource use, there are implications of intense cropping systems for use of labor, production of food and income, and distribution of both input use and food output through the year.[2] These are factors in choice of cropping patterns and crop/animal systems by farm families, and further expand the complexity of how the total resource base is used over time and through space on the farm.

B. Efficiency of Light, Nutrient, and Water Use

Planting two or more crops sequentially in the field each year allows greater total light interception than would be possible with either crop alone, unless one crop had an extremely long growth cycle. More efficient use of light comes from overlapping two or more crops by planting a second crop when the first is flowering or filling grain. This system is called relay cropping or relay intercropping. Most intensive use of light by intercrops comes when two or more species are planted near the same time. The most efficient intensive intercrops appear to be mixtures of dissimilar plant types such as tall maize or sorghum with low-growing legumes such as peanut, cowpea, or drybean. The taller cereal takes advantage of full sunlight with a unique process of carbon assimilation, while the understory legume or other crop makes relatively efficient use of lower light intensities. Thus, the two crops make more efficient total use of this resource while they are actively growing in the field than could a single homogeneous crop species. These aspects of light use were described and reviewed by Donald,[29] Marshall and Willey,[30] and Trenbath.[31-33]

Nutrient availability for crop growth and development is critical for successful economic production. Certain levels of nitrogen and other nutrients are available as internal resources from the organic and mineral fractions of the soil. Their availability depends on rainfall and temperature as well as soil physical, chemical, and biological properties. Each of these factors is influenced to some degree by management and cropping pattern — but none so much as soil organic matter. Sanchez[34] reviewed the fertility relationships of multiple crop systems. Some of the specific management practices which will influence fertility include stand densities of crops in mixtures, rotations and green manures, and which plant products are removed from the field.[35] An intensive intercrop through much of the year has potential to increase organic matter and subsequent nitrogen available for future crops. But the small farm family may remove more from the field than the grain, e.g., using lower maize leaves for cattle feed and stalks for fuel, straw for bedding, or whole plant residues for other purposes. In this case, intercropping may result in greater total extraction of nutrients from the soil reserve and actually reduce soil organic matter. Thus, the nutrient balance is dependent on more than just the cropping pattern employed on the farm. There is a continuing debate about whether legumes actually produce nitrogen which is used by associated cereals during the same cropping season. There is no doubt about the availability of some of this fixed nitrogen for subsequent crops.[36]

Water use by crops is primarily a function of how much rain falls during the cropping season and is trapped and stored and made available during the crop growth cycle. In many regions of the tropics, year-round favorable temperatures allow planting at any time, yet actual dates of planting crops depend on the arrival of seasonal rainfall. In Northern Nigeria and in some coastal areas of Central America, a long-season sorghum is planted with other cereals and legumes. The legumes use rain which falls during the principal cropping season, while the sorghum develops slowly at first and uses residual moisture after harvest of the early crops. Following one or two crops of rice in the Philippines, residual moisture is used by a short-cycle "catch crop" such as soybean or mungbean planted into the rice paddy. Sometimes the legume is planted before rice is harvested. Cassava often is planted in the rainy season together with short-cycle cereals and other crops, and continues its growth and development through the next dry season with residual moisture. This popular root crop can be stored in the ground for many months with little loss of food value, and cassava is a staple carbohydrate source in much of the lowland tropics where small farmers maintain subsistence agriculture.

C. Pest Protection in Multiple Species Systems

Reports on the incidence of weeds, insect pests, and pathogens in multiple cropping systems compared to monoculture are varied, yet many point out advantages of the diversified, multi-species systems used by small farmers. The recent review by Altieri and Liebman[37] concentrates on weed and insect populations, since these appear to be most affected by more complex cropping patterns. They suggest that traditional systems "are based on countless distinct crop varieties and crop mixtures that were selected by farmers over thousands of years for resistance to pests and other characteristics. These varieties and mixtures are grown using agricultural practices that usually enhance cultural and biological pest control". In fact there are no doubt numerous mechanisms of pest management which are operating in traditional systems which should be studied and preserved during any change to new component technology.

Weed problems may represent the major limiting constraint in many small-scale farms, seriously reducing crop yields in both monoculture and intercropping. Labor needed to control weeds at the critical time of early crop development may limit size of area planted in some regions. Successful management of weeds without expensive herbicides involves level of soil fertility and rainfall pattern, choice of crops and rotations, method of tillage and cultivation, insect prevalence, and labor available at key times in the season.[38] An entire range of cultural practices imposed by the farmer can influence weed populations and pressures; these include crop species and their relative proportions, dates of planting and total density, and the spatial organization of intercrop components. Herbicides and cultivation techniques for weed control are much better known than their biological control counterparts.[37] Since intercropping creates an intense interaction among crops for growth resources, it is useful to manipulate systems so that the crops planted compete favorably with unwanted weeds for these growth factors.

Mechanisms of insect control by intercrops have been described by Hasse and Litsinger[39] as that interfere with insects finding the right host and those that interfere with insect development and survival. Some of these mechanisms are physical: insects don't encounter a pure stand of the crop they desire and get lost looking for the next plant. Other mechanisms are chemical: attractant odors from host plants are diluted by the presence of other plants in the mixture. A number of management practices have been shown to influence the incidence of insect pests. In Colombia, Altieri et al.[40] found that maize planted 3 to 4 weeks before beans reduced leafhopper population in beans by 66% compared to planting the two crops on the same day. In many areas, maize is planted before beans by small farmers, and this may be the rationale. Yet in the same studies, maize fall armyworm was reduced by 88% when the beans were planted earlier. Choice of optimum dates for planting the two crops may depend on which insect is most likely to occur during a particular planting season or in a specific area. Weed populations may have a profound effect on insect prevalence, but this relationship is poorly understood in the tropics.[37]

Plant pathogen incidence appears to be reduced by greater diversity in the cropping pattern, although there is less information available about diseases. Altieri and Liebman[37] suggest that different crop mixtures may have fewer disease problems by "delaying the onset of disease, reducing spore dissemination, or modifying . . . humidity, light, temperature, and air movement" in the mixed crop canopy. Use of multiline cereals to reduce incidence of devastating diseases has been suggested by plant breeders[41,42] and used with success in a few practical crop improvement programs. Some of the control of unwanted species, especially weeds, may be due to chemical suppression or allelopathy.[43] Much more research is needed to understand why multiple crop systems seem to have fewer problems with crop pests, and then these mechanisms can be used to design new approaches using internal resources and careful management.

D. Statistical Measures of Efficiency

Outlined above are some of the differences between monoculture and multiple cropping in use of resources (light, water, nutrients) and in system-based control of plant pests. How can these differences be quantified? To begin, most subsistance farmers keep few records in the sense of input purchases and exact yields and income. Instead there is an intuitive sense of what has worked in the past and which years had adequate food and income from the cropping patterns on the farm. Researchers find these measures inadequate to compare alternative systems, and thus seek a numerical index or other guide to how systems perform. Such indices help in the comparison of different options and serve as a measure of biological, economic, or nutritional productivity.

One of the most frequently employed measures is the "land equivalent ratio" (LER) which evolved from the relative yield total (RYT) described by de Wit and van den Bergh.[44] Use of this measure was reviewed by Willey[45,46] in his classical review articles on multiple cropping. Briefly, the LER is the land required under monoculture to produce the same yield as the area will produce with intercropping. For example, if maize produces 5000 kg/ha in monoculture and 4000 kg/ha when planted with beans, and beans produce 1000 kg/ha in monoculture and 600 kg/ha when planted with maize, the LER would be: 4000/5000 + 600/1000 = 1.40.

In the example, one hectare of sole planted maize would produce 5000 kg and 1 ha of beans would produce 1000 kg; 2 ha of the crops planted together would produce 8000 kg of maize and 1200 kg of beans. Thus, the intercrop system is 40% more efficient than the composite of the monocrops. This index is the one most widely used in the literature to describe intercropping in comparison to sole cropping or monoculture. There have been studies of the distributions of LERs[47,48] which conclude that these indices can be analyzed using standard techniques if the divisors (monocrop yields) are kept constant. There is always a question about which monocrop yields to use in a calculation, yields from the same experiment or comparison on-farm, yields with the same densities or fertilizer levels, or average monocrop yields from the region. Although most researchers have found the method useful, it does have limitations.

Any type of index or pooling of data leads to some loss of information. One of the drawbacks of the LER method is the lack of detail on how individual components perform and how the system can be manipulated to improve yields and performance. This led to development of the "effective LER" by Mead and Willey[49,50] to guide changes in proportions of the crops and eventual choices of the most desired products from the system. Another question about the LER is the lack of a time dimension. Since many intercrop systems are used in the tropical latitudes where temperature allows crop growth during much of the year, there is a need to include the time factor as well as crop intensity in deriving a measure of efficiency. The index has been called the "area-time equivalency ratio", and this takes into account the time that each component crop is in the field.[51] Other measures of intensity include the "cropping index", i.e., the number of crops grown per year in a given field × 100. The "income equivalent ratio" is calculated in the same manner as the LER, and other measures are the "calorie index" and "monetary index" suggested by Gomez and Gomez.[20] These various statistical indices and measures for analysis mean little to the small farmer who is unlikely to keep careful records of inputs and outputs for various enterprises on the farm. The mathematical tools do help the researcher to put numbers onto systems and see whether new alternatives really do produce greater yields, food value, or economic return. It is critical to assess the farm family's goals and be sure that whatever measure is used in research and validation of technology is consistent with the perceived needs of the family. Otherwise, the statistical measures of efficiency may be irrelevant for a practical application of new methods and varieties.

IV. BIOLOGICAL STABILITY WITH MULTIPLE SPECIES

In a recent book on agroecology, Altieri[52] describes our present farming methods as agroecosystems which have converted wide areas of biologically diverse ecosystems into narrow genetic plantings suited for human needs. He describes crop monoculture as "simplification ... to increase the proportion of solar energy fixed by the plant communties that is directly available to (humans). The net result is an artificial ecosystem which requires constant human intervention for its operation". As farming systems have moved away from the diversity found in natural ecosystems and in early cropping patterns,[1] they have lost some of the stability which Murdoch[53] ascribes to biological diversity and complex structural arrangements found in nature. Multiple cropping systems provide an environmentally rational way to reverse this trend.

A. Diversity Range in Cropping Systems

There is a wide range of diversity represented in today's cropping systems.[54] The most genetically homogeneous situation is a single cross hybrid of maize or sorghum where every plant is the product of a cross between the same two parents, and those parents are highly inbred and uniform. Slightly less uniform are the three-way crosses and double crosses (four parents) of maize which are common products of many breeding programs and form the bulk of seed distribution in national programs for crop improvement in developing countries. Varieties of self-pollinated crops are often products of seed increase from a single plant or family of closely-related plants. These varieties of small grains or legumes are highly uniform, but gain some diversity through mutations and mixtures during the seed multiplication phase and perhaps through extremely low levels of outcrossing.[55,56] One proposed way to reduce their genetic uniformity while maintaining some agronomic constancy is the formation of multiline varieties which would be resistant to a range of diseases, for example.[41,42]

Small farmers who plant beans in the coffee zone in Colombia often intercrop with maize, taking advantage of the maize stalks as support for climbing beans. They may mix several types of beans; bush and climbing types, several colors of seed, and different maturities add diversity and some stability through the range of unpredictable climatic conditions during that season, especially variation in rainfall. At the extreme of this conscious diversity in planted cropping systems is the multi-species mixture shown in Figure 1 from Nigeria.[14] These cropping systems could be characterized as a spectrum ranging from minimum to maximum genetic diversity as shown in Figure 2.[54] The figure also illustrates parallels with natural ecosystems.

B. Natural Systems Diversity

Figure 2 illustrates a series of natural ecosystems and the range of diversity which exists from minimum genetic variation in geothermal pools and northern forests to maximum diversity in tropical rain forests. In the fairly constant conditions and uniform niches of a thermal pool, specific algae adapt to a certain depth in the pool and predominate there. A northern forest has more diversity, but the short growing season and relatively cool temperature favor a few well-adapted species and do not promote rapid species change. In a tropical climate, in contrast, the seasonal

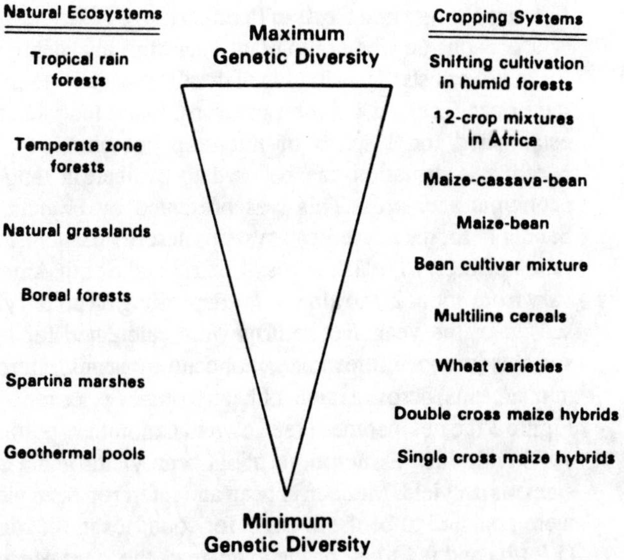

FIGURE 2. Spectrum of genetic diversity in natural ecosystems and in cropping systems. (From Smith, M. E. and Francis, C. A., in *Multiple Cropping Systems*, Macmillan, New York, 1986, 219. With permission.)

changes in rainfall and humidity, the high temperatures, and the variability in soils and fauna all promote rapid genetic changes in the plant community.

A diverse plant community such as the tropical rain forest is characterized by a number of dynamic processes which occur quickly and further stimulate change. These include energy flow through the system, cycling of nutrients, and capture and use of rainfall.[57,58] These factors all combine to create a wide range of ecological niches, each favoring a different type of plant organism, and as a whole promoting great diversity of plant life. It is this complex system which is most adaptable and most stable in the natural environment. This is a model in which plant to plant interactions can be studied and defined, and the principles applied to potential new combinations of crop species.[59]

C. Research Data on Biological Stability

It is generally assumed or suggested in ecological literature and recent papers on multiple cropping systems that diverse ecosystems are more stable than simple monoculture cropping systems. In a review, Goodman[60] concluded that "observational confirmation of the diversity-stability hypothesis (has) never materialized." Thus there is a lack of experimental evidence on quantified stability in different cropping systems, complicated by the lack of an agreed upon yardstick or definition with which to measure stability.[49] C. A. S. Hall from Cornell University (personal communication) contends that more diversity in a system increases stability only if food chains are connected "in parallel", where each species has several sources of growth factors. If these food chains or nutrient cycles are connected "in series" then even a diverse cropping system or natural ecosystem could be drastically affected by loss or decline of just one component and thus be very unstable.[54] Thus, there is much conjecture and assumption about stability of natural systems and cropping patterns, but only limited quantitative data.

V. ECONOMIC STABILITY IN MULTIPLE CROPPING

According to Lynam et al.,[61] intercropping systems are not more consistently profitable than monoculture systems. They contend that "profitability depends on relative prices, costs, and the degree of complementarity or competition of the activities; hence, other factors affecting profitability are specific to site, time, and input level". Over 2 years of maize/bean studies in Colombia, Sanders and Johnson[62] found monoculture more profitable the first year and an intercrop of the two species more profitable the second year. Thus, small farmers believe that "in intercropping systems net income advantages appear to be secondary to risk reduction, particularly in farm systems where subsistence is an important objective".[61]

A. Economic and Nutritional Diversity

A farm family can diversify crops and enterprises either by planting a range of different species in different sole cropping fields or by intercropping those same species in the same fields. Crop diversification could be defined as "planting various crops on the same farm without them being interplanted, (a strategy which) would reduce yield and price uncertainty, as the weather and economic conditions would have different effects on the different crops".[61] There is evidence that diversification has increased stability of income from crops in semiarid regions of India.[63] The main objectives of crop diversification on a farm may be to take advantage of specific adaptation of crops to different soil types or variations in topography which would favor one species over another. In most regions, crops have specific adaptation to contrasting seasons due to differences in rainfall, temperature, or other climatic changes through the year.[64] These are all ways to diversify cropping strategies, and do not necessarily involve intercropping. In a region with year-round growing conditions there may well be multiple cropping, in this case two crops per year on the same field although they are not grown together.

Intercropping is used to advantage where there are interactions among crop components and between crops and environment which favor more efficient resource use. These details have been explored in a previous section. It may be more effective use of total rainfall, soil nutrients, or solar energy, or may be a management approach to controlling weeds, insects, or plant pathogens. Three potential explanations may describe why small farmers intercrop:[61,65]

1. The more variable the production conditions and the more buffering is needed within the system, the more likely intercropping will be used by farmers.
2. The greater the two or more crops complement each other when grown in close association, the more likely these systems will be used.

3. The more the family depends on farm output for most of the food supply, the more likely crops will be grown in some mixed cultural system and the more variable the species mix will be.

Nutritional diversity has been an important factor through the centuries of development of cropping systems.[1] There is evidence that diets in some areas were better balanced before the introduction of improved technology and transportation. In the highlands of Guatemala, for example, the traditional cropping system of intercropped maize and beans provided two key staples in the diet. Maize is deficient in lysine and tryptophan, two essential amino acids, while beans are deficient in the sulfur-containing amino acids methionine and threonine. When the diet consisted of about 75% maize (10% protein) and 25% beans (20 to 22% protein), about half of the protein came from each source and there was an excellent balance of essential growth factors. With the introduction of higher yielding maize varieties, a ready market for both beans and maize, both the cropping and marketing patterns began to change. Some maize was planted in monoculture to achieve higher yields and incomes, and increasing prices for beans caused these to be sold as well through the newly developed road system. The consequence was a reduction of bean protein in the diet relative to maize consumed, and a less favorable nutritional balance (R. Bressani, INCAP, Guatemala, personal communication). It is critical to examine all potential impacts before introducing an apparently better technology.

B. Variances in Production and Income

Yield stability can be measured in different ways, and this presents a confusing story in the literature. For this reason it is important to define stability before analyzing and evaluating data to determine relative differences in "stability" between cropping systems or among varieties or hybrids. The simplest approach is to look at relative variation among treatments. This statistic is the coefficient of variation that puts the standard deviation in the scale of the mean.[49] Mead further points out that paired comparisons of intercrop with monocrop across locations is more valuable than a comparison of relative variation between the treatments without such pairing.

Francis and Sanders[66] evaluated 20 maize/bean trials in Colombia, comparing monoculture of each crop with the intercrop. The variation in yield was lowest in the intercrop and highest in monocrop beans. The same was true for net income: the probability of at least breaking even in the three systems was 0.92 (intercrop), 0.65 (maize monocrop), and 0.80 (bean monocrop). In a parallel study 2 years later with sorghum/pigeonpea systems in ICRISAT, Rao and Willey[67] analyzed 94 trials in India and found similar results. Their probabilities of at least breaking even were 1.00 (intercrop), 0.95 (sorghum monoculture), and 0.91 (pigeonpea monoculture). These careful studies by researchers confirm one of the reasons why small farmers persist with intercrop combinations on their farms.

C. Price Ranges and Costs of Production

Decisions on whether to plant intercrops also depend on the relative costs of production of the alternative systems and the prices received for the two crops. Once the yields are established for a series of intercrop combinations, this baseline information can be used to evaluate a range of economic scenarios. This was presented by Francis and Sanders[66] for the maize/bean systems described above. Price ratios of bean to maize across Central and South America vary from about 2:1 to almost 7:1 depending on country and season of the year. Net returns were calculated for three systems (monoculture maize, monoculture beans, intercrop maize/beans) across a range of bean to maize price ratios. In Figure 3 the net incomes are shown in Colombian pesos for 1978 with three assumptions about bean yields and a constant maize yield. Monocrop bean and intercrop bean yields were assumed to be the average for countries in the region (1.2 t/ha and 0.4 t/ha), or the average of the 20 experiment station trials (3.0 t/ha and 1.2 t/ha), or the highest yields achieved in one of the trials (4.0 t/ha and 2.0 t/ha). Intercrops were more profitable than monoculture beans at all price ratios up to 7:1 assuming farmer yields in the region, up to 4:1 assuming average experiment trial yields, and up to 3:1 assuming the highest experimental yields. Similar results were found when different assumptions were made about costs of production. It is clear from these data why small farmers continue to plant intercrops of these two important sources of food and income.

D. Food vs. Export Crops

Enough Food, a recent overview of agricultural development based on regenerative agriculture techniques,[4] describes the growing current awareness of many governments about the need for a high degree of self-reliance in basic food production. Although our conventional wisdom in economics describes comparative advantage and economies of scale as determinants of where each crop should be grown, there are other factors to be considered in an imperfect and volatile international economy. Emphasis on export crops and the foreign exchange they generate has led to reduced domestic food availability in some countries. A dependence on the world market and foreign food aid makes a country highly vulnerable to fluctuations in currencies, harvest success in other parts of the world, competition from other importers, and political decisions by both importer and exporter country governments. Given this reality of international politics and economics, it is good planning to increase self-reliance in basic foods. National governments can encourage greater domestic food production by realistic local price supports for food crops instead of export crops, by financial incentives and loan programs, and by setting the right import and export policies. There is also an obligation to international lenders and bilateral assistance program planners to give emphasis to food crops.

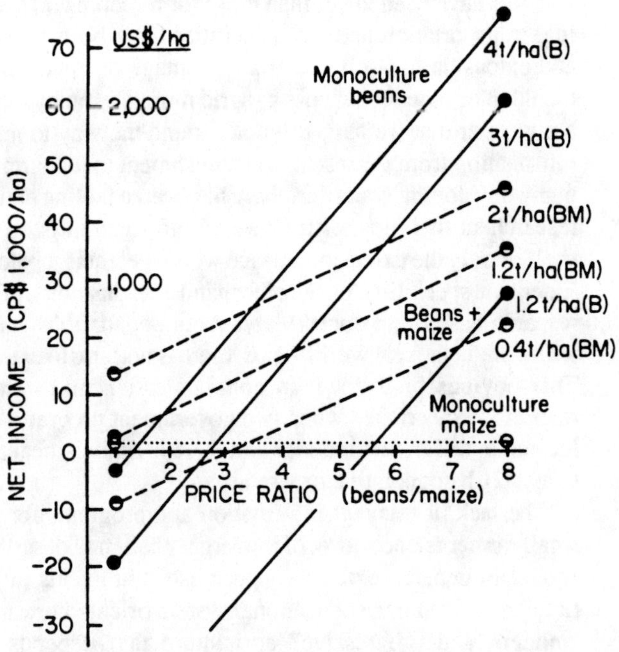

FIGURE 3. Net income from three cropping systems at different field bean/maize price ratios with different field bean yields; B = monoculture beans, BM = Bean/maize intercrop; high, medium, and low levels of bean production included. (From Francis, C. A. and Sanders, J. H., *Field Crops Res.*, 1, 319, 1978. With permission.)

VI. FUTURE PERSPECTIVES FOR MULTIPLE CROPPING

Each of the recent books and workshops on multiple cropping has given some perspective on the future of these complex systems in the developing world.[15,18,20,28] These ideas are especially important to scientists and development administrators who make decisions on allocation of personnel and resources in the research and extension organizations that have the potential to reach small farmers. Continued focus on monoculture cropping, high technology solutions to world food challenges, and large scale commercial enterprises will mean further denial of the importance of a large share of the world's agriculture. It is important to explore some of the biological realities that are coming out of today's research, and to see how this should influence the research agendas of the international and national programs which deal with food crops. In addition, there are currently a number of institutional constraints which make it difficult to both improve and extend information about multiple cropping. Each of these questions is explored.

A. Biological Advances and Understandings

As more information is accumulated about the biology of both natural and agricultural systems, it is easier to understand the basic interactions among species and between plants and environment that contribute to productivity and stability. Especially when chemical inputs are removed from some of the more simple of cropping patterns, a number of biological processes are enhanced and can be studied in detail. The integrated nature of specific biological processes in a crop community could be called the biological structuring of that system.[68]

There are both linear and cyclical changes that occur over time as a part of this structuring. The conscious management decisions that farmers make in designing appropriate crop rotations to take maximum advantage of the natural potentials could be called the "progressive biological sequencing" of cropping systems.[69] When these several fields or enterprises, each built on biological principles, are brought together and interact closely within one family's operation, this could be called the "integrative farm structuring" of the total farm system. These concepts provide a conceptual framework on which further study of component technology and total systems can build.

Some of the specific processes that are under study, and that appear to contribute efficiency to biological systems, include nutrient cycling, nitrogen fixation, weed population cycling, disease suppression, and insect population control. Much is yet to be learned about these interactions in complex cropping systems, since most research has been done on monocultures and with high levels of pesticide and fertilizer input. It will take time to learn enough about the details of biological management of systems to be able to build workable recommendations for this alternative approach.

B. Research Agendas for the International Centers

The network of International Agricultural Research Centers began primarily as an effort in plant breeding to bring improved genetic potential to the world's growers of food crops. The IARCs quickly became involved in active training programs for research and extension specialists from national programs, and further expanded their vision with the formation of teams in multiple cropping or farming systems work. There have been major advances in the improvement of yield potential of major food crops, and some of the "green revolution" technology has been successful where land was fertile and inputs were available.[70] Yet most of this technology has been unavailable or unaffordable to the majority of the world's small farmers.[4] It's time that the centers focus greater resources on the development of low-input, sustainable alternatives.[71]

Some of the varieties and populations developed in the strong genetic improvement projects bring valuable traits to crops which are somewhat "scale neutral" in their application. Resistance or tolerance to insect or disease problems or ability to withstand drought are all useful characteristics for the small farmer as well as the large operator. These yield-protecting traits allow reduction of inputs or more stable production under stress conditions, and thus contribute to the objectives of the small farmer. But too often there have been unfavorable traits associated with the new, high-yielding varieties: differences in taste or flavor in rice, changes in cooking requirements of opaque-2 maize, differences in color and taste of dry beans. There is a need to concentrate more on consumer preferences and specific needs of local groups. Further, there is a critical need to carefully test new varieties under a range of prevailing conditions in the area where those genetic materials are to be used by farmers. Much of this testing needs to be carried out in marginal conditions, low fertility, natural rainfall, and prevalent cropping systems in the zone. The farming systems or multiple cropping programs have gone up and down since their energetic beginning more than 15 years ago. These integrative research activities need to be an integral part of each center's operation. Efforts at participatory development of technology, with center staff working closely with key farmers as well as national program scientists, need to be accelerated.

C. Research Agendas for National Programs

Priorities in national programs often are determined in response to the most serious perceived needs of the country's agriculture, and this may be closely linked to a national development plan. Too often these plans appear to be a "top down" approach to transform agriculture and the entire rural scene, and may not take into consideration the immediate food and income needs of small farmers. Research planners and specialists in the experiment stations often are too far removed—culturally, economically, physically—from the small farm sector to really determine what is needed to help families meet their food needs. For example, research emphasis and production credit programs to sustain yields of export crops such as cotton and groundnut in West Africa may have been perceived at the national level as a valid strategy over the past decade. During the recent drought years the exports of these crops scarcely declined while there was a serious shortfall in basic food grains. The latter had to be imported at high cost, and some countries became more dependent on distant suppliers for their basic food subsistence. This is not a sustainable approach to agriculture and development in today's volatile world.

More appropriate for national programs is a focus on the primary food crops which sustain people and bring them a modest income.[4] The OTA study of low-resource agriculture in Africa released in mid 1988 (*Enhancing Agriculture in Africa: A Role for U.S. Development Assistance*, Office of Technology Assessment, U.S. Congress, Washington, D.C.), outlines the problems that face Sub-Sahelian countries today and calls for a research strategy based on efficient exploitation of internal resources on the farm. There are severe soil fertility constraints, and a critical need for some outside (external) inputs to overcome these production bottlenecks. But these need to be met in a low-cost manner and the productivity of soils built up to where they can be sustained with proper choice of crops and cropping patterns. This is an emphasis which national research programs can bring to their experiment stations and extension activities over the next decades.

D. Importance of Extension Activities

Agricultural extension has been viewed by many as the "poor relative" of research through the past several decades. Budgets have been lower than those for research and teaching, transportation and other facilities have been lacking, and extension has suffered from an image of lower status than that of other endeavors in agriculture's official sector. No wonder that we have often not found the way to move information from the research establishment to the farm! In many developing countries, there has been a feeling among researchers that most of the needed information is on the shelf, if only the extension service would get their job done. Extension specialists, on the other hand, feel that much of the research has not produced viable recommendations which will work in the real world: Thus, there is nothing to extend. This obvious breakdown in communication and mutual respect is exacerbated when two government programs are located in different ministries and there is limited incentive to establish cooperative programs.

The lack of relevant information and programs for the small farmer is one subset of the larger challenge described above. In general, extension specialists and agents in the field are trained in an educational system oriented toward a "modern and progressive" agriculture that depends on chemical fertilizers, pesticides, and mechanization. This model has been successful in other parts of the world, they are taught, and the best route to a successful agriculture is to emulate that model. Importance of women in agricultural production has been virtually ignored in many extension programs, with an absence of female specialists in the system and no effective ways to reach the majority of farmers. There is limited exposure to the principles of biological interactions in cropping systems and such practices as crop rotation, green manure crops, interplanting of legumes, and multiple cropping systems. With a research organization dedicated to work on high-technology approaches and strong emphasis on export crops, and an extension service which is poorly equipped to get to the field with relevant information, it is not surprising that the official sector often is poorly prepared to reach the small farmer.

Fortunately, there is an emerging awareness of the importance of small farmers, many of whom are women, and their contribution to national food production. This comes at a time when research on the biological interactions in cropping systems is also accelerating and there is greater concern about how internal resources can be used more effectively to reduce production costs and make systems more sustainable. There is growing appreciation and improved perspective in both the international centers and the national research/ extension programs about the impacts of the green revolution on agricultural production. Decision makers realize that these high-technology practices that depend on imported production inputs are unavailable to most farmers who are outside the mainstream and who cannot afford to purchase them. The next decades will see a greater emphasis on technology that is relevant for the small farmer, and that can help establish a more sustainable agriculture and food supply.

REFERENCES

1. **Plucknett, D. L. and Smith, N. G. H.,** Historical perspectives on multiple cropping, in *Multiple Cropping Systems,* Francis, C. A., Ed., Macmillan, New York, 1986, 20.
2. **Francis, C. A.,** Rationality of new technology for small farmers in the tropics, *Agric. Hum. Values,* 2, 54, 1985.
3. **Wortman, S. and Cummings, R. G., Jr.,** *To Feed this World,* Johns Hopkins University Press, Baltimore, 1978.
4. **Francis, C. A. and Harwood, R. R.,** *Enough Food,* Rodale Institute, Emmaus, PA, 1985.
5. **Pinchinat, A. M., Soria, J., and Bazan, R.,** Multiple cropping in tropical America, in *Multiple Cropping,* Spec. Publ. 27, Papendick, R. L, Sanchez, P. A., and Triplett, G. B., Eds., American Society of Agronomy, Madison, WI, 1976, 51.
6. **Castillo, M.,** Algunos sistemas de producción agrícola en Guatemala, in *Conferencia sobre Sistemas de Producción Agrícola para el Trópico,* Centro Agronómico Tropical de Investigación y Ensenanza, Informe Final, CATIE, Turrialba, Costa Rica, 1974.
7. **Soriña, J.,** Sistemas de producción bajo condiciones ecológicas en América Latina, con enfasis en el mejoramiento de la agricultura tradicional de pequenos productores, in *Consulta de Expertos sobre Investigacioñ Agrícola en América Latina,* FAO, Panama, 1975.
8. **Rao, M. R.,** Prospects for sorghum and pearl millet in the cropping systems of Northeast Brazil, in *Proc. Farming Systems Workshop,* CIMMYT/ ICRISAT/INTSORMIL, El Batán, Mexico, 1984.
9. **Alas L. M.,** Breve descripción del sistema de producción del pequeño agricultor en El Salvador, in *Conferences sobre Sistemas de Producción Agrícola para el Trópico,* Centro Agronómico Tropical de Investigación y Enseñanza, Informe Final, CATIE, Turrialba, Costa Rica, 1974.
10. **Francis, C. A.,** Variety development for multiple cropping systems, *Crit. Rev. Plant Sci.,* 3, 133, 1986.
11. **Gutierrez, U., Infante, M., and Pinchinat, A.,** *Situación del Cultivo de Frijol en America Latina,* Boletin Informe ES-19, CIAT, Cali, Colombia, 1975.
12. Instituto Interamericano de Ciéncias Agrícolas, Reunión Técnica sobre Programación de Investigación y Extensión en Frijol y Otras Leguminosas de Grano para América Central, Publ. Zn, 112-65, Bol 2, IICA, Turrialba, Costa Rica, 1969.
13. **Francis, C. A., Flor, C. A., and Temple, S. R.,** Adapting varieties for intercropping systems in the tropics, in *Multiple Cropping,* Spec. Publ. 27, Papendick, R. I., Sanchez, P. A., and Triplett, G. B., Eds., American Society of Agronomy, Madison, WI, 1976, 235.
14. **Okigbo, B. N. and Greenland, D. J.,** Intercropping systems in tropical Africa, in *Multiple Cropping,* Spec. Publ. 27, American Society of Agronomy, Madison, Wisconsin, 1976, 63.
15. **Steiner, K. G.,** *Intercropping in Tropical Smallholder Agriculture: with Special Reference to West Africa,* German Agency for Technical Cooperation (GTZ), D-6236, Eschborn, Germany, 1982.
16. **Nye, P. H. and Greenland, D. J.,** *The Soil Under Shifting Cultivation,* Commonwealth Agric. Bur. Tech. Commun. 51, Harpenden, U.K., 1969.
17. **Jameson, J. D.,** *Agriculture in Uganda,* Uganda Government Ministry of Agriculture and Forestry, Oxford University Press, London, U.K., 1970.
18. **Beets, W. C.,** *Multiple Cropping and Tropical Farming Systems,* Westview Press, Boulder, CO, 1982.
19. **Perkens, D. H.,** *Agricultural Development in China,* Aldine, Hawthorne, New York, 1969.
20. **Gomez, A. A. and Gomez, K. A.,** *Multiple Cropping in the Humid Tropics of Asia,* IDRC, Ottawa, Canada, IDRC-176e, 1983.
21. **Harwood, R. R. and Price, E. C.,** Multiple cropping in tropical Asia, in *Multiple Cropping,* Papendick, R. I., Sanchez, P. A., and Triplett, G. B., Eds., Spec. Publ. 27, American Society of Agronomy, Madison, WI, 1976, 11.
22. **Hardy, F.,** *Suelos Tropicales, Pedologia Tropical con Énfasis en América,* Herrero Hermanos Sucs., S. A., Mexico, 1970.
23. **Norman, D. W.,** Rationalizing mixed cropping under indigenous conditions: the example of Northern Nigeria, *J. Dev. Stud.,* 11, 3, 1974.
24. **Rodale, R.,** Internal resources and external inputs: the two sources of all production needs, Workshop Rept.: Regenerative Farming Systems, USAID, Washington, D.C., Published by Rodale Institute, Emmaus, PA, 3, 1985.
25. **Francis, C. A. and King, J. W.,** Cropping systems based on farm-derived, renewable resources, *Agric. Syst.,* 27, 67, 1987.
26. **Gilbert, E. H., Norman, D. W., and Winch, F. E.,** *Farming systems research: a critical appraisal,* MSU Rural Devel. Papers, No. 6, Michigan State University, Lansing, 1980.
27. **Andrews, D. J. and Kassam, A. H.,** The importance of multiple cropping in increasing world food supplies, in *Multiple Cropping,* Papendick, R. I., Sanchez, P. A., and Triplett, G. B., Eds., Spec. Publ. 27, American Society of Agronomy, Madison, WI, 1976, 1.
28. **Francis, C. A., Ed.,** *Multiple Cropping Systems,* Macmillan, New York, 1986.
29. **Donald, C. M.,** Competition among crop and pasture plants, *Adv. Agron.,* 15, 1, 1963.
30. **Marshall, B. and Willey, R. W.,** Radiation interception and growth in an intercrop of pearl millet/groundnut, *Field Crops Res.,* 7, 141, 1983.
31. **Trenbath, B. R.,** Resource use by intercrops, in *Multiple Cropping Systems,* Francis, C. A., Ed., Macmillan, New York, 1986, 57.
32. **Trenbath, B. R.,** Plant interactions in mixed crop communities, in *Multiple Cropping,* Papendick, R. I., Sanchez, P. A., and Triplett, G. B., Eds., Spec. Publ. No. 27, American Society of Agronomy, Madison, WI, 1976, 129.
33. **Trenbath, B. R.,** Light-use efficiency of crops and the potential for improvement through intercropping, in *Proc. Int. Workshop on Intercropping,* Willey, R. W., Ed., ICRISAT, Hyderabad, India, 141, 1979.
34. **Sanchez, P. A.,** *Properties and Management of Soils in the Tropics,* Wiley & Sons, New York, 1976.
35. **Oelsligle, D. D., McCollum, R. E., and Kang, B. T.,** Soil fertility management in tropical multiple cropping, in *Multiple Cropping,* Papendick, R. I., Sanchez, P. A., and Triplett, G. B., Eds., American Society of Agronomy, Special Publ. No. 27, Madison, Wisconsin, 275, 1976.
36. **Barker, T. C. and Francis, C. A.,** Agronomy of multiple cropping systems, in *Multiple Cropping Systems,* Francis, C. A., Ed., Macmillan, New York, chap. 8, 161, 1986.
37. **Altieri, M. A. and Liebman, M.,** Insect, weed, and plant disease management in multiple cropping systems, in *Multiple Cropping Systems,* Francis, C. A., Ed., Macmillan, New York, 1986, 183.
38. **Bantilan, R. T., Palada, M. C., and Harwood, R. R.,** Integrated weed management. I. Key factors affecting crop-weed balance, *Philipp. Weed Sci. Bull.,* 1, 14, 1974.
39. **Hasse, V. and Litsinger, J. A.,** The influence of vegetational diversity on host finding and larval survivorship of the Asian corn borer, *Ostrinia furnacalis,* IRRI Saturday Seminar, Entomology Dept., Los Baños, Philippines, 1981.
40. **Altieri, M. A., Francis, C. A., van Schoonhoven, A., and Doll, J. D.,** A review of insect prevalence in maize *(Zea mays)* and bean *(Phaseolus vulgaris)* polycultural systems, *Field Crops Res.,* 1, 33, 1978.
41. **Borlaug, N. E.,** The use of multilineal or composite varieties to control airborne epidemic diseases of self-pollinated crop plants, in *Proc. 1st International Wheat Genetics Symposium,* University of Manitoba, Winnepeg, Canada, 12, 1959.
42. **Jensen, N. F.,** Intra-varietal diversification in oat breeding, *Agron. J.,* 44, 30, 1952.
43. **Gliessman, S. R.,** Allelopathic interactions in crop/weed mixtures: applications for weed management, *J. Chem. Ecol.,* 9, 991, 1983.
44. **de Wit, C. T. and van den Bergh, J. P.,** Competition among herbage plants, *Neth. J. Agric. Sci.,* 13, 212, 1965.

45. **Willey, R. W.**, Intercropping — its importance and research needs. I. Competition and yield advantage, *Field Crops Abstr.*, 32, 1, 1979.
46. **Willey, R. W.**, Intercropping — its importance and research needs. II. Agronomy and research priorities, *Field Crops Abstr.*, 32, 73, 1979.
47. **Oyejola, B. A. and Mead, R.**, Statistical assessment of different ways of calculating land equivalent ratios (LER), *Exp. Agric.*, 18, 125, 1981.
48. **Oyejola, B. A.**, Some statistical considerations in the use of the land equivalent ratio to assess yield advantages in intercropping, Ph.D. thesis, University of Reading, Reading, UK, 1983.
49. **Mead, R.**, Statistical methods for multiple cropping, in *Multiple Cropping Systems*, Francis, C. A., Ed., Macmillan, New York, 1986, 317.
50. **Mead, R. and Willey, R. W.**, The concept of a land equivalent ratio and advantages in yields from intercropping, *Exp. Agric.*, 16, 217, 1980.
51. **Harwood, R. R.**, *Small Farm Development*, Westview Press, Boulder, CO, 1979.
52. **Altieri, M. A.**, *Agroecology: the scientific basis of alternative agriculture*, Div. Biol. Control, University of California, Berkeley, 1983.
53. **Murdoch, W. W.**, Diversity, complexity, stability, and pest control, *J. Appl. Ecol.*, 12, 795, 1975.
54. **Smith, M. E. and Francis, C. A.**, Breeding for multiple cropping systems, in *Multiple Cropping Systems*, Francis, C. A., Ed., Macmillan, New York, 1986, 219.
55. **Allard, R. W.**, *Principles of Plant Breeding*, John Wiley & Sons, New York, 1960.
56. **Briggs, F. N. and Knowles, P. K.**, *Plant Breeding Methods*, Reinhold, New York, 1965.
57. **Hart, R. D.**, Ecological framework for multiple cropping research, in *Multiple Cropping Systems*, Francis, C. A., Ed., Macmillan, New York, 1986, 40.
58. **Odum, E. P.**, *Fundamentals of Ecology*, 3rd ed., W. B. Saunders, Philadelphia, 1971.
59. **Gliessman, S. R.**, Plant interactions in multiple cropping systems, in *Multiple Cropping Systems*, Francis, C. A., Ed., Macmillan, New York, 1986, 82.
60. **Goodman, D.**, The theory of diversity-stability relationships in ecology, *Q. Rev. Biol.*, 50, 237, 1975.
61. **Lynam, J. K., Sanders, J. H., and Mason, S. C.**, Economics and risk in multiple cropping, in *Multiple Cropping Systems*, Francis, C. A., Ed., Macmillan, New York, 1986, 250.
62. **Sanders, J. H. and Johnson, D. V.**, Selecting and evaluating new technology for small farmers in the Colombian Andes, *Mountain Res. Dev.*, 2, 313, 1982.
63. **Walker, T. S., Singh, R. P., and Jodha, N. S.**, Dimensions of farm-level diversification in the semi-arid tropics of rural south India, Economics Program Prog. Rept. 51, ICRISAT, Patancheru, India, 1983.
64. **Stoop, W. A., Pattanayak, C. M., Matlon, P. J., and Root, W. R.**, A strategy to raise the productivity of subsistence farming systems in the West African semi-arid tropics, in *Sorghum in the Eighties*, House, L. R., Mughogho, L. K., and Peacock, J. M., Eds., ICRISAT, Patancheru, India, 1982, 519.
65. **Flinn, J.**, Agronomic considerations in cassava intercropping research, in *Intercropping with Cassava: Proceedings International Workshop*, Weber, E., Nestle, B., and Campbell, M., Eds., Trivandrum, India, Int. Dev. Res. Center, Ottawa, Canada, 1979.
66. **Francis, C. A. and Sanders, J. H.**, Economic analysis of bean and maize systems: monoculture versus associated cropping, *Field Crops Res.*, 1, 319, 1978.
67. **Rao, M. R. and Willey, R. W.**, Evaluation of yield stability in intercropping studies on sorghum and pigeon pea, *Exp. Agric.*, 16, 105, 1980.
68. **Harwood, R. R.**, The integration efficiencies of cropping systems, in *Sustainable Agriculture and Integrated Farming Systems*, Michigan State University Press, East Lansing, MI, 1984.
69. **Francis, C. A., Harwood, R. R., and Parr, J. F.**, The potential for regenerative agriculture in the developing world, *Am. J. Altern. Agric.*, 1, 65, 1986.
70. **Dalrymple, D. G.**, Development and spread of high-yielding rice varieties in developing countries, Bureau of Science & Technology, U. S. Agency for Int. Dev., Washington, D.C., 1986.
71. **Francis, C. A.**, Research and development strategies for a sustainable agriculture, USAID International Centers Day, Washington, D.C., Oct. 21, 1987.

16 INDIGENOUS SOIL MANAGEMENT IN THE LATIN AMERICAN TROPICS: NEGLECTED KNOWLEDGE OF NATIVE PEOPLES

Susanna B. Hecht

I. INTRODUCTION

During the last 20 years, more than 20 million ha of lowland forests have been converted to other land uses in the Amazon basin.[1] Most of this land eventually ends up in cattle pastures which rapidly become degraded and unproductive. The potential long term global impacts of this process of deforestation and degradation include changes in the global carbon balance, atmospheric moisture recycling, and also declines in hydrological resources, and genetic diversity. While these effects are speculative and the subject of much debate, the most consistently documented impact of deforestation is the degradation of soil resources once the nutrient flush from conversion of forest to pasture or agriculture is over. The main forms of regional agriculture, cattle raising and annual cropping that follow forest conversion are very unstable and declining soil fertility is frequently identified as a factor in agricultural failure.[2-7] As soil nutrients become exhausted, and the costs of weeding become more arduous, farmers and ranchers abandon old areas and clear new ones and thus create an ever-expanding front of forest removal and land degradation.

This paper focuses on two main themes that relate to soil management issues in small farm development. First it discusses the indigenous vs. modernization approaches to soil resource management in Amazonian research and development strategies by focusing on the agricultural practices of a native group in the Brazilian Amazon, the Kayapo Indians. Kayapo systems are structurally compared with the dominant agricultural systems that prevail in Amazonia. Then, the outcomes of Kayapo, colonist and livestock systems are compared in terms of soil fertility and yields. These results show the greater yields and sustainability of the Kayapo land management. While indigenous systems are complex, the principles that underlie them are not. Native land management models could inform land resource management approaches in a fuller way than they have so far.

II. AMAZON SOILS AND RESEARCH APPROACHES: FROM TRANSNATIONAL TO TRIBAL PARADIGMS

The dynamics of agricultural and pasture failure in the Amazonia are extremely complex and are not uniquely determined by soil parameters.[8,9] What is clear, however, is that Amazonian soils for the most part are extremely poor, and soil constraints are severe for many crops grown under current cropping systems. As Table 1 indicates, more than 90% of Amazonian soils are deficient in phosphorus and nitrogen, about 75% are deficient in potassium, have serious problems of aluminum toxicity, or low levels of calcium and magnesium. Only about 4% of the soils of the Amazon basin exhibit no major agricultural constraints. The problems inherent in conventional crop-soil management in areas with such poor soils have given tropical zones a reputation for "fragility". This perception ignores the resiliency of many tropical forest formations, and overlooks the fact that indigenous Amazonian populations have developed complex systems of agriculture and intensive soil management that have been able to overcome these difficulties.

The dramatic nature of soil degradation after forest conversion is recognized in much of the regional research on tropical agronomy. Research related to soil management is a central focus of most agronomic research stations in the humid tropics, and indeed more than 60% of the research budgets involve surveying, mapping and classifying soils, and fertilizer and management trials. This research, along with climatic data and germplasm selection, is viewed as essential for developing the scientific basis for technical transfer of improved crops and cropping systems. In the main, researchers argue that soil constraints, considered to be the most critical agronomic limitation, can be overcome by the application of fertilizers and other modern inputs.[10-12] One of the main justifications of conventional soil research is the idea that if production systems, especially annual production systems, could be stabilized through the use of modern inputs, then the migration and deforestation linked to cropping failures would be diminished. Hence the overwhelming preponderance of fertilizer and varietal trials in tropical agricultural research stations.

A. The Yurimaguas Model

The most famous Amazonian example of this approach is that developed by Sanchez's group in Yurimaguas.[13-15] Following the U.S. Land Grant College model of agronomic research and development, soil and crop management strategies focus on target crops or soil problems, with minimal reference to knowledge and land use practices developed by local populations. This kind of approach has generated useful information on crop-soil dynamics for specific tropical zones, but as a strategy it concentrates primarily on what scientists have been able to learn on research stations through the application of western scientific techniques to rather narrowly defined pedological/agronomic problems. The overriding response to the soil management problems has emphasized agronomic techniques rooted in temperate zone agricultural intensification models that consist principally of fertilizer applications.[16-19] In approaching soil issues in this manner, a large body of local knowledge about soil potential and management has been ignored, and the social and economic contexts in which most agriculturalists in the humid tropics must function is often overlooked.

There are a number of reasons why an approach based on soil management which focuses on chemical fertlizers such as the Yurimaguas model is open to question. Some of these issues have been outlined elsewhere,[20,21] but include agronomic reasons such as (1) soil nutrient imbalances, and micronutrient difficiencies that cannot be easily monitored by most Amazonian peasants; (2) pest outbreaks that can reduce yields regardless of soil management; (3) erosion problems; and (4) physical changes in soil properties. Institutional problems include: (5) availability of inputs at the proper time; (6) access to inputs; and (7) adequate quality of inputs. Finally, market factors such as: (8) adequate return; (9) affordable transport costs; and (10) affordable credit are also important. Broader structural questions that impinge on the use of such technologies may involve problems of land title. Peasants with insecure title may not want to invest in costly, and for them, often risky technologies when they may be expelled from their land. The very low incomes of most Amazonian inhabitants, about U.S. $1,000 and less mean that the $250 to 300 costs for fertilizer per hectare represent almost one third of an average household's income for producing commodities whose prices are often controlled, whose marketing is difficult and costly and whose production in Amazonia is quite risky. While credit is usually proposed as the solution to this impasse, only about 4% of Amazonian peasants receive credit. The inherent riskiness of annual crops often reduces the use of high technology packages[22] and the opportunity costs of investment of labor and cash in annual cropping systems limit the adoption of such crop/soil technologies. The Yurimaguas approach has undergone several modifications over the years and increasingly incorporates practices that are less input demanding.[23]

Some analysts are beginning to argue that tropical land use models should be based on land management methods developed by local populations and presumably more closely integrated into the dynamics of tropical ecological systems, and the needs and constraints of local peoples.[24-26] Researchers increasingly recognize the role of native populations in the development of ecologically sound, productive land uses. However, the amount of research on indigenous soil management techniques, and how well they perform in the tropics is a research question that has received little attention in spite of the enormous budgets devoted to tropical pedology and agronomy. One might well ask why such a central issue in tropical development has so systematically ignored local knowledge systems. A bag of fertilizer is not inherently a "bad" technology for Amazonian soil management but it is not the only one available. The real question is what are the range of technical options for the Amazon. Existing native practices could suggest useful hypotheses for land management that could be tested under experiment station conditions. Indigenous systems could be used as a springboard for incorporating the best of both systems. It is a great irony that perhaps the finest studies of Amazonian agriculture, the Yurimaguas group, and the team headed by Denevan and Padoch[27] have interacted so little.

B. The Kayapo Model*

The characteristics of the the Kayapo and Yurimaguas systems are outlined in Table 2. This table shows the rich array of techniques and strategies for managing relatively low fertility soils and the points of intersection between the modernization and traditional model. The Kayapo system includes a soil taxonomy, selection for varietal diversity, and complex planting pattern in space and time. Figure 1 shows an idealized version of the cropping pattern, often called concentric ring planting.[28,29] This pattern involves spatial segregation of some crops relay planting, and successional

* The Kayapo data were derived from yield measurements of crops in the field, household harvests, and informant estimates. Because Kayapo planting and harvests are continuous and our field presence was not, the numbers cited are probably underestimates. Colonist production data was derived from field interviews, and estimates from the Conceicao de Araguaia IBGE office, as well as generalized estimates on colonist agricultural productivity derived from government and academic literatures. Livestock data is derived from field research. The livestock and colonist field work was undertaken 25 km from Redencao in the direction of Gorotire. The Kayapo work has been underway since 1984.

TABLE 1
Main Soil Constraints in the Amazon under Native Vegetation

Soil constraint	Million hectares	% of Amazon
Phosphorus deficiency	436	90
Aluminum toxicity	353	73
Drought stress	254	53
Low potassium reserves	242	50
Poor drainage/flood hazard	116	24
High phosphorus fixation	77	16
Low cation exchange capacity	64	13
High erodibility	39	8
No major limitations	32	7
Slopes of over 30%	30	6
Laterite hazard if subsoil exposed	21	4

Source: Sanchez (1987).

TABLE 2
Agriculture Formations of the Kayapo

Village gardens
1. Household gardens
2. Swidden plots (Puru)
3. Successional plots derived from Swidden
4. Grave sites
5. Marantaeae gardens on hill slopes (Krai kam puru)

Planting associated with travel
6. Trails between villages and gardens
7. Trails between villages
8. Hunting/trekking trails
9. Planting around old camp sites

Forest planting
10. Natural ecosystem gaps (Ba kre ti)
11. Man-made gaps
12. Plantations in mature forest (viz Euterpe, Bertholetia)
13. Fruit groves "in memory of the dead"

Cerrado planting (Apete)

management techniques. Soil conserving practices are incorporated within this physical and temporal structure. These include the use of multiple cropping systems, crop rotations, crops like sweet potato whose vines cover the ground, concentrated tillage and soil turning in high use areas, direct additions of nutrients in the form of applications of ashes, mulches, residues, dung and enriched soils, transfering forest litter, composting, and controlled periodic, in-field burning. There are clearly points of similarity in the Yurimaguas/Kayapo systems in the use of crop residua, relay planting, nutrient additions and short fallows. The active "arsenal" of the Kayapo agricultural system is very rich and requires no purchased inputs.

There is a complex intellectual system that underlies the native management of soil resources, the ensemble of which is "ethnopedology". Ethnopedology includes the study of native land classification systems, management techniques and their variations and how the practical and theoretical knowledge is developed, expanded, encoded and reproduced. Land uses reflect some implicit or explicit assessment of the relative capabilities of soils and practical techniques ranging from crop selection to soil management to address these capabilities. Folk soil taxonomies are widespread (although generally underinvestigated) and generally correlate well with discernable quantitative differences among soil types.[28-33] Crop genetic selection and experimentation is impressively widespread in Amazonia.[37-39] An emerging body of ethnobiological work in Amazonia suggests that native populations developed complex systems of resource management that are ecologically sustainable,

FIGURE 1. Schematic view of Kayapo concentric ring agriculture.

yield well, and generate levels of income that exceed the regional average.[34-36]

The study of indigenous soil management has many of the advantages of classic soils research: (l) soil properties can be more precisely specified than vegetation processes; (2) edaphic data from one area can illuminate the probabilities of land use outcomes on similar sites more clearly than vegetation data; (3) soil data are particularly powerful tools for evaluating and comparing the impact of land management over time. In addition, insights from millenia of tropical land management experience that has survived the test of time and the vaguaries of environment can help inform new strategies and testable hypotheses. Given the power of this form of analysis, it is rather surprising that so little serious attention has been paid to native land management techniques. If Kayapo techniques can sustain productivity on a given site by maintaining soil fertility, or increase output through soil modification and crop management, these techniques could serve as the foundation for sustainable agricultural models for small scale farmers in Amazonia.

In the next section, Kayapo agricultural yields and soil dynamics are compared with the two dominant regional land uses: livestock and colonist agriculture.

III. COMPARING KAYAPO AGRICULTURE WITH COLONIST AGRICULTURE AND LIVESTOCK IN THE EASTERN AMAZON

The Gorotire Kayapo inhabit a region characterized by a complex geology and geomorphology in southern Para, Brazil. Lying at the interface between the Precambrian Brazilian shield and more recent metamorphic and sedimentary formations, several major soil orders are found within short distances. Most Kayapo agriculture is carried out on four main soil types: a high fertility alfisol, a relatively high fertility ultisol, a low fertility ultisol and a low fertility oxisol. Because the four soil orders managed by the Kayapo are similar to more than 80% of the Amazonian soils, the principles, the techniques and the impacts of Kayapo management could have wider implications for tropical soil management in the Amazon Basin.

The Kayapo designate 14 types of land use as "agriculture" (See Table 3). These various land uses are complex, and include ceremonial planting, reforestation, and trek gardens, as well as swidden plots. The Kayapo practice concentric ring/crop segregation agriculture based on sweet potatoes, manioc, yams, and perennials, periodically intercropped with maize, beans, cucurbits, introduced rice, and numerous other minor crops and ritual plants. Kayapo swiddens stay in active root crop production for about 5 years, and continue to contribute these products at reduced levels for as long as 11 years. This is an exceptionally long production period before an agricultural plot goes into fallow. This protracted production period is a function of eight main factors:

1. A mixture of short and long cycle cultigens and cultivars
2. Sequential harvesting and replanting
3. Root crop cultivars well adapted to fire (including fire tolerant manioc, sweet potatoes, yams, marantaceaes)
4. Systematic, differential, periodic burning within the agricultural field for the entire production sequence
5. Mulching
6. Nutrient additions
7. Agricultural structure
8. Manipulated fallows

TABLE 3
Kayapo and Yurimaguas Production Systems

	Kayapo	Yurimaguas
Soil classification	Yes	Yes
Clearing	Slash & burn	Slash & burn
Crop diversity	High	Low
Medium cycle crops (2—4 year production periods)	Yes	No
Arboreal species	Yes	Rarely
Polyvarietal planting	Yes	No
Plant structure	Concentric ring	Pure stand monocultures
Nutrient inputs	Yes — Ash, mulch, termite nests, litter, palm fronds	Yes — 30 kg N, 22 kg P, 48 kg K
Residue return	Yes — rice, corn stover, banana leaves, vines of yam, sweet potato, manioc peelings	Yes — Rice and cowpea stover
Cultivation practices		
Intercropping	Yes	Rarely
Relay planting	Yes	Yes
Mulching	Yes	No
Continuous planting	Yes	No
In-field burning	Yes	No
In-field mulch pits	Yes	No
Weed control		
Manual	Yes	Yes
Fire	Yes	No
Mulch	Yes	Sometimes
Allelopathy	Possibly	No
Scandent crops	Yes	No
Herbicides	No	Yes
Fallow	5—10 years	Yes — kudzu 1—2 years

Informants indicate that concentric zones facilitate the creation or manipulation of in-field microvariability that "fine tunes" soil nutrients to crop demands. Considering the dominant physical and chemical properties of the regional soils, slight lateral changes in those properties can be strongly reflected in the growth and productivity of most annual and bienniel crops. Concentric field architecture permits the use of specific soil nutrient management techniques in a controlled manner. For example, the frequent use of in-field burning throughout the agricultural cycle requires that particular crops be separated from others to control nutrient additions, and to minimize fire damage. Mulch application, specific nutrient additions, soil aerating, etc. can be more effectively practiced when crops are spatially segregated.

TABLE 4
Comparison of the Structure of Kayapo, Colonist and Livestock Production Systems

	Kayapo	Colonist	Livestock
Clearing	Slash & burn	Slash & burn	Slash & burn
Clearing Size	About 1 ha	2—5 ha	Up to 20,000 ha
Planting patterns			
Cropping zonation	Yes	Rarely	No
Continuous cropping	Yes	Sometimes	Yes
Continuous planting	Yes	Rarely	No
Relay cropping	Yes	Yes	No
Monocropping	No[a]	Often	Usually
Intercropping	Yes	Sometimes	No
Polyvarietal crops	Yes	Rarely	No
Arboreal species	Yes	Rarely	Rarely
Number of cultivated species in field	10—42+	5—10	1—5
Harvest pattern	Continuous	Pulsed	Pulsed
Soil conservation practices that maintain soil fertility	Yes, mulching, controlled infield burning, use of nutrient acculating species (legumes, maximiliana) application of soil fertility elements, use of crops with scandent habits, complex coplanting	Rarely	Rarely, occasional use of legumes
Main crops	Sweet potato, yams, manioc, maize, musa, beans, squash	Rice, manioc	*Panicum, Brachiaria*
Labor	40 md/ha	25 md/ha	4.5 md/ha

[a] Parts of the Kayapo system are monocultured in the concentric field, but the entire field is rarely monocropped.

A. Comparing Land Uses in the Eastern Amazon

The dominant forms of land use in the Amazon are pasture and short cycle agriculture, which are notorious for their lack of sustainability and low rates of economic return. The features of these systems are described elsewhere in the literature and will not be elaborated here except in summary form (See References 45 to 52). The features of Kayapo agriculture are outlined and compared with these two other land uses in Table 4. There are real structural differences in these production systems ranging from field pattern, crop species diversity, total time in production, use of arboreal species, and harvest patterns, to degree of agricultural integration into the larger economy. The table demonstrates a gradient of management intensity, ecological complexity and declining labor allocation per hectare. Because they are so different the production comparison focuses on production yields and proteins.

The edible harvest and protein yields for each of the land uses are outlined in Table 5 for periods of 5 and 10 years. The Kayapo yields per hectare over 5 years are roughly 200% higher than colonist systems, and 175 times that of livestock. Colonist agriculture rarely continues beyond 5 years, hence there is no comparison between the Kayapo and colonist in the 10-year period. However, Kayapo and livestock comparisons show that animal production is a mere 700 kilos of animal to more than 84 tons of Kayapo product.

TABLE 5
Production in Kilograms and Proteins per Hectare of Kayapo Colonist and Livestock Systems Over Time

Production system	5 years	10 years
Kayapo		
Production of all crops[1] (kg/ha)	61,750	84,050
Protein (kg/ha)[2]	1,248	1,704
Colonist		
Production of all crops[3]	21,800	—
Protein	602	—
Livestock		
Production (kg/ha)	350	700
Protein[2,4]	105	210
	(63)	(126)

[1] Estimates based on in-field measurement area harvest weights, household harvests, and informant estimates. The crops here include sweet potato, yam, manioc, plantains and bananas, maize, beans, squash, peanuts. Many other minor crops such as *Colocassia* and other tubers, papayas, watermelons, peppers, mangos, pineapples, are not in the calculation.

[2] Protein estimates derived from Wu Leung (1961), Dufour (1987), USDA (1981).

[3] Based on average yields of rice, maize and manioc in colonist agriculture in Amazonia.

[4] Assumes that virtually the entire animal, including the hide, is consumed, of which roughly 30% is protein. If the animal is dressed out it weighs about 60% of its liveweight. The protein yield is indicated in parentheses.

The data follow the same trends when protein production is analyzed. Kayapo protien yields from vegetable sources are roughly double those of colonists and more than ten times the protein production for the entire animal. The protien per 100 g of beef is roughly 30 g (USDA). If the dressed out animal, (that is one with hide, bones and offal removed, usually about 40% of live animal weight) is used as the basis of the analysis, the kilos of pure protien produced over 5 years are a scandalous 63 kilos/ha. In 10 years, using these calculations 1 ha of pasture has produced less than a ton of meat, and slightly more than 100 kilos of protein, or roughly 5% of the protein generated by the Kayapo system. Incidently, Kayapo gardens in their later phases become habitat for animals such as agouti, peccaries and deer, and thus producers of animal protein during the fallow.[45] This protien production is not taken into account in this analysis.

The Kayapo system is based on root crops, especially sweet potatoes which are very productive in the tropics and also drought tolerant so that they can continue production even during the dry season. The sheer volume of production assures carbohydrate adequacy and with minor supplements, protein sufficiency. While these crops are often reviled for their low protein contents,[46] nutritional studies by Heung[47] on adult Yami tribesmen show that diets were nutritionally adequate when subjects are 2.5 kilos of sweet potatoes a day. Several studies in New Guinea show that protein content varies significantly between cultivars,[48] and that intestinal flora of some groups of New Guinean sweet potato eaters may have been able to fix nitrogen.[49]

TABLE 6
Changes in Soil Fertility Elements in Kayapo, Colonist, and Livestock Systems

	Forest	Year one	Year five	Year ten
Kayapo				
Ph	4.7	5.4	5.6	5.4
N%	0.05	0.07	0.03	0.06
P ppm	1.0	5.0	3.0	3.16
K*	0.17	0.37	0.23	0.33
Ca*	0.75	1.55	1.31	1.90
Mg*	0.31	0.89	0.97	1.67
Colonist				
Ph	4.8	5.4	5.4	—
N%	0.12	0.10	0.06	—
P ppm	1.2	6.0	1.0	—
K*	0.12	0.32	0.09	—
Ca*	1.90	2.1	1.30	—
Mg*	0.34	0.59	0.42	—
Livestock				
Ph	4.7	5.5	5.2	5.0
N%	0.10	0.07	0.06	0.06
P ppm	2.0	7.0	2.0	1.0
K*	0.10	0.17	0.10	0.05
Ca*	1.3	1.7	0.92	0.64
Mg*	0.42	0.65	0.60	0.30

Data derived from soil samples taken at Gorotire (Kayapo), Nixdorf Fazenda located near Rendencao (livestock), and Colonist agriculture of squatters on the Nixdorf Ranch. Colonist and livestock agriculture sampled in 1982.

B. Soil Effects

The next issue is what impact this high production has on soil properties compared with colonist and livestock systems. Soil samples were taken on sites with similar soil charcteristics, in this case dystrophic paleudults. Adjacent forest sites were used as "controls" and samples were collected on areas in the first year of production, fifth year of production and tenth year. Sampling areas were roughly 1 ha in size, and the samples were collected randomly. Ten samples were taken per "treatment".

Table 6 shows several clear trends. First, pH tends to improve with burning, and this effect persists over time in all the systems, mainly as the result of decomposition of larger tree boles. In the Kayapo case, higher pHs are maintained for a longer time, probably due to the continual in-field burning, the cooking within the fields themselves, and importing wood for these cooking fires. Nitrogen levels are very low in all three systems, but the importance of low nitrogen is less pronounced in Kayapo agriculture because of its emphasis on root crops rather than grains for most of the production cycle. Rice requires about 23 kg of nitrogen per ton, while manioc and sweet potato remove 3.7 and 4.3 kg per ton, respectively, and require little nitrogen for good production.[58] Phosphorus levels are low in all the soils but Kayapo production maintains higher levels of P over time. Potassium is a very labile element, it is easily leached in tropical conditions and one that is closely associated with productivity in root crops like manioc and sweet potatoes. The use of high K mulches such as *Maximiliana* leaves, crop residues, and cooking ash additions[59] maintains K levels high to compensate for the production losses. This element stays at levels equal to the first year of production because of these practices. Calcium and magnesium levels are also maintained over time, in spite of the crop off take because of the soil management techniques of the Kayapo. Kayapo land management produces more product and protein at less environmental cost than the livestock or colonist systems. While the labor demands are greater in this system it does assure the reproduction of subsistence and tropical forest. Tropical soils are difficult, but not impossible to manage. The Kayapo have much to teach us about how this can be done. There are no big mysteries to Kayapo agricultural systems, they incorporate the features of fine agricultural practices.

The Kayapo have developed several soil management techniques that are of special interest for small farm develeopment. What is of interest is the array of potential inputs, mulches and cropping patterns that produce the high yielding sustainable systems.

Those involved in small farm development might approach the question of soil management in a different way than they have to date. The basic step is really to build on what is already known, and to test native knowledge in controlled circumstances. What are the folk classifications and soil lore found in the area? How well do these coincide with "western knowledge" and are there "hidden logics" (plant disease control, fertility enhancement, etc) behind particular practices?

Fire is one of the most powerful tools that tropical agriculturalists have, and yet we know very little about the subtleties of fire management in native agricultural systems. While the current approach to fire in tropical agroecosystems borders on a "Smokey the Bear" fire suppression approach, there are arrays of burning technologies that may be extremely appropriate in some environments, particularly when associated with fire tolerant cultivars.

Mulching and alternative input systems are necessary for sustainable tropical systems. Local systems can provide valuable clues to what species are most likely to be useful, and how, when and in what form they should be used.

Finally, developers should look at the landscapes and agricultural systems around them and ask what it takes to keep each running, and what features of each system could potentially contribute to the others. The Kayapo, and rural populations don't just manage agricultural fields, they manage landscapes.

IV. CONCLUSION

This paper emphasizes two main points. First, indigenous knowledge systems and the agriculture on which they are based are rich in management information about nutrient poor tropical soils. Second, these systems are better producers of calories and proteins than any of the alternatives without damaging the resource base.

The real question becomes why do some production systems prevail over others. The livestock story has been the subject of a spate of recent articles,[51-56] and colonist migrations are a central feature of the Amazon's 20th century history. These land uses and the deforestation associated with them are driven by an array of social and economic factors ranging from government policy, credit incentives, land speculation, and rural violence among others. The political economies of accumulation and subsistence associated with the new occupation of the Amazon drive the destructive and low productivity patterns that we see. These patterns are related to processes that have very little to do with questions of technological adequacy, in spite of the discourse that posits production failure and land degradation as a central driving force behind deforestation. For this reason, neither the Yurimaguas nor indigenous models will be very effective at alterating this particular deforestation process.

This does not make exploration of sustainable management systems a trivial exercise. There are technological questions about land management in the Amazon that could increase the production or at least maintain it a bit longer. However, hundreds of millions of dollars have been funneled into surveys and experiments which have not made colonists agriculture more stable, or livestock more productive. At the same time, the total budget for exploring indigenous Amazonian soil management which is based on locally available inputs, and whose principles are straight forward, has recieved virtually no support. Why this critical source of knowledge has been systematically ignored requires explanations, few of which have to do with the welfare of Amazonian populations.

Finally, the populations which have created the cultivars and sustainable land resource management techniques are under extraordinary pressures. Our society pays for libraries, universities, and research facilities of all types. It should be prepared to protect the producers of the sustainble Amazon land use systems and the cultures that support them. No investment in R and D is likely to have greater return.

ACKNOWLEDGMENTS

Funds for this project were generously provided by the Wenner-Gren Foundation, Resources for the Future, the U.S. Fullbright, and World Wildlife Fund. The Kayapo field work was carried out with Darrell Posey through the Kayapo Project of the Museu Goeldi and the Federal University of Maranhao. This paper is modified from one presented at LASA's 1988 Fragile Lands Symposium.

REFERENCES

1. **INPE,** Relatorio de Atividades do Projeto IBDF-INPE 1987, Sao Jose dos Campos, Brazil. 1987.
2. **Sanchez, P.,** *Management and Properties of Tropical Soils*, John Wiley & Sons, New York, 1976.
3. **Moran, E.,** *Underdeveloping the Amazon*, University of Indiana Press, Bloomington, 1982.
4. **Hecht, S.,** Deforestation in the Amazon Basin: magnitude and soil nutrient effects, *Studies in Third World Societies*, 13(1), 61—101, 1982.
5. **Hecht, S.,** Environment, development and politics: the livestock sector in the Eastern Amazon, *World Development*, 13(6), 663—685, 1985.
6. **Buschbacher, R.,** Tropical deforestation and pasture development, *Bioscience*, 36(1), 22—28, 1986.
7. **Fearnside, P.,** *Human Carrying Capacity in the TransAmazon Highway*, Columbia University Press, New York, 1986.
8. **Hecht, S.,** *Reanalysing colonist attrition*, forthcoming, *World Development*.
9. **FAO,** *Polonoreste Agricultural Evaluation*, Rome.
10. **Sanchez, P.,** *Management of Acid Soils in the Humid Tropics*, Acid Soils Inaugural Workshop, Brasilia, Brazil, 1985.
11. **Sanchez, P. and J. Benites,** Low input cropping systems for acid soils of the humid tropics, *Science*, 238, 1521—1527, 1987.
12. **Sanchez, P., Bandy, D., Villachica, J., and Nicolaides, J.,** Soils of the Amazon and their management for continuous crops production, *Science*, 216, 812—827, 1982.

13. **Sanchez, P.,** Management of Acid Soils in the Humid Tropics, Acid Soils Inaugural Workshop, Brasilia, Brazil, 1985.
14. **Sanchez, P., Bandy, D., Villachica, J., and Nicolaides, J.,** Soils of the Amazon and their management for continuous crops production, *Science*, 216, 812—827, 1982.
15. **Sanchez, P. and J. Benites,** Low input cropping systems for the tropics, *Science*, 238, 1521—1527, 1987.
16. **Falesi, I.,** *Ecossistema de Pastagem Cultivada*, CPATU, Belem, Brazil.
17. **Sanchez, P., Bandy, D., Villachica, J., and Nicolaides, J.,** Soils of the Amazon and their management for continuous crop production, *Science*, 216, 812—827, 1982.
18. **Sanchez, P. and Benites, J.,** Low input cropping systems for the tropics, *Science*, 238, 1521—1527, 1987.
19. **Sanchez, P.,** *Management of Acid Soils in the Tropics*, paper presented at the Acid Soils Network Inaugural Workshop, Brasilia, Brazil, 1985.
20. **Fearnside, P.,** Rethinking continuous cultivation in Amazonia, *Bioscience*, 27(3), 209—213, 1987.
21. **Altieri, M.,** *Agroecology*, Westview Press, Boulder, CO, 1987.
22. **Scott, J.,** *The Moral Economy of the Peasant*, Yale University Press, New Haven, 1978.
23. **Sanchez, P. and Benites, J.,** Low input cropping systems for acid soils of the tropics, *Science*, 238, 1521—1527, 1987.
24. **Altieri, M.,** *Agroecology*, Westview Press, Boulder, CO, 1987.
25. **Clay, J.,** Indigenous models of development and forest management, Cultural Survival Report #27, Cambridge, 1988.
26. **Alcorn, J.,** Development policy, forests and peasant farms, *Econ. Bot.*, 38, 389, 1984.
27. **Denevan, W. and Padoch, C., Eds.,** *Swidden-Fallow Agroforestry in the Peruvian Amazon*, Advances in Economic Botany, New York Botanical Garden, 1988.
28. **Stocks, A.,** Candoshi and Cocamilla swiddens in eastern Peru, *Hum. Ecol.*, 11(1) 69—84, 1983.
29. **Beckerman, S.,** A note on ringed fields, *Hum. Ecol.*, 12(2) 203—206.
30. **Johnson, A.,** Ethnoecology and planting practices in a swidden agricultural system, in *Indigenous Knowledge Systems and Development*, Brokenshaw, D., Warren, D., and Werner, O., Eds., 49-67, U.S.A. Press, Washington, DC, 1982.
31. **Christanty, L.,** Shifting cultivation and tropical soils, in *Traditional Agriculture in Southeast Asia*, Marten, G., Ed., Westview Press, Boulder, CO, 226—239, 1987.
32. **Behrens, C.,** The relation between Shipibo and western soil classification, *Am. Ethnol.*, in press, 1988.
33. **Hecht, S. B. and Posey, D.,** Preliminary analysis of Kayapo soil management, *Adv. Econ. Bot.*, 7, 179, 1989.
34. **Moran, E.,** Developing the Amazon, University of Indiana Press, Bloomington, 1982.
35. **Toledo, V. and Barrera-Bassals, J.,** *Ecologia & Desarrollo Rural en Patzucuaro*, Instituto de Biologia, Mexico, 1984.
36. **Carter, J.,** *New Lands, Old Traditions*, University of Florida, Gainesville, 1969.
37. **Chernela, J.,** Os cultivares de mandioca na area do Uaupes, *Suma Etnologica Brasileira*, Vozes, Sao Paulo, 1986.
38. **Kerr, W. and Posey, D.,** Additional notes on Kayapo agriculture, *Interciencia*, 9(6), 392—400, 1984.
39. **Boster, J. S.,** Classification, cultivation, and selection of Aguaruna cultivars of *Manihot esculenta*, *Adv. Econ. Bot.*, 1, 34—48.
40. **Anderson, A., Gely, A., Strudwick, J., Sobel, G., and Pinto, M.,** Un sistema agroflorestal na varzea do estuario amazonico, *Acta Amazon.*, 15(1), 195—224, 1985.
41. **Denevan, W. and Padoch, C.,** Swidden fallow agroforestry in the Peruvian Amazon, *Adv. Econ. Bot.*, Vol. 5, 1988.
42. **Hecht, S. B. and Schwartzman, S.,** The Good, the Bad and the Ugly: Extraction, Colonist Agriculture and Livestock in Comparative Perspective, GSAUP working paper. 1988.
43. **Hecht, S. B. and Posey, D.,** Preliminary analysis of Kayapo soil management, *Adv. Econ. Bot.*, in press, 1988.
44. **Clay, J.,** Indigenous models of agriculture and forest development, Cultural Survival Report #27, 1988.
45. **Redford, K. and Richards, J.,** The game of choice, *Am. Anthropol.*, 89(3), 650—667, 1987.
46. **Gross, D.,** Protein capture and cultural development in the Amazon Basin, *Am. Anthropol.*, 77, 526—549, 1985.
47. **Hueng, D.,** Nutritive value of sweet potato, International Symposium on Sweet Potato, Taiwan, 1983.
48. **Hayward and Nakikus,** Protein, Energy and Nutrition in Papua New Guinea. Papua New Guinea Food Crops Conference, Port Moresby, New Guinea, 1981.
49. **Heung and Lee, N.,** Evidence of No Intestinal N. Fixation in Sweet Potato Eaters, *Am. J. Clin. Nutr.*, 32, 1741, 1979.
50. **Hecht, S. B. and Posey, D.,** Preliminary analysis of Kayapo soil management, *Adv. Econ. Bot.*, 7, 179, 1989.
51. **Buschbaker, R.,** Tropical deforestation and pasture development, *Bioscience*, 36(1), 22—28.
52. **Browder, J.,** The social costs of rainforest destruction, *Interciencia*, 13(3), 115—120, 1988.
53. **Hecht, S. B.,** Cattle Ranching Development in the Eastern Amazon: Evaluation of a Development Policy, Ph.D. thesis, University of California, Berkeley, 1982.
54. **Hecht, S. B.,** Environment, Development and Politics; the Livestock Sector in the Eastern Amazon, *World Development*, 13(6), 663—685, 1985.
55. **Hecht, S. B., Norgaard, R., and Possio, G.,** The Economics of Cattle Ranching in Eastern Amazonia, *Interciencia*, 47(1), 25—85, 1988.
56. **Repetto, R.,** *The Forest for the Trees?* WRI, Washington, 1988.

The Dynamics of Traditional Agriculture: Case Studies

Mountain Environments

17 CROP DEVELOPMENT IN CENTERS OF DOMESTICATION: A CASE STUDY OF ANDEAN POTATO AGRICULTURE

Stephen B. Brush

I. INTRODUCTION

Cradle areas of domestication of the world's crops have played an important role in several generations of agricultural research. The areas where different crops originated have been identified, crop origins have been dated, the processes of domestication that developed a few wild species into cultivars have been generally defined, and similarities among these areas have been recorded.[1] Cradle areas are typified by environmental diversity and by marked seasonality of available moisture. Many of them are associated with mountains. They were originally identified because of the diversity of the crops that are grown within them.[2] It has been shown that the productivity of crops is lower in their cradle areas than outside.[3] Finally, virtually all are characterized by being economically impoverished and technologically traditional, in other words underdeveloped.

Although thousands of plants have been domesticated and used by humans over the past 8000 years, many crops and crop varieties have been dropped from the inventory of human diets. This trend may be accelerating with increasing urbanization, commercialization, and communication. It is also affected by the technology available to farmers. As more and more people depend on fewer plant species and varieties, the relative importance of the remaining staples increases dramatically.[4] This dependence is compounded by the decreasing portion of the human population directly engaged in food production. Our greater dependence on a few crops and fewer farmers has made the task of conserving the genetic resources found in cradle areas of domestication urgent.[5]

During the last 5 decades, a worldwide effort has been undertaken to collect and conserve crop genetic resources. A network of regional, national and international crop genetic resource centers has been created, and this network is estimated to hold a significant percentage of the germplasm of major crops.[6] This effort has been directed at gene bank conservation and has played relatively little heed to the processes of conservation and change that occur in the farming systems where the genetic resources originated. Research is now underway in several cradle areas of domestication, to study farmers' roles in generating and maintaining genetic resources,[7] including an emphasis on the farming system associated with crop genetic diversity. This chapter presents a case study of the farming system of the potato (*Solanum* spp.) in its cradle of domestication, the central Andes of Peru,[8] and examines the dynamics of potato production in a single valley on Peru's eastern Andean escarpment. This is the Tulumayo Valley whose headwaters rise in the Huatapallana glacier east of the city of Huancayo.

This research is important for small farm development in two ways. First, it documents how small-scale farmers manage their farming system in a marginal agroecological zone. Of particular relevance here is the understanding of decision-making regarding the choice of technology, exemplified by decisions to grow improved vs. traditional potato varieties. This choice is at the heart of all "green revolution" development strategies. Second, this research documents a case of farmer conservation of genetic resources. Such conservation can be seen as an important complement to off-site conservation in genetic resource centers.

II. ZONATION ALONG AN ANDEAN TRANSECT

The highly variable mountain environment of the Andean area can be divided by different criteria into ecological and agricultural production zones.[9-11] Besides latitude and longitude, the primary determinants of zonation are altitude, slope, soils and orographic features. A transect across the central Andes following the 12° S parallel eastward from

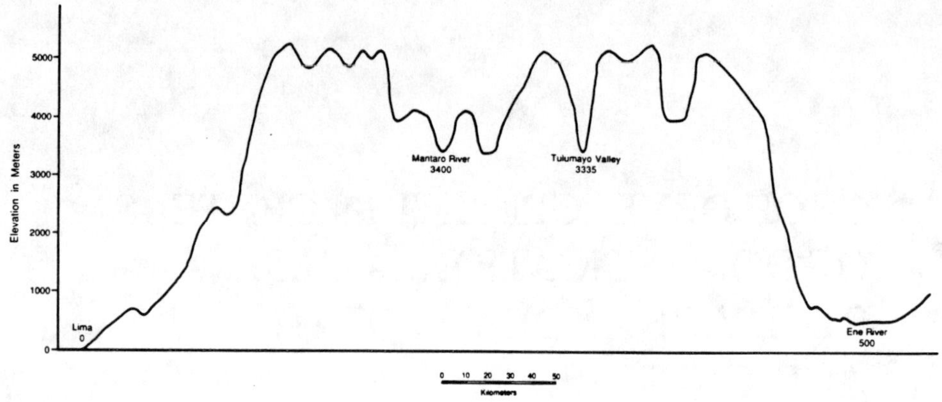

FIGURE 1. Cross section of the Peruvian Andes (12° South Latitude).

Lima (77° W) to the Ene River (74° W) demonstrates the major environmental and agricultural zones of the Andean region (Figure 1). This transect crosses five general environmental and land-use zones: coast, western escarpment, high plateau, inter-Andean valley, and eastern escarpment. Within each of these, however, there are numerous local zones which are derived from micro-environmental features and determine actual land-use.

At Lima, the climate is dominated by cold ocean currents. It is dry year round, and agriculture is only possible through irrigation with waters from the Rimac River. This agriculture is capitalized, without regular fallow and organized around private farms. It emphasizes truck crops for the large urban market: vegetables, maize *(Zea mays)*, flowers, and manioc *(Manihot esculenta)*.

Above 2000 m altitude, rainfed agriculture begins to appear. Cereals, especially maize and barley *(Hordeum* spp.), are the major crops in the lower elevations of rain-fed agriculture. In some locations maize is produced on irrigated bench terraces which date from the Inca period. This agriculture is less intensive than the coastal type, relying on fewer purchased inputs, with more frequent fallow and destined for consumption as well as the market. Communal activity and control begin to complement individual farm management, for instance in the operation of irrigation systems.

At between 3000 and 4000 m, tuber agriculture predominates. Besides the potato, three other Andean tubers are cultivated (oca — *Oxalis tuberosa*, mashua — *Tropaeolum tuberosum*, and ollucu — *Ullucus tuberosus)*. Here agriculture begins to shift to a greater subsistence orientation, purchased inputs are used less, and agriculture is less intensive than at lower elevations with short cropping cycles and regular fallow. Land-use in the upper elevations (>3600 m) is extensive in that it involves a short cropping sequence (1 to 3 years) and a long fallow (3 to 15 years). Communal activity and control in the agricultural cycles is significant, for instance in the ownership of land and the scheduling of the agricultural calendar. At these altitudes and above, herding becomes increasingly important, and above 4100 m, herding replaces agriculture. Much of the herding is done on communally managed pastures or on pastures managed by collectives that succeeded private estates which were expropriated during the Peruvian land reform (1969 to 1974).

Crossing the broad Mantaro Valley (3400 m) an inter-Andean pattern is evident. The orographic effects of the high *cordilleras* create a zone with an extended dry season. Maize and other cereals are grown on irrigated lands in the valley bottom, and agriculture is fairly intensive as it is geared for the large urban market of Huancayo. Tuber agriculture and communal activity become more and more important as one climbs above 3500 m.

Descending from the central *cordillera* along the eastern slopes of the Andes, moist narrow valleys are intensively cultivated in tuber crops. The intensity of this production, however, reflects dense populations and subsistence production as well as market production of potatoes. Mid-valley slopes are terraced through the use of hedgerows, by intentional slope modification and by the formation of linchets. Maize predominates on the valley floors above 2000 m, and coffee *(Coffea arabica)*, sugar *(Saccharum officinarum)*, coca *(Erythroxylon coca)*, fruits, and manioc are major crops below this. With the eradication of malaria and road construction along the Amazon periphery, the lower valleys of the eastern slopes have become important areas of migration and colonization for cash cropping, grazing livestock, and gold mining by Andean peoples. Along the western Amazonian rim, islands of colonization are surrounded by vast expanses of forest.

Probably 90% of the people living along this transect live in urban centers. This is skewed from a more normal distribution by including Lima (>5,000,000) and Huancayo (350,000). It does, however, reflect the increasing trend toward urbanization of the Andean population. Approximately 65% of Peru's population is now urban, and roughly 25% of the country's people live in the Lima area.[12] The highlands are nevertheless still predominantly rural and very important to the nation's economy. An estimated 65% of the country's total rural population live between 2000 and 4000 m.[13] This rural population is comprised overwhelmingly of small-scale farmers, with 77% of the farm units owning less than 5 ha.[14] The highlands produce one half of the food consumed in Peru.[15]

This transect shows the importance of three factors: relation to rain-bearing winds (longitude), altitude, and proximity to urban centers. Equally important in the actual land-use patterns, but concealed by the large scale of the transect, are local features of climate, soil, topography, and socio-economic features, such as presence or absence of roads or *haciendas* (estates). Research at the regional level reveals local zonation in cropping patterns and the production of single crops. The Mantaro Valley region can be easily divided into three major agroclimatic zones (low-intermediate-high) according to altitude, humidity, and relation to urban centers.[16] Besides different environmental conditions, these three zones are characterized by different crop inventories, crop varieties, tillage practices, reliance on purchased inputs, degrees of commercialization, and land tenure patterns.

Farming systems in the Andes reflect both these macro- and micro-regional features. Out of the tremendous local variation of agroclimatic zones, four major agricultural zones can be distinguished in the central Andes, that is between 2000 and 4800 m:

1. A cereal zone between 2000 and 3000 m where maize is important, followed by European cereals (esp. barley) and then by Andean crops (e.g., Kiwicha — *Amaranthus caudatus)*
2. A tuber zone between 3000 and 4000 m dominated by potatoes but also a zone where some other minor Andean crops (e.g., tarwi — *Lupinus mutabilis* and quinoa — *Chenopodium quinoa)* are grown
3. A pasture zone where Andean camelids (llama and alpaca) and European animals (especially sheep and cattle) are grazed
4. A lower zone along the eastern slopes (<1500 m) where coca, fruit, manioc, coffee, and sugar cane are grown.

A great deal of anthropological research has been done on the use of these zones and the relations between them.[17-19] This research has emphasized the goal of the Andean people to manage their vertical landscape in such a way as to gain access to these zones, either directly, through migration, or by trade. Both the past and present of Andean agriculture, economics, and politics can be understood by reference to the principle of complementarity between these zones, as Andean people manage their labor and other resources to gain access to different ecological zones. This complementarity certainly relates to the need to use land and labor resources efficiently and to manage risk in a highly variable environment. Thomas,[20] for instance, suggests that no single zone is capable of providing adequate nutrition for the Andean people. Among seven villages in Southern Peru, Figueroa[21] found a household average of 6.1 crops cultivated on 13.2 fields. These households owned 10 head of livestock (7 sheep, 2 cattle, 1 pig) plus 7 minor animals (fowl and guinea pigs). Our research in the Tulumayo reflects a similar pattern, but it also reflects the overwhelming importance of potatoes in Andean agriculture, as two thirds of all cultivated plots are in potatoes.

The present distribution of the Andean farming population and its land tenure stresses the direct control of the first three zones by Andean farm communities. The most typical pattern is community ownership of vertical transects from narrow valley floors, up steep slopes to high rolling pastures. These transects include cereal producing lands at their base, tuber zones at mid-slope, where most of the population lives, and pasture land above the tuber zone. Individual households within these communities own parcels at different elevations and have rights to keep animals on communal pastures. Communities have legal ownership of these lands and exercise different degrees of control over their land at different altitudes. Intensively cultivated lower valleys are least controlled by community mechanisms, while the higher tuber zones and pasture zones are often under direct community control.

III. FARMING SYSTEMS OF THE ANDEAN POTATO

Although complementarity is a principle that can be broadly used over several production zones, it is also an important principle within single zones and for single crops. This is especially evident for the Andean region's most important crop, the potato, and for the tuber zone in general. On closer inspection, there is diversity within this single zone which is comparable to the total diversity of Andean agriculture. As with the overall system, the diversity of potato agriculture can be understood as complementarity between land and labor, relating to environmental, socio-economic and cultural features. The efficient use of resources and the management of risk are equally important in understanding the dynamics of this single cropping system and in explaining the diversity of the crop. Besides efficiency and risk management, Andean cultural patterns, such as social organization, cuisine and ritual, are significant in determining the zonation of potato agriculture.

A. The Diversity of Andean Potatoes

The current biological taxonomy of cultivated potatoes recognizes eight species of the crop.[22] All of these are grown in the Andes, in comparison to only one in North America *(S. tuberosum)*. These divide into four ploidy levels ($2n = 24$ to $5n = 60$). The world collection of potatoes at the International Potato Center has some 13,000 named accessions, and geneticists there estimate that there are approximately 4000 distinct clones or varieties in this collection. Andean farmers are well aware of this diversity, employing a systematic folk taxonomy to classify the potatoes, and using different types of selection and exchange in maintaining their diverse collections.[23]

The diversity of the potato crop can be attributed to three major factors: (a) environmental diversity, (b) crop evolutionary history, and (c) human selection. The highly dissected intermontane valleys of the Andes provide numerous micro-environments and isolating mechanisms. Although most species and ploidy levels are cosmopolitan in the central Andes, most varieties and certain species are localized. A clear example of this is the bitter diploid *S. ajanhuiri* Juz. et Buk. that is located almost exclusively in the Lake Titicaca basin.[24] Like other bitter species (*S. x curtilobum* Juz. et Buk. and *S. x juzepczukii* Buk.) this species is grown at high elevations (3800 to 4100 m) because it is frost resistant. These bitter species are processed by freeze-drying and washing to remove the bitter alkaloids. The diploids, *S. stenotomum, S. phureja, and S. goniocalyx* are more common at lower altitudes. The most diverse and cosmopolitan of all cultivated potato species is the tetraploid *S. tuberosum* subsp. *andigena*. Ethnobotanical research[25,26] indicates that over 80% of the total Andean potato crop is *andigena*, while Ochoa's survey of collections in the Andean area between 1966 and 1972 reveal that 75% of all samples (accessions) are of this subspecies.[27]

Potatoes have been cultivated for at least the last 6000 years in the Andes, and this long period has offered the opportunity for hybridization and selection leading to new varieties. It has also offered an adequate time frame for the evolution of numerous species of pests and pathogens. Although one major pathogen, the late blight fungus *Phytophthora infestans*, is exogenous to the Andes, most of the pests and pathogens are native. These include numerous viruses, fungi, bacteria, and insects. This large pest community may be one reason why there is so much potato diversity, although horizontal resistance has not been demonstrated for the crop.

Human selection plays a critical role in creating and maintaining diversity. This role is the identification, selection, maintenance and exchange of potato germplasm, and these functions relate to a more general culture of potato farming. Murra[28] noted that there is far less status or ritual accorded to the major Andean staple, potato, than to maize. Unlike other crops in their cradle areas, for instance rice in Asia, "potato culture" in the Andes is less visible. Nevertheless, such a culture may be defined by the many human activities surrounding this crop: the knowledge system of the crop (e.g., naming of varieties, potato pests and pathogens), production practices (e.g., selection, tools and tillage), and processing (e.g., storage and culinary practices). Viewed in this manner, the potato has as rich and complex culture associated with it as any other Andean crop. The potato is fundamental to both the household and regional economies of the Andes, and it is, therefore, a critical unit in the analysis of the farming system at any level.

IV. FARMING IN THE TULUMAYO VALLEY

A case study of potato farming system in the Tulumayo Valley in central Peru illustrates the complexity of Andean agriculture. This valley is located on the eastern escarpment of the Andes, 30 km east of the Mantaro Valley and three hours from Huancayo over an all weather road. The population of the valley is approximately 13,000. Most of the population lives in nucleated settlements, including two large towns, Comas (1200 people) and Cochas (700 people), and numerous villages and hamlets. Potato cultivation is found between 4100 and 2800 m in the valley. A collection of some 200 named varieties (526 accessions) from six villages yielded approximately 62 biologically distinct varieties.[29] These varieties include all four ploidy levels and all of the cultivated *Solanum* species and subspecies, except *S. ajanhuiri*. Farmers grow mixed collections ("*chalo*") for consumption, bitter varieties ("*shiri*") for processing by freeze drying and washing to make *chuño*, and selected native varieties and improved, white varieties for market and consumption.

A. Household Production and Community Functions

As in most small-scale farming economies of the world, the primary unit of production and consumption in Andean potato farming is the household. Here, the primary decisions in the allocation of land and labor are made. Before recent times, production on small-scale farms was for consumption, and resources for production were almost entirely local. Today, consumption is still a major goal of household production, but commercial production is increasingly important.

Kinship is bilateral, and households tend to be nuclear with a married couple and their dependent children living together. People from the extended kinship network may also occupy the same compound. The average household has 4.5 persons. Only inscribed and resident members of the peasant community may own land. Membership is gained through birth, marriage, and long-term residence. As long as residence in the community is continued, parcels are held indefinitely by nuclear families. They can be passed to other members through inheritance, exchange, sale or, after abandonment, by community allocation. Thus, there is no land market in the valley that might bring absentee renters or landlords. Although the community may retain *de jure* ownership of the land and prohibit sales to outsiders, the household is the *de facto* owner of land. It is this unit that decides what crops and varieties are to be grown, the tillage system to be used, the amount and organization of labor to be employed, the type and amount of purchased inputs to be used, and how to utilize the harvest.

Families own an average of 11.4 fields, plant 7.3, and keep 4.1 in fallow. Out of the 7.3 fields planted, 4.9 are in potatoes and 2.4 are in other crops. Potatoes are followed by maize and then barley in frequency. Besides crop production, most families are engaged in other activities to produce food, fiber, and income. Many families engage in livestock production, especially of cattle and sheep, on communal pastures.

Most families have members, usually the husband, who sell their labor off-farm and frequently outside of the Tulumayo Valley. Figueroa's[30] survey data from villages in central and southern Peru suggest that 45 to 70% of household income is comprised of goods produced and consumed by the household, with an average value from six communities of 51%. Wage migration accounts for only 11.2% of the family income of Figueroa's sample.[31] This compares with the Cajamarca region, where Deere and deJanvry[32] found that 46.7% of household income came from farm activities and 33.8% from wage labor. Although the Tulumayo data does not have specific household income data, my estimate is that it approximates these cases. Whereas many families report some source of off-farm income, most of this is irregular. The production of commercial potatoes is a predominate feature of virtually every household, and 72% of the potatoes produced in the Tulumayo Valley are sold. Both Deere and deJanvry and Figueroa show that wage labor and migration are more important to households with less land, a feature also evident in the Tulumayo. Here, the most common way to earn off-farm wages is to migrate to the eastern *montaña* around Satipo to work in the coffee harvest.

The use of extensive fallow and difference in altitude, slope, drainage, and soils greatly reduce the actual amount of land under cultivation. The use of multiple fields for potato production is typical of the extreme fragmentation of land tenure that has been noted for the Andes. In the community of Irpa Chico, Bolivia, for instance, Carter and Mamani report that families may own 90 different parcels.[33] Figueroa's survey of seven villages in the Peruvian Andes shows household's owning from 8.5 parcels in a low altitude village in Huancavelica Department to 84.4 parcels in a village on the *altiplano* in Puno Department.[34] Out of the total parcels surveyed in our study, 56% were in production and the rest in fallow. The Tulumayo numbers are comparable to these: 11.4 owned, 64% under production.

This land fragmentation derives from several factors: from risk management in a highly variable environment, the inheritance system, communal mechanisms to manage agriculture, and strategies to produce different types of potatoes under different cultivation practices. Risks derive from both abiotic and biotic factors. Drought, frost, erosion, late blight and animal damage are the most frequently cited risks, but there are numerous other limiting factors, pests and pathogens that threaten the crop. Crop and variety diversity is one method to manage risk, and field dispersion is another. Many ethnographies mention the aim of farmers to have parcels in different places so as to take advantage of different slope, soil, and drainage conditions. Moisture seems to be especially significant, as farmers try to locate fields in areas where ground water is available in dry years. In wet years, however, drainage may be a problem in these areas, and additional labor may be required to produce a crop.

The inheritance system is through both males and females, and the norm is for elderly people to divide their land among all children who remain in the village. The creation of new households thus joins dispersed fields from both husband and wife. Additionally, fields may be acquired by trade or purchase within the community.

Communal activity also contributes to the ownership of numerous, dispersed parcels. Highland communities from central Peru through Bolivia divide their lands into two general types, land which is individually managed and land which is managed communally.[35] Carter and Mamani's detailed study of Irpa Chico suggest that the land allocated to each type is roughly equivalent.[36] Communal management is given different terms in different areas, *laymi, manay, suerte, sector, aynuqa,* but the management principles are generally the same. This is a system of sectoral fallow that is managed by the community. Community lands, usually in the higher altitude zones, are divided into several sectors, generally seven. Each sector is cultivated in sequence, starting with potatoes for one or two years, followed by other tuber crops or barley, and returned to fallow after 3 or 4 years of cultivation. Households have *de facto* rights to plots in each of the sectors, and may own several plots in any one sector. In Irpa Chico, households own an average of 21 plots in communally managed lands.[36]

Communal management is exercised in the rotation of sectors and in determining agricultural calendar. The community decides which sectors to open for cultivation and grazing and when cultivation and the harvest should begin. A key factor in this management is to provide single blocks of fallow fields for pasturing animals. The community enforces the sectoral rotation by penalizing animal owners whose animals are responsible for crop damage within the regular rotation. Families who break this rotation by cultivating their individual plots out of sequence are not afforded this protection, and they must individually fence, guard their plots, and seek damages without community help. Apart from determining the rotation and scheduling the calendar, the community role in managing crop production is minimal.

The final cause for land fragmentation is the economic strategies of households to produce different crops and varieties. Apart from risk management mentioned above, there are two economic strategies that contribute to field dispersion. One of these is to use family labor most efficiently. Golte[37] demonstrates that polycropping spreads out labor demand throughout the year, while monocropping concentrates it. Looking at the Arequipa Department in southern Peru, Golte argues that a monocrop of maize would use only 36% of the available labor, potatoes 42%, maize plus potatoes 58%, and 83% for a combination of maize, potatoes, onions and squash.

Besides crop production, we must also be cognizant of the need for the numerous other tasks and activities required of small-scale farming households: livestock production, artesanry, communal and religious duties, temporary labor migration, and so forth.[38] During recent field work, I observed numerous families working both day and night in their fields, approximately 18 h/d. By day they were plowing potato fields and by night harvesting barley fields. Farming in the Andes is always difficult work, requiring strenuous physical labor, numerous skills, and a knowledge of many crops, soils, and environmental features. Conventional images of idle or underemployed peasants are greatly at odds with this reality.

B. Zonation of Potato Production

Field dispersion in potato production may allow farmers to use labor more efficiently or to minimize risk, but Tulumayo farmers describe this under a different rationale, to take advantage of different micro-zones to produce different crops and varieties for different purposes. Two principal types of production are found, *hatun tarpuy*, "big harvest", and *mahuay tarpuy*, "early harvest". *Hatun tarpuy* is generally confined to higher areas and to the major subsistence crop, potatoes. In some Andean areas it now includes commercial crops, such as barley production for beer. *Mahuay tarpuy* is generally a more intensive land-use system, often dependent on irrigation, and more closely associated with the use of purchased inputs and commercial production. Like the other system, *mahuay tarpuy* has secondary crops, e.g., maize, besides potatoes.

Andean research has emphasized the concept of altitudinal zonation as an organizing principle.[39-41] While the distinction between *hatun tarpuy* and *mahuay tarpuy* is based on cultivation at different altitudes, differences in crop type, types and levels of inputs, degree of intensification, the role of the community, and the intended use of the crop are equally important to altitude. The idea of system, whether it be defined as a production, farming or cropping system, is perhaps more useful than zone to convey the idea that many things other than altitude are involved.

The *hatun tarpuy* in the Tulumayo Valley is most strongly associated with subsistence potato production, using land in the communal management zone, relying on certain types of potatoes, and employing certain tillage practices. The amount of land under communal management varies from village to village, but most lands in the upper zones (3400 m) have some degree of sectoral fallow. In these lands, a form of minimum-till cultivation is practiced, known locally as *tikpa*. Potatoes are planted in untilled ground, by placing seed tubers in holes opened by the footplow *(chakitaklla)*. A handful of fertilizer (dung, guano, or chemical) and/or pesticide may be thrown into the hole along with the tuber. After the tuber has sprouted in 4 to 6 weeks, the sod is turned and mounded against the young plant, creating long furrows that customarily run vertical to the contour for drainage and for ease of working with the footplow and harvest hoe *(raukana)*. More fertilizer and insecticide may be added at this step. A second cultivation may be used midway through the growing season, and fungicides to control late blight are applied several times. The growing season is eight to nine months. No irrigation is used in the upper zone.

Three types of variety selection are practiced in the

TABLE 1
Production of Different Potato Types by Zone

	Chalo	Native/ commercial	Improved
% total potato crop	12	26	62
% in *hatun tarpuy* zone	57	70	26
% in *mahuay tarpuy* zone	43	30	74
% sold	45	61	80
% total potato sales	6	24	70

Tulumayo Valley: (a) for mixed native varieties *(chalo)*, (b) for specific native varieties, and (c) for improved varieties. The latter two are associated with commercial production and with the use of higher levels of purchased inputs, although these also are used in the first. Table 1 indicates the relative portions of each of these selection strategies, their use by households, in the two production zones, and the portion of each that is sold.

The selection strategy most often associated with consumption and with *hatun tarpuy* is for diversity and random mixing in the field. The local term for this is *chalo* or "mixed colored" potatoes. Diversity is appreciated for its culinary quality, not for its agronomic virtues. Native potatoes yield less than improved varieties, and they are regarded as having less resistance against the principal risk factors: frost, drought, and late blight. Culinary quality is, however, an essential criteria in this subsistence crop. The one agronomic advantage that these mixed native collections seem to have is that they are more tolerant of virus infection. The result of this infection, besides lower yields, is small and deformed tubers. Farmers observe that improved varieties, imported from outside of the valley, degenerate after a few years, requiring them to purchase new seed. Native varieties, although equally infected by virus, do not seem to degenerate or to require such frequent replenishing of seed. Native seed is, however, rotated through plots at different altitudes, and it is exchanged between farmers.

While *chalo* cultivation is associated with consumption, certain native varieties and improved varieties are selected for commercial purposes. Roughly a half dozen native varieties are grown because of their high market price, and most of these commercialized native potatoes are grown in the upper zone. Improved varieties have been used for the past 30 years in the Tulumayo Valley. While most of these are grown at lower altitudes in the valley, some are produced in the upper zones. They are appreciated for their high yielding potential and resistance to drought, frost, and late blight. But, they are regarded as culinary inferiors to native varieties and are produced principally for market. This valley is the source of seed for some coastal and highland commercial producers who like to buy seed tubers of improved varieties at high altitudes. This preference may be due to lower aphid infestation and related virus infection. Like *chalo* in the upper zones, these two types are grown under the *tikpa* system, although often at higher levels of inputs.

Mahuay tarpuy refers to both a production zone and a system of production that are distinct from the *hatun tarpuy*. *Mahuay tarpuy* is the production of early potatoes, often under irrigation. It occurs at lower altitudes (2500 to 3400 m), during a different cultivation cycle, and using different types and levels of inputs. The major potato producing cycle (*hatun tarpuy*) coincides with the rainy season, November to May, and is entirely rain-fed. Potatoes are harvested in July, a period when the potato market is depressed by a temporal glut. *Mahuay tarpuy* potatoes are planted before the rainy season and must depend on irrigation during significant portions of their growth cycle. They are harvested in December and January, when the potato market is relatively strong.

In distinction to production in the higher zones, the *mahuay tarpuy* zone is more intensive in several aspects besides irrigation. Land is kept under production for longer periods, without the three or more year fallow period of the higher elevations. Rather than the extensive, no-till method of *tikpa*, fields in this zone are prepared by thoroughly plowing the soil, frequently with oxen, and removing all other vegetation before planting. Planting is done in carefully prepared beds, and relatively greater levels of purchased inputs (seed, fertilizer, fungicide, and pesticides) are used. The higher temperatures and higher humidity of lower elevations may favor the development of fungal diseases, especially late blight, and the use of fungicides. Indeed, *mahuay tarpuy* was not an important potato producing system until the advent of more fungus resistant varieties and fungicides as well as the introduction of precocious (3 month) varieties.

V. INTENSIFICATION OF ANDEAN AGRICULTURE

A common lament in Peru is that the peasant economy of the Andes is stagnant and incapable of providing either enough employment or food for the nation's rapidly expanding population. The swollen shanty towns of Lima and backward highland villages are suggested as vivid evidence of this fact. Figures presented by Gonzalez de Olarte[42] show that the annual rate of national increase in the value of agricultural production was only 0.9% between 1970 and 1974 and 0.4% between 1975 and 1977. For the highlands, potato production is a more useful guide to the performance of small farm agriculture. Using national statistical sources, Fano and Scott[43,44] suggest that highland potato productivity per hectare actually declined by 0.1% between the periods 1948/50 and 1977/79. Gonzalez de Olarte[45] indicates that agricultural production in the central highlands declined by 2.4% between 1970 and 1977. In a general review of Peruvian agriculture, Caballero concludes that "the Andean peasant economy is maintaining itself at a level of simple reproduction, at least during the last two decades: output, yields, accumulated capital, incomes, and population have grown little or not at all".[46]

The weight of evidence from various sources is that agriculture in the highlands has been relatively stagnant for the past 40 years. This is in spite of major technological innovations, including the breeding of improved crops that are higher yielding, the invention of pest control chemicals, and the availability of chemical fertilizers. It is also in spite of major capital investments in the highlands, including the construction of roads and irrigation systems, the development of regional universities and agricultural research services. Finally, this stagnation occurs during a period of rapid population growth and the development of large regional and national markets.

How do the potato farmers of the Tulumayo compare to this image of stagnation? Simply put, Tulumayo potato farmers seem to dramatically contradict it. They have intensified their production for commercial purposes by emphasizing *mahuay tarpuy* cultivation, and they have adopted a broad range of innovations, especially improved varieties and agrochemicals. Our on-farm interviews indicate that 1985 yields were 10.2 t/ha in the *hatun tarpuy* and 11.4 t/ha in the *mahuay tarpuy*. The overall average of these two production systems in the Tulumayo Valley is 10.3 t/ha. These yields are almost three times as large as yields reported in a national rural household survey in 1984. This national survey indicated that average potato yields in 1984 were 3.7 t/ha overall in the highlands, with 5.0 t/ha on irrigated plots and 3.6 t/ha on rain-fed plots.[47]

The apparent discrepancy between the intensification of potato production by Tulumayo farmers and the overall performance of potato agriculture in the central highlands between 1948 and 1984 might be explained in several different ways. It may reflect the inadequacy of the data, either because the aggregate data are inaccurate or that the case study is atypical. Aggregate data on small-scale farming in isolated communities is notoriously suspect. The practice of asking farmers to estimate the area of their fields is open to question, and it is possible that farmers underreport their production to official census takers. On-farm research by social scientists and agronomists seems to corroborate this by revealing significantly higher yields than those reported in the official statistics. This suspicion is magnified by the difficulty of research in remote and rugged terrain. Pro-urban bias may also contribute to underreporting the contribution of small-scale farms in distant Andean valleys.[48]

The discrepancy between general reports of highland agricultural stagnation and our evidence of intensification in the Tulumayo may also reflect differentiation between farming regions. The importance of agroclimatic zones as a contributor to agricultural performance is widely recognized. This has been well documented for potato production in the Andes.[49] Although the Tulumayo could be extraordinary, it is similar to the least productive zones of the adjacent, well studied Mantaro Valley. It is a high altitude area, dominated by small-scale farms, and a significant percentage of its potato production is for consumption. These characteristics suggest that it is not atypical of other Andean communities. Its productivity also suggests this, when compared to detailed on-farm research results rather than survey data.

Measuring productivity is exceedingly difficult in Andean agriculture because of the irregularity of fields, uneven conditions between different zones, and the lack of record keeping of small-scale farmers. Detailed case studies suggest that higher yields are in fact attained than aggregate statistics indicate. Horton et al.[50] report figures of 13.0 t/ha for improved varieties and 10.1 for native varieties for the Mantaro Valley region, 30 km west of the Tulumayo. A 3-year study among small-scale producers in Ayacucho reports yields of 8.8 and 8.9 t/ha in two different communities.[51]

VI. CONCLUSIONS

These detailed case studies corroborate the evidence from the Tulumayo study area that technological change, intensification of land use and increasing productivity are common to small-scale Andean farmers. This agricultural change is prompted by several factors. The valley's population has grown, but at below the national rate, by increasing 22% between 1961 and 1981 (1.1%/year). Regional population growth, however, has been significant. Nearby cities and towns, Huancayo, Jauja, and Concepcion, have grown rapidly in the last three decades and are now important market centers for regional produce. Infrastructure, especially roads and other apparatus of marketing, has been greatly expanded, and the inputs necessary for agricultural change are available. Our research shows that most households have taken advantage of these conditions to increase their productivity. Finally, agrarian reform has been instituted in Peru, and the allocation of individual parcels to previous tenants should have a positive effect on productivity.[52] Several communities in the Tulumayo Valley received land expropriated during the agrarian reform. Caballero notes that there is no evidence of decapitalization, which might depress productivity, in Peruvian agriculture either before or after the agrarian reform.[53]

Limits to agricultural change in the Tulumayo Valley are also evident. Stable or even declining labor is a major constraint. Off-farm employment and agricultural labor in the *montaña* draw down the labor available for agriculture in the Tulumayo Valley, and there is no available or appropriate mechanization to replace human labor. The use of oxen is restricted by available pasture, and pasture area is reduced as farmers shorten fallow cycles. Farmers are very articulate in their complaints about the risks and potential unprofitability of investing in new technology. Purchased inputs and transportation costs comprise the majority of all costs. Although farmers may be willing to see the return for their labor decrease under unfavorable market conditions, they are reluctant to risk decapitalization or indebtedness by overexpansion. They compete in the same urban markets as large producers and regions with better infrastructure and access to markets. They are also producing for a market where new urban tastes for rice and bread may lower the demand for potatoes. Even if they succeed in producing a good crop, they are likely to be undercut by other producers or to have prices fall because of inadequate demand.

Finally, the unprofitability of small-scale agriculture in the Andes discourages change. Economic analysis shows that highland people in general receive a disproportionate share of Peru's national wealth, and that rural wages are significantly below urban ones. Caballero, for instance, reports that in 1961, rural wages in the highlands were only 23% of Lima's, and in 1972, they fell to 18%.[54] It is also clear that the small-scale farm sector of the highlands receives an even lower proportion of urban wages. The result is an exodus from the farm to earn cash wages, both on a permanent and temporary basis. Whereas some capital from this wage migration is invested in the rural area, only a small portion is invested in agriculture. Figueroa shows that a high proportion (45%) of rural income is spent on food, but this expenditure seems to reflect preference for products of urban origin (salt, sugar, cooking oil, noodles, flour) rather than an inability to produce enough food.[55]

Despite these constraints, intensification of potato agriculture in the Tulumayo Valley has proceeded, and a large proportion of the production is now commercial. How has this affected the production of native varieties? While estimates of the base of potato germplasm before intensification are unavailable, the current status of native varieties gives us a glimpse of the process of the loss of germplasm. This loss results from environmental causes such as frost, drought pathogens, and pests, but human selection for commercial varieties (native or improved) would seem to outweigh these causes as the single most important factor. A majority of the potato plots are sown with highly selected varieties rather than the random mixture of native ones. Nevertheless, conservation of native varieties can be detected. This occurs primarily in the upper, *hatun tarpuy*, zone and in fields planted for consumption. There is a positive correlation between the amount of *chalo* planted and the number of varieties reported in household inventories. Households in the village of Maranioc on the Tulumayo Valley's southern side, maintain an average of 21.5 varieties, while an average of only 10.4 varieties is found on the northern side.

The Andean farmers have used their complex and rugged land for several millennia to feed themselves, their families and urbanites. By using different zones, crops, varieties and agricultural technology, the Andean farmer has been able to provide relatively stable subsistence production for several millennia. Except in the highest altitudes (>3800 m) and during extraordinary environmental crises, such as the 1982 drought associated with the *el niño* phenomenon, Andean technology has been successful. In recent decades, however, Andean farmers have faced new pressures, from the region's highest population and from new consumption patterns that require income as well as subsistence. Generally, Andean agriculture has not been successful in meeting these new demands. Peru is a food deficit nation, and there has been a mass exodus of the rural population to cities in search of wage labor. The inability to mechanize agriculture, the small scale and dispersion of plots, *de jure* community ownership of land, the cost-price squeeze between inputs and farm produce value, and high environmental heterogeneity and risk are some of the many factors that influence the lack of capital accumulation and rapid change in Andean agriculture. In contrast to this generally gloomy picture, however, farmers of the Tulumayo Valley have introduced technology and new production methods to increase yields and boost incomes. At the same time, they have conserved their Andean heritage in such practices as selecting mixed *chalo* potato varieties and planting them according to the minimum tillage *tikpa* method.

ACKNOWLEDGMENTS

Research for this paper was supported by the National Science Foundation (BNS 8416724) and by the University of California, Davis. My co-investigators, Enrique Mayer and Cesar Fonseca provided important insight and advice for the background of this paper.

REFERENCES

1. **Hawkes, J. G.**, *The Diversity of Crop Plants*, Harvard University Press, Cambridge, 1983.
2. **Vavilov, N. I.**, The origin, variation, immunity and breeding of cultivated plants, *Chron. Bot.*, 13, 1, 1951.
3. **Jennings, P. R. and Cook, J. H.**, Centres of origin of crops and their productivity, *Econ. Bot.*, 31, 51, 1977.
4. **Wilkes, G.**, Current status of crop plant germplasm, *Crit. Rev. Plant Sci.*, 1(2), 133, 1983.
5. **Plucknett, D. L., Smith, N. J. H., Williams, J. T., and Anishetty, N. M.**, Crop germplasm conservation and developing countries, *Science*, 220, 163, 1983.
6. **Lyman, J.**, Progress and planning for germplasm conservation of major food crops, *Plant Genetic Resources Newsletter*, 60, 3, 1984.
7. **Brush, S. B.**, A second generation of crop genetic resource conservation, unpublished paper delivered to symposium on Genetic Diversity and Food Security: A Global Perspective, annual AAAS meeting, Chicago, 1987.
8. Funding for this research was provided by the National Science Foundation (BNS 8644843).
9. **Troll, C.**, The codilleras of the tropical Americas. Aspects of climate, phytogeographical and agrarian ecology, in *Geo-Ecology of the Mountainous Regions of the Tropical Americas*, Band 9, Troll, C., Ed., Ferd. Dummlers Verlag, Bonn, Germany, 1968, 15.
10. **Brush, S. B.**, *Mountain Field and Family: The Economy and Human Ecology of an Andean Valley*, University of Pennsylvania Press, 1977.
11. **Mayer, E. and Martel, C. F.**, *Sistemas Agrarios en la Cuenca del Rio Cañete*, Impreso Onern, Lima, 1979.
12. **Murphy, E. M.**, World population: toward the next century, *Population Reference Bureau*, 1, 1985.
13. **Figueroa, A.**, *Capitalist Development and the Peasant Economy*, Cambridge University Press, Cambridge, 1984.
14. **Caballero, J. M.**, *Economía Agraria de la Sierra Peruana, Instituto de Estudios Peruanos*, Lima, 1981.
15. **Caballero, J. M.**, *Economía Agraria de la Sierra Peruana*, Instituto de Estudios Peruanos, Lima, 1981.
16. **Mayer, E.**, *Land Use in the Andes. Ecology and Agriculture in the Mantaro Valley of Peru with Special Reference to Potatoes*, International Potato Center, Lima, 1980.
17. **Murra, J. V.**, *Formaciones Economicas y Politicas del Mundo Andino*, Lima, Inst. Estudios Peruanos, 1975.
18. **Lehmann, D., Ed.**, *Ecology and Exchange in the Andes*, Cambridge University Press, Cambridge, 1982.
19. **Masuda, S., Shimada, I., and Morris, C., Eds.**, *Andean Ecology and Civilization*, University of Tokyo Press, Tokyo, 1985.
20. **Thomas, R. B.**, *Human Adaptation to a High Andean Energy Flow System*, 7, Anthropology, The Pennsylvania State University, University Park, 1973.
21. **Figueroa, A.**, *Capitalist Development and the Peasant Economy*, Cambridge University Press, Cambridge, 1984.
22. **Hawkes, J. G.**, Biosystemics of the potato, in *The Potato Crop*, Harris, P. M., Ed., Chapman & Hall, London, 1978, chap. 2.
23. **Brush, S. B., Carney, H. J., and Huaman, Z.**, Dynamics of Andean potato agriculture, *Econ. Bot.*, 35, 70, 1981.
24. **Johns, T. and Keen, S. L.**, Ongoing evolution of the potato on the altiplano of Western Bolivia, *Econ. Bot.*, 40(4), 409, 1986.
25. **Brush, S. B., Carney, H. J., and Huaman, Z.**, Dynamics of Andean potato agriculture, *Econ. Bot.*, 35, 70, 1981.
26. **Jackson, M. T., Hawkes, J. G., and Rowe, P. R.**, The nature of solanum x chaucha juz. et buk., a triploid cultivated potato of the South American Andes, *Euphytica*, 26, 775, 1977.
27. **Ochoa, C.**, Potato collecting expeditions in Chile, Bolivia and Peru, and the genetic erosion of indigenous cultivars, *Crop Genetic Resources for Today and Tomorrow, International Biological Programme*, 2, 167, 1975.
28. **Murra, J. V.**, Rite and crop in the Inca state, in *Culture in History: Essays in Honor of Paul Radan*, Columbia University Press, New York, 1960, 393.
29. Analysis was done at the International Potato Center, Lima, by general protein electrophoresis, described by Stegeman, H. & V. Loeschcke, *Index of European Potato Varieties*, Biol. Bundenanst. Land-und Forstw., Berlin, 1976.
30. **Figueroa, A.**, *Capitalist Development and the Peasant Economy in Peru*, Cambridge University Press, Cambridge, 1984, 42.
31. **Figueroa, A.**, *Capitalist Development and the Peasant Economy in Peru*, Cambridge University Press, Cambridge, 1984, 45.
32. **Derre, C. D. and de Janvry, A.**, A conceptual framework for the empirical analysis of peasants, *Am. J. Agric. Econ.*, 61(4), 602, 1979.
33. **Carter, W. E. and Mamani, P. M.**, *Irpa Chico. Individuo y Comunidad en la Cultura Aymara*, Libreria — Editorial "Juventud", La Paz, Bolivia, 1982.
34. **Figueroa, A.**, *Capitalist Development and the Peasant Economy in Peru*, Cambridge University Press, Cambridge, 1984.
35. **Orlove, B. S. and Godoy, R.**, Sectoral fallowing systems in the central Andes, *J. Ethnobiol.*, 6, 169, 1986.
36. **Carter, W. E. and Mamani, P. M.**, *Irpa Chico. Individuo y Comunidad en la Cultura Aymara*, Libreria — Editorial "Juventud", La Paz, Bolivia, 1982.
37. **Golte, J.**, *La Racionalidad de la Organizacion Andina*, Instituto de Estudios Peruanos, Lima, 1980, 43.
38. **Brush, S. B.**, The myth of the idle peasant: employment in a subsistence economy, in *Peasant Livelihood*, Halperin, R. and Dow, J., St. Martin's Press, New York, 1977, 60.
39. **Murra, J. V.**, *Formaciones Economicas y Politicas del Mundo Andino*, Instituto de Estudios Peruanos, Lima, 1975.
40. **Mayer, E.**, Production Zones, in *Andean Ecology and Civilization*, Masuda, S., Shimada, I., and Morris, C., Tokyo University Press, Tokyo, 1985, 45.
41. **Gade, D. W.**, *Plants, Man and the Land in the Vilcanota Valley of Peru*, 6, Dr. W. Junk B.V., the Hague, 1975.
42. **De Olarte, E. G.**, *Economias Regionales del Peru*, Instituto de Estudios Peruanos, Lima, 1982.

43. **Fano R., H. F.,** *Cambio Tecnologico y Tendencias de la Produccion de Papa en la Region Central del Peru 1948—1979*, Univ. Nacional Agraria La Molina, Lima, 1983.
44. **Scott, G. J.,** *Mercados Mitos e Intermdiarios*, Universidad del Pacifico, Lima, 1985.
45. **De Olarte, E. G.,** *Economias Regionales del Peru*, Instituto de Estudios Peruanos, Lima, 1982.
46. **Caballero, J. M.,** Agriculture and the peasantry under industrialization pressures: lessons from the Peruvian experience, *Latin American Research Review*, Vol. XIX (2), 19, 1987.
47. Peru, Instituto Nacional de Estadistica, *Encuesta Nacional de Hogares Rurales, Resutados Definitivos*, Documento No. 5, Ministerio de Agricultura, Lima, 125, 1986.
48. **Caballero, J. M.,** Agriculture and the peasantry under industrialization pressures: lessons from the Peruvian experience, *Latin Am. Res. Rev.*, Vol. XIX (2), 3, 1987.
49. **Horton, D.,** Potato farming in the Andes: some lessons from on-farm research in Peru's Mantaro Valley, *Agri. Syst.*, 12, 171, 1983.
50. **Horton, D., Tardieu, F., Benavides, M., Tomassini, L., and Accatino, P.,** *Technologia de la Producción de Papa en el Valle del Mantaro, Peru*, Centro Internacional de la Papa, Lima, 1980.
51. Universidad Nacional San Cristobal de Huamanga, *Diagnostico Tecnico Agropecuario de las Communidades Campesinas de Arizona, Qasanqay Ayacucho*, Instituto Internacional de Cooperacion para la Agricultura, Lima, 1983.
52. **King, R.,** *Land Reform: A World Survey*, Westview Press, Boulder, CO, 1977.
53. **Caballero, J. M.,** Agriculture and the peasantry under industrialization: lessons from the Peruvian experience, *Latin Am. Res. Rev.*, 19, 3, 1984, 10.
54. **Caballero, J. M.,** *Economía Agraria de la Sierra Peruana*, Instituto de Estudios Peruanos, Lima, 1981.
55. **Figueroa, A.,** *Capitalist Development and the Peasant Economy in Peru*, Cambridge University Press, Cambridge, 1984.

The Dynamics of Traditional Agriculture: Case Studies

Mid-Arid Environments

18 THE ECOLOGY OF TRADITIONAL AGROECOSYSTEMS IN AFRICA

Roger A. Kirkby

I. INTRODUCTION

Africa has evolved over the past 3000 years a large number of local systems adapted to agricultural production in a wide range of agroecological conditions. This chapter is concerned primarily with semi-arid tropical areas having dry seasons of 2.5 to 7.5 months that cover 12 million square kilometers or 60% of Africa's land area.[1]

II. AFRICAN FARMING SYSTEMS

The indigenous crops of Africa, such as sorghum, millets, yam, cowpea, and coffee, have been supplemented by introductions over the past 1000 years; most notable among exotic crops that have become widely incorporated into farming systems are banana, cassava, the Asian yams, sweet potato, maize, rice, and beans. Ethnic diversity and geographical isolation have increased the range of strategies used by farmers in satisfying subsistence needs. However, most systems still show that a farmer's main concern is for the reliability of production from primary carbohydrate sources.

The principal physical and biotic constraints faced by farmers in this region are inadequate rainfall, poor soils, and a wide array of pest problems. In the following sections traditional practices related to the management of these ecological determinants of Africa's farming systems are discussed in relation to ecology. This is followed by a discussion of modifications to systems, otherwise designed on ecological principles, that have been introduced by farmers as a result of labor constraints, a principal socioeconomic determinant of these systems. Finally, formal R and D programs for the improvement of small farm systems are reviewed in brief.

A. Variation in Rainfall

Rainfall in most of the semi-arid tropics of Africa is associated with the movements of the Inter-Tropical Convergence Zone, and amounts of rain and length of the rainy season decrease with distance from the equator. This is seen in the West African annual rainfall belts, which follow lines of latitude and show a decline in average annual rainfall of approximately 140 mm for every 100 km traversed in a northerly direction. In Eastern Africa this pattern is modified by the equatorial highland topography, which tends to produce two rainy seasons.

Year-to-year variation in rainfall produces coefficients of variation typically around 30% in the 300- to 900-mm rainfall belt. Greater variation at drier sites is associated largely with differences in the number of rain days per year.[2] Crop ecological conditions are made more difficult by a concentration of rain in a few storms that often last several days, separated by dry spells of varying length. Weather prediction is particularly difficult due to spatial variation of discrete, conventive storms. Farmers' management of agroecosystems is therefore conditioned both by expected seasonal rainfall and by the variation in rainfall patterns. Similarities among cropping patterns are seen in widely separated locations that fall within the same climatic zone, local modifications being made according to differences in soils, altitude (temperature), and local socioeconomic circumstances such as population density.

1. Effect of Average Season Length upon Cropping System Design

Farmers generally are adept at selecting a range of crop species and varieties that make full use of the average growing season for the area, and new varieties are fitted in where they are ecologically well adapted.[3]

a. Cropping Patterns for Long Seasons

Opportunities for double cropping with early maturing varieties are confined to areas with long seasons exceeding about 200 days, as with maize and beans in Burundi.[4] Relay cropping and mixed cropping with simultaneous sowing of a long and a short season crop are normal in areas having at least 120 days of reliable rainfall. In Mali a traditional mixture of a 100—120-day maize crop with a long cycle pearl millet variety of 140—155 day maturity utilizes 85% of the available annual moisture. Efficient use of water in intercropping may be due to rooting patterns and to microclimatic effects within the canopy, the taller component acting as a windbreak that raises the relative humidity around the understorey crop and reduces its transpiration.[5] Mixtures of a tall cereal and a low-growing legume also improve interception and utilization of light. In mixtures of crops having different maturity periods, the slower growing component is normally relatively drought tolerant as in the sorghum/millet example from Mali.

b. Cropping Patterns for Seasons of Intermediate Length

Farmers employ a wider range of options in managing crop production under rainfall regimes between 80 and 120 days. In West Africa, where cereal/cowpea cropping systems are most widespread, local practice is to manage carefully the complex interaction among land types (dry uplands, moist lower slopes, wet lowlands), cropping patterns (by selecting maize, sorghum, or pearl millet as the cereal component), and cultural practices (including planting densities, sowing dates, and fertility management).[6] Maize, having the least tolerance to dry conditions, is confined to valley bottoms and lower slopes; the importance of this crop is increasing as new, earlier maturing varieties confer a greater margin of safety (and consumers are attracted to the crop by its ease of preparation). Pearl millet is sown on lighter and drier soils, particularly on upper slopes, while sorghum is intermediate in tolerance and position in the landscape.

c. Cropping Patterns for Short Seasons

Mixed cropping with simultaneously maturing species is common where seasons are very short (less than 80 days). However, yield increases due to intercropping are less common because interspecific competition for moisture is acute.[7] Traditional intercropping patterns are generally less complex, and sole cropping is frequently practiced.

Planting densities are low, often less than 15,000 plants per hectare. It should be noted, however, that there are other reasons for the common use of low seeding rates by resource-poor farmers, including the investment represented by the seed itself. This is seen particularly in peanuts, which have a large seed size and are difficult to store without loss of viability. Low seeding rates minimize plant-to-plant competition, increase the yield per plant and the returns on seed invested,[8] and favor local varieties having an indeterminate or prostrate growth habit.

d. Early Planting

Planting as soon as the rains start enables crops to make full use of the available rainfall and early flush of nutrients. In the high yielding highland environment of western Kenya, the potential yield of maize is reduced by 600 kg/ha for each week's delay in sowing, due in this case also to slower initial growth as soil temperatures are reduced after the start of the rains.[9] In the drier region of northwestern Zimbabwe, where cultivation with oxen is not possible due to the presence of trypanosome-carrying tsetse fly, hoe planting of sorghum at the start of the rains is speeded up by digging planting holes before the rains start. In eastern Kenya a proportion of the maize crop is dry planted before the rains using a traditional form of minimum tillage; once the ground is softened by rain, other fields are more completely tilled by ox ploughs.

e. Increasing the Water Supply

Water microcatchments are constructed each season for sorghum planting over wide areas of Somalia, where the crop is grown on only 250 mm rain per season. Microcatchments, varying in size with field slope, increase water penetration into the soil profile and reduce erosion caused by runoff of storm water. Small-scale supplementary irrigation using water channelled from uplands is another feature of dry areas such as the rainshadow zone on the western footslopes of the Pare mountains in Tanzania.

2. Managing Rainfall Variation within an Agroecosystem

a. Taking Advantage of Soil Variation

Specific management of each soil type along a toposequence, typically comprising light, drier soils at the top of the slope and a gradation down to a heavy, wetter and more fertile soil in the valley, provides some overall stability of production. In a relatively dry year only the heavy land yields well; labor may then be concentrated even upon low-lying patches within a field, leaving unweeded other crop areas that are not expected to repay the effort.[10]

The indigenous land classification system of the Wasukuma of northwestern Tanzania includes the concept of a viable farm unit as being an assemblage of fields that sample the soil types of the area.[11] The Kisii who inhabit a nearby area of Kenya strive to maintain small but unfragmented holdings oriented often as a narrow strip down the steep hillsides, a strategy that also combines the different characteristics of soils distributed along a catena.[12] An extreme example of this strategy is the dry season cultivation of wetlands, valley bottoms and flood plains, a common practice wherever these microenvironments occur and analogous to the concept of "verticality" in the integration of Andean ecological systems.[13]

b. Time of Sowing

Successive sowings of maize on farms throughout Southern Africa ensures that a reasonable overall harvest is obtained. Simultaneous sowings at the start of the rains would give the highest yield when averaged over several years, but a midseason drought that happens to catch the crop at flowering, its most susceptible stage, would cause unacceptable yield loss in some years.[14] However, staggered planting is not practiced where the season is generally too short to sustain later sowings.

Risk of poor germination of sorghum in the very erratic rainfall regime of Somalia is avoided in the second season by ratooning much of the first season's crop; local varieties appear to have been selected for propensity to regrow.

c. Adjusting to Seasonal Conditions

A striking case of farmers' adjusting their management as the season unfolds is seen in the Ethiopian highlands, where 900 mm of rainfall are distributed unevenly within a single season of 6 months. The principle crops are sorghum (late maturing local varieties), maize, faba beans, wheat, teff (*Eragrostis tef*), and chickpea, which are planted, in that order, between April and September. Whenever a dry spell interrupts planting, a reduced area is cropped to that species and additional land is planted later to a subsequent crop in the sequence. Teff, being the staple cereal as well as a cash crop, is not entirely replaced but failure of early sowings lead to increased use of a quicker maturing but non-preferred (red seeded) local variety kept for this purpose. Failure of early crops may also lead to a quick maturing minor crop, such as common bean, becoming prominent in a particular year. Clearly this soundly based strategy makes central planning by government more difficult.

d. Intercropping for Stability

Although the potential of intercropping for intensifying production through more complete exploitation of physical resources has received more attention from researchers, resource-poor farmers probably are more influenced by the relative stability of yields from intercropping. Intercropping rainfed rice in maize is a widespread practice from Sudan to Malawi, while sorghum/millet intercropping is stable for similar reasons in the drier areas of West Africa: whether the season is wetter or drier than the average, one crop is favored climatically and may then take a disproportionate share of resources.

e. Surviving Crop Failure

Agricultural strategies for ensuring food supply in the event of serious crop failure include the integration of livestock into dryland farming systems, and crop storage. Livestock, which are less dependent upon a single season's rainfall, utilize as fodder the leaves and stems of annual crops that have failed to set grain. Full integration of livestock and crop production, involving production of fodder crops and reciprocal benefits for crop production through animal traction and manuring, is found in the highlands of Madagascar, Kenya, and Ethiopia. Lower levels of integration are common in West Africa, where dual-purpose crops, such as indeterminate, leafy varieties of cowpea, are grown to encourage pasturing of livestock.[15,16]

The long-term storability of finger millet (*Eleusine coracana*) is one of the reasons for the persistence of this relatively low-yielding crop in the rapidly intensifying small-scale agriculture of southern Zimbabwe. Cassava has been incorporated into systems of drought-prone areas as a famine-relief crop, the bitter varieties being storable *in situ* in the ground.

III. THE MAINTENANCE OF SOIL FERTILITY

Soils in much of Africa outside the volcanic highlands of Eastern Africa are older, more weathered and less fertile than those of other tropical regions of the world. Although inadequate rainfall often appears more striking as an ecological factor in agricultural productivity, a growing body of experimental evidence suggests that productivity of many semi-arid areas is more limited by the availability of plant nutrients.[17]

Present-day agroecosystems may be differentiated into shifting cultivation systems and those that are permanent or semi-permanent. Both types have served farming populations in the past and have been sustainable over long periods until the recent rapid increases in population growth rates in Africa (now 3.2% overall, and 4.0% in some countries) overtook the indigenous evolution of the systems themselves.

A. Shifting Cultivation

In shifting cultivation, smaller trees and shrubs are cut down in the dry season and burned. A mixture of crops is sown into the ash with a minimum of cultivation, and annual cropping continues for several years until soil fertility declines to the point at which low crop yields and the increasing competitiveness of weed growth justify the clearing of a new plot.[18] Originally a 25-year period would be allowed for regeneration of deep-rooted trees, including some such as *Combretum* spp. which are resistant to bush fires. These trees accumulate nutrients that are released rapidly upon clearing and burning. Burning also tends to produce a liming effect upon the soil, renders it friable and easier to cultivate, but also leaves it more susceptible to erosion. A fast-growing crop makes best use of mobile nutrients, particularly nitrogen, that are quickly leached beyond crop rooting depth.

Among the many variants on the general strategy of shifting cultivation are the *chitemene* systems of northern Zambia, also found in parts of southern Sudan, which utilize land at two levels of intensity. A soil of inherently low fertility is enhanced temporarily by concentrating nutrients from an outfield area into an infield area only one fifth to one tenth as large. Trees are lopped and carried to the infield area for burning, and the first of two seasons' crops is sown into the ash circle. Although the ratio of exploited land to cultivated land varies with soil fertility, a regeneration requirement of at least 20 years allows sustainable population density to reach only about 20 persons per square kilometer.[19]

B. Semi-permanent and Permanent Cultivation

Of increasing importance are the various forms of ring cultivation system, in which a permanently cropped compound area near the house is surrounded by semi-permanent cultivation in outlying fields. Permanent cultivation near the house is maintained by applications of farmyard manure, household refuse, and through conservation methods made possible by tree crops — soil shading to protect from erosion and to reduce the rate of organic matter decomposition, mulching with banana leaves, and so on. Outlying fields are maintained in semi-permanent cultivation by alternating a few years of cropping with a similar or longer period of fallow.

C. Maintaining Fertility in Semi-permanent Cultivation

Farmers use various techniques to augment natural regeneration of fertility, especially as population increase reduces the period that a field can be left out of production.

Zimbabwean small farmers increase manure availability by enclosing animals at night (often against wild animals) and feeding stored crop residues. In western Ethiopia and Madagascar manure is concentrated by moving a grazing and feeding enclosure progressively across a field where a subsequent fertility-demanding crop such as potato is to be grown. Both methods show evolution parallel to that of Andean integrated crop/livestock systems.[20]

In areas where the subsoil is more fertile due to pH value, organic matter and clay contents, termite mounds are used preferentially for cropping and may be spread to cover a wider surface. In a higher rainfall area of eastern Zaire, farmers select quickly decomposable annual weed species to store nutrients, which include the flush of nitrates released into the soil as organic matter decomposes at the start of the rainy season (the Birch effect).[21] Early weed growth is cut, dried, and stored, and applied as green manure to coffee or banana plants later in the season when soil nitrogen has been depleted by leaching.[22]

Differences among crops in their responsiveness to fertility are commonly exploited in cropping sequences in semi-permanent cultivation. Cassava is often established in the final season in association with a shorter-term crop and is left to complete its development as bush fallow takes over the land. Minor tropical crops can become important locally; for example, niger seed (*Guizotia abyssinica*) which is well adapted physiologically to frequently waterlogged, cool conditions of vertisols in the Ethiopian highlands.[23]

D. Maintaining Permanent Cultivation

Except on particularly fertile soils, in high rainfall areas or around compounds, permanent cultivation usually requires an association with trees for maintenance of soil fertility. A particularly widespread system in savanna areas involves the use of the leguminous tree, *Acacia albida*. A population density of 400 persons per square kilometer is sustained in the eastern Ethiopia highlands under conditions of moderate soil fertility and a semi-arid climate (500 to 900 mm annual rainfall) by an *Acacia*/sorghum/maize/beans association. Young *Acacia* trees are allowed to remain during annual ploughing, and cereals are planted under the grown trees, which shed their leaves during the rainy season and therefore do not compete for light. Average yields of cereals directly underneath *Acacia* trees were found to be increased from 1.5 to 2.4 t/ha. This increase of 55% was attributed to improved moisture retention due to higher organic matter, improved drainage in heavy clay soils permeated by tree roots, and to nitrogen fixation by the trees. Additional benefits derived from the trees are fuelwood lopped every 4 to 5 years (amounting to $1 m^3$/ha/year) and an annual crop of pods as a dry-season protein supplement for cattle and sheep.[24] Proportionally larger increases in cereal yields have been recorded in West Africa under less fertile conditions.

Compound farms and gardens represent a more highly integrated and intensive system. Ecological diversity is high, and typically includes fast- and slow-growing trees (which may have fruiting, fuelwood, religious, and other functions) with bananas, coffee bushes, annual and perennial food and spice crops, and fodder grasses. This species mixture ensures a more variable quality and timing of decomposition and a greater buildup of soil organic matter. Uptake of nutrients is also stabilized by the differing requirements and properties of the root systems of component species. Overlapping canopies protect the soil from erosion, promote the development of surface organic matter by lowering temperatures, and act as windbreaks.

Compound gardens often serve the special functions of a testing ground for new plants introduced by the farmer[25] and for educating children in agriculture.[26] They therefore contribute to the longer-term maintenance and development of an ecosystem that is planned for stability achieved through diversity. While competition among species is the norm and selectivity may be shown in the application of manure (preferentially applied to bananas in the case of Rwandan homegardens),[27] a similar level of overall complementarity would not be expected from a random conglomeration of species. The importance of banana in quickly providing shade, mulch, and fruit probably underlies the eastern Zaire practice of using this crop as the pioneer in establishing a permanent compound on an unfavorable location, i.e., on a hillside prone to drought.

E. Effects of Population Growth on Soil Fertility

Increasing pressure on land has stimulated the development and intensification of semi-permanent and permanent systems. Sustained by techniques outlined above and often by the adoption of inorganic fertilizers, productivity has increased with higher cropping indices and intercropping complexity, and labor-intensive methods. Fertile but difficult-to-work valley bottom soils are being brought into cultivation; on the smallest farms on Kenyan roadside verges, lower leaves are stripped from maize to increase the yields of associated beans and to help feed an animal.

These systems are not developing fast enough, however, to satisfy the needs. In too many areas, fuelwood demands reduce the density of trees; manure availability declines as fallows give way to more crops; shifting cultivation spreads to marginal soils, erosion increases, and soil fertility is mined faster than it can be restored; cassava, adapted to exhausted soils, is cultivated extensively.

IV. MANAGEMENT OF INSECTS, DISEASES, AND WEEDS

A. Agronomic Practices

Although farmers tend to attribute the losses caused by small insect species and most diseases to their climatic determinants, some traditional agronomic practices have evolved at least partly for pest management reasons.

Burning cut vegetation in the dry season helps control cutworms[28] and nematodes in topsoil. Ugandan farmers, in sowing their principal bean crop towards the end of the rains so as to minimize disease infections, are aware that potential yields will be reduced by soil moisture stress. Low seeding rates reduce humidity within the crop canopy and discourage development of certain diseases, while up to 750,000 bean plants per hectare are used in dry and relatively disease-free areas of the Ethiopian Rift Valley to suppress weed growth without weeding. Grazing, cutting, or removing maize stalks after harvest reduce populations of diapausing larvae that initiate the following season's infestation.[29]

B. Crop Diversity

Common intercropping patterns drastically reduce the incidence of cowpea foliar insect pests,[30] and pearl millet may respond by compensatory tillering if intercropped sorghum is killed by shootfly attack. However, crop diversity does not always reduce pest losses. The migratory and polyphagous African bollworm, *Heliothis armigera*, is stabilized by year-round vegetation in Uganda and rarely becomes a serious pest. On the other hand, marked dry seasons in Tanzania cause annual buildup of successive generations during the rainy season and a higher pest incidence where cotton is intercropped with maize, an alternative host.[31]

Disease reduction is one reason for farmers' use of varietal mixtures, which may reduce the proportion of plants susceptible to a particular pathogen or race. In Uganda the proportions of the components in a local bean mixture are adjusted for early or late sowing to ensure an optimum mix of high-yielding susceptible types with resistant types having a lower potential yield in the absence of disease.

Some control of the parasitic witchweed, *Striga hermonthica*, is achieved by rotating maize or sorghum after a non-susceptible trap crop of cotton, beans or peanuts, which, under certain conditions, induce lethal germination of witchweed seeds.

C. Genetic Resistance of Crop Varieties

Local crop varieties often have some tolerance to prevalent pests and diseases as a result of repeated exposure and selection; tillering ability in local sorghums and millets is an effective recovery mechanism after stem boring, provided rainfall is adequate. Conscious selection against diseased bean seeds is practiced by Rwandan farmers.[32] However, levels of genetic resistance present in local varieties are often inadequate.

D. Adaptive Behavior in Pest Control

Timing and frequency of weeding by farmers generally show that an adaptive strategy is employed to minimize labor outlay. A similar approach has been demonstrated in the adoption of fixed-interval spraying against cotton and coffee insects in Tanzania, where small farmers have reduced the recommended number of sprays to match observed pest incidence and/or expected yield potential of the crop.[33]

V. FARM LABOR CONSTRAINTS

Small farmers in Africa differ most from their counterparts in Asia and Latin America in the low productivity of labor. This situation is due in part to the low fertility of many African soils, to the strong seasonality of African farming systems in semi-arid areas and to the continued use of hoe cultivation in many areas.

Labor is still a more critical constraint than land shortage in much of Africa. Communal and reciprocal labor arrangements are common, and make more efficient use of the family labor that is available, while the importance of task differentiation by gender is discussed elsewhere in this volume.

Since clearing land with hand tools is the most laborious component of shifting cultivation systems, the indigenous development of stable, semi-permanent systems has had a strong incentive. The decision to abandon or to fallow a particular field is triggered by the laboriousness of hoe weeding as weed species adapted to low fertility come to dominate annual crops.[34] Concern for optimizing the returns to labor also account for the widespread practice of choosing to locate a new field on an easily cultivated sandy soil, rather than on a heavier and more fertile valley-bottom soil. Unfortunately, it is this logical short-term response by each individual subsistence farmer that has led to overcultivation of the more fragile soils, including sodic soils which are highly erodible once disturbed.

Most semi-arid annual cropping systems experience a labor bottleneck at the start of the rainy season. Under these conditions a basic sequence of priorities usually is followed in preparing and planting the land, to ensure subsistence: food crops usually take precedence over cash crops such as cotton, and different crops are planted according to soil type in a definite sequence that may be altered if the start of the rains is much delayed.[35]

Labor requirements for food production are better spread through the year in permanent cultivation systems with tree crops. Production intensity is greatest close to the house, saving time and effort, and is lowest in distant fields.

The gradual spread of animal traction, an indigenous

practice only in Ethiopia, during the past 70 years has relieved the seasonal labor bottleneck in land preparation in some semi-permanent systems based on fallows. As animal traction is not helpful in bush clearance, its use generally has not been adopted by farmers who practice shifting cultivation.[36] In areas such as the drier zone of southern Zimbabwe, farmers having oxen are able to plant more of their land earlier and obtain higher cereal yields.[14] In Sukumaland, Tanzania, the benefits from ox plowing derive from an increase in the area of heavy soil that is cultivable and in the total production that can be obtained by one family; yields per hectare are lower because tillage is less effective than the traditional hoe ridging system (designed to bury plant debris under the ridge where it can be used by the following crop) and because hoe weeding is made more difficult by broadcast sowing.[37] In this example, ox plowing appears to be economically viable but environmentally inappropriate.

The compromises forced upon hoe cultivators by their lack of power for tillage and for weeding are perhaps the aspects of traditional African agriculture that have been most difficult for traditionally educated Western agriculturalists to appreciate. In spite of the application of techniques such as those mentioned above, yields of many crops are reduced by late planting or by inadequate control of weeds. Farmers are obliged to make priorities. In the highlands of Ethiopia, the staple food crop, teff, is usually free of weeds despite its being a slow-growing, broadcast-seeded, small cereal and an extremely poor competitor. Weed control is achieved by the stale seedbed technique, using several plowings early in the season, followed after crop germination by hand pulling. Unsurprisingly, the food legumes — faba beans, lentils, chickpeas, and common beans — that complement this system are rarely weeded and, although they sustain yield loss to weed competition, they are less affected than teff would be.

Weeding frequency is a little less critical for many traditional varieties, including those grown as dual-purpose grain/fodder crops. These are vegetative in growth habit, with a low index of grain to total dry matter yield. Examples are the prostrate habit in legumes and tall, leafy, or tillering characters in cereal crops. Plants of this type tend to be competitive not only with weeds but also with others of the same species, and consequently are sown at relatively low densities.

VI. SMALL FARM RESEARCH AND DEVELOPMENT

Agricultural research in sub-Saharan Africa started with the establishment of research stations during the period 1900 to 1950. Most research programs were, and still are, organized to improve a crop commodity, priority being given initially to export crops and, in the 1960s post independence era, to staple cereal crops.

Production of crops such as cotton expanded rapidly in many countries, although the contribution of research was often to facilitate area expansion despite the emphasis given on stations to yield-increasing technology. New cotton varieties resistant to leafhoppers and bacterial blight were used by the Wasukuma people in opening up new land, leading to an eightfold increase in Tanzanian cotton sales between 1950 and 1966;[38] this development delayed to the present decade the pressing problem of declining soil fertility in densely populated areas of Sukumaland. In a similar semi-arid area of Uganda, where considerable indigenous development of farming systems had occurred between 1910 and 1938 with the adoption of cotton, of ox plowing, and of root crops into the diet, later government interventions have had little effect upon food crops productivity. As in other case studies, the introduction of ox plowing exacerbated the weeding labor bottleneck and reduced yields per hectare.[39] The development of effective, affordable ox-drawn weeders adapted to local circumstances is a key to the intensification of low-input semi-permanent cultivation systems and remains today the most urgent requirement in the field of mechanization.

Commodity-based research in West Africa also has had limited impact. Only 25% of the cereals area is estimated to be sown with new varieties and fertilizer use is less than 2 kg/ha, due to low and very variable response rates and to shortages of foreign exchange and of credit facilities for farmers.[40,41] Plant breeding, which has been estimated to comprise about one half of all tropical agricultural research, has often proved to be more site specific than had been expected.[42] Extension packages, including agronomic and plant protection components designed on research stations to realize high yield potentials from new varieties, all too often have performed poorly in farmers' less fertile fields or have proven difficult for them to apply.

Overcoming these limitations of traditional research programs has been the objective of farming systems research (FSR), introduced in most countries during the past 10 years. Understanding farmers' production constraints and priorities through surveys and other techniques is enabling commodity research programs, a convenient and efficient organizational form for interdisciplinary research, to adjust their objectives and design criteria to fit wider needs. For example, a maize variety for Burundi must be not only higher yielding but also no later in maturity than traditional varieties, to permit associated legume crops to yield well.[4] Characters that stabilize yields and that will be especially important in dry areas include drought tolerance, seedling vigor and resistances to pests or diseases.[40] Technology development is no longer considered complete until successful testing has been conducted under the control of representative farmers and adoption has started.

A wider range of attributes in new crop technology is now recognized as being necessary for small farmers managing dynamic agroecosystems. An early-maturing, erect cowpea variety has proven successful in semi-arid regions of Nigeria because it added a new strategy, based on double cropping, to farmers' traditional cereal/cowpea intercropping.[3] Its adoption should not be expected, however, to lead to rapid replacement of existing varieties and practices, which may be improved through specific development of other technology.

The incorporation of a farming systems perspective and on-farm research into commodity research programs is a necessary but insufficient condition to the development of more sustainable and productive systems for low fertility conditions.[42] New farming systems are also required, starting probably with the replacement of natural fallows with selected plants, mostly herbaceous legumes, that are effective in quickly restoring soil chemical and physical condition.

Fallow plants need to be selected for rapid growth to produce a closed canopy, high biomass production, deep rooting and nitrogen fixation ability; they need to be easy to clear and to yield attractive byproducts, such as stakes or fodder. Grazing would reduce the direct benefits but could improve crop/livestock integration by filling the dry season feed shortage. Selected indigenous species may be useful for adaptation to difficult conditions such as acid soils. Pilot scale results under relatively cool, humid conditions at Nyabisindu, Rwanda included a fourfold increase in subsequent maize yields through substituting a 10-month legume fallow.[43]

The critical change to stable permanent cultivation could be encouraged by developing adapted tree/crop systems analagous to the indigenous *Acacia albida* model. Alley cropping development by the International Institute of Tropical Agriculture in humid areas of Nigeria uses lopped branches as mulch for associated annual crops, and provides a means of recycling nutrients leached from the A horizon and of bringing up newly weathered material. In drier areas, competition for moisture between agroforestry components and protection of young trees from browsing animals pose new research problems.[44]

The most labor-intensive form of agroforestry, permanent multistorey cropping, might be reached by a stepwise progression from planted fallows.[44] Added fertilizer or manure may still be necessary to sustain this system, analagous to intensive compound gardens, on poorer soils.

VII. CONCLUSIONS

These new hopes for overcoming Africa's food crisis will require at least as much site-specific research as more traditional programs have employed, and with more attention to ecological aspects. Willing incorporation of farmers into the research process can tap local knowledge and assist design and evaluation, but this participation will need to be more effective and longer term than has been the case in most FSR and rural development projects.

REFERENCES

1. **Okigbo, B. N.**, Land Use and Production Potentials of African Savanna, in *Ecology and Management of the World's Savannas*, Tothill, J.C. and Mott, J.J., Eds., Commonwealth Agricultural Bureaux, Farnham Royal, 1986.
2. **Dennett, M. D.**, Variation in rainfall-the background to soil and water management in dryland regions, *Soil Use Manage.*, 3, 47, 1987.
3. **Ashraf, M., Balogun, P., and Jibrin, A.**, *A Case Study of On-Farm Adaptive Research in the Bida Agricultural Development Project—Nigeria*, OFR Bull. No. 1, International Institute of Tropical Agriculture, Ibadan, Nigeria, 1985, 29.
4. **Zeigler, R. S.**, Application of a systems approach in a commodity research programme: evaluating Burundi highland maize, *Exp. Agric.*, 22, 319, 1986.
5. **Steiner, K. G.**, *Intercropping in Tropical Smallholder Agriculture with Special Reference to West Africa*, German Agency for Technical Cooperation, Eschborn, West Germany, 1982, 303.
6. **Stoop, W. A.**, Agronomic management of cereal/cowpea cropping systems for major toposequence land types in the West African savanna, *Field Crops Res.*, 14, 301, 1986.
7. **Jones, M. J.**, Soil water and crop production in Botswana, *Soil Use Manage.*, 3, 74, 1987.
8. **Taylor, B. R.**, Groundnut agronomy research in southeast Tanzania, in *Proc. Regional Groundnut Workshop for Southern Africa*, ICRISAT, Patancheru, India, 1985, 99.
9. **Law, R. and Cooper P. J. M.**, The effect and importance of soil temperature in determining the early growth rate and final yield of maize in Western Kenya, *E. Afr. Agric. and Forest. J.*, 41, 189, 1976.
10. **Walker, T. S. and Jodha, N. S.**, How small farm households adapt to risk, in *Crop Insurance for Agricultural Development: Issues and Experience*, Hazell, P., Pomareda, C., and Valdez, A., Eds., Johns Hopkins University Press, Baltimore, 1985, chap. 2.
11. **Uchendu, V. C. and Anthony, K. R. M.**, *Agricultural Change in Geita District Tanzania*, East Africa Literature Bureau, Nairobi, Kenya, 1974, 1.
12. **Uchendu, V. C. and Anthony, K. R. M.**, *Agricultural Change in Kisii District Kenya*, East Africa Literature Bureau, Nairobi, Kenya, 1975, 1.
13. **Murra, J. V.**, El "control vertical" de un máximo de pisos ecológicos en la economía de las sociedades andinas, in *Vista de la Provincia de Leon de Huánuco (1562), Iñigo Ortiz de Zúniga, II*, Murra, J.V., Ed., Hermilio Valdizán University, Huánuco, Peru, 1972, 429.
14. **Collinson, M. P.**, Eastern Africa farming systems and some indications of research priorities, in *Proc. Eastern Africa-ACIAR Consultation on Agricultural Research*, Australian Centre for International Agricultural Research, Nairobi, Kenya, 1983, 187.
15. **Powell, J. M.**, Yields of sorghum and millet and stover consumption by livestock in the subhumid zone of Nigeria, *Trop. Agric. (Trinidad)*, 62, 77, 1985.
16. **McIntyre, J. and Gryseels, G.**, Crop-livestock interactions in sub-Saharan Africa and their implications for farming systems research, *Exp. Agric.*, 23, 235, 1987.
17. **Ahn, P. M.**, Soil factors affecting rainfed agriculture in semi-arid regions with particular reference to the Sahel zone of Africa, in *Proc. Int. Symp. Rainfed Agriculture in Semi-Arid Regions*, Cannell, G.H., Ed., University of California, Riverside, 1977, 128.
18. **Nye, P. H. and Greenland, D. H.**, *The Soil under Shifting Cultivation*, Commonwealth Agricultural Bureau, Farnham Royal, 1960, 1.
19. **Allan, W.**, *The African Husbandman*, Oliver and Boyd, Edinburgh, 1965, 1.
20. **Cornick, T. R. and Kirkby, R. A.**, *Interactions of Crops and Livestock Production in the Generation of Technology in Sloped Areas*, Cornell International Agriculture Mimeograph 90, Cornell University, Ithaca, NY, 1981, 1.
21. **Birch, H. F.**, Mineralization of plant nitrogen following alternate wet and dry conditions, *Plant and Soil*, 20, 43, 1964.

22. **Fairhead, J.**, Conceptions des Paysans pour Différencier des Varietés au Kivu Nord, Zaire, presented at Regional Workshop on Production and Improvement of Beans in the Great Lakes States, Kigali, Rwanda, November 18 to 21, 1987, 1.
23. **Abebe, M.**, Ecophysiology of niger (*Guizotia abyssinica* Cass.), in *Oil Crops Niger and Rapeseed/Mustard: Proc. Third Oil Crops Network Workshop*, Omran, A., Ed., International Development Research Centre, Ottawa, 1987, 230.
24. **Poschen, P.**, An evaluation of the *Acacia albida*-based agroforestry practices in the Hararghe highlands of Eastern Ethiopia, *Agrofor. Syst.*, 4, 129, 1986.
25. **Okigbo, B. N.**, Broadening the food base in Africa: The potential of traditional food plants, *Food Nutr.*, 12, 2, 1986.
26. **Nankumba, J. S.**, The socio-economic aspects of small gardens near smallholder farmers' houses in Malawi — a case study, *Res. Bull.* (Bunda College of Agriculture, Malawi), 10, 1, 1979.
27. **Balasubramanian, V. and Egli, A.**, The role of agroforestry in the farming systems in Rwanda with special reference to the Bugesera-Gisaka-Migongo (BGM) region, *Agrofor. Syst.*, 4, 271, 1986.
28. **Ruthenberg, H.**, *Farming Systems in the Tropics*, 3rd ed., Oxford University Press, London, 1980, 1.
29. **Usua, E.J.**, Observations on diapausing larvae of *Busseola fusca*, *J. Econ. Entomol.*, 60, 1466, 1967.
30. **Jackai, L. E. N. and Daoust, R. A.**, Insect pests of cowpeas, *Annu. Rev. Entomol.*, 31, 95, 1986.
31. **Southwood, T. R. E. and Way, M. J.**, Ecological background to pest management, in *Concepts of Pest Management*, Rabb, R.L. and Guthrie, F.E., Eds., N.C. State University, Raleigh, 1970, 6.
32. **Trutmann, P.**, personal communication, 1987.
33. **Cox, P. G.**, The Organisation of User Recommendations and Pesticide Distribution in Tanzania, AAU Discussion Paper 9, Overseas Development Institute, London, 1982.
34. **Moody, K.**, Weeds and shifting cultivation, *Pest Artic. News Summ.*, 21, 188, 1975.
35. **Matlon, P. J.**, Local Varieties, Planting Strategies, and Early Season Farming Activities in Two Villages of Central Upper Volta, West Africa Economics Progress Report 2, ICRISAT, Upper Volta, 1980.
36. **Pingali, P., Bigot, Y., and Binswanger, H. P.**, *Agricultural Mechanization and the Evolution of Farming Systems in Sub-Saharan Africa*, The World Bank, Washington, D.C., 1987, 216.
37. **Kjoerby, F.**, *Problems and Contradictions in the Development of Ox-Cultivation in Tanzania*, Research Rep. No. 66, Scandinavian Institute of African Studies, Uppsala, Sweden, 1983, 163.
38. **Percy, H. C.**, Factors affecting cotton production in western Tanzania, *Cotton Grow. Rev.*, 52, 253, 1975.
39. **Carr, S.**, *The Impact of Government Intervention on Smallholder Development in North and East Uganda*, A.D.U. Occasional Paper No. 5, School of Rural Economics, Wye College, University of London, 1982, 88.
40. **Matlon, P. J. and Spencer, D. S.**, Increasing food production in Sub-Saharan Africa: environmental problems and inadequate technological solutions, *Am. J. Agric. Econo.*, 66, 671, 1984.
41. **Eicher, C. K. and Baker, D. C.**, *Research on Agricultural Development in Sub-Saharan Africa: A Critical Survey*, MSU International Development Paper 1, Michigan State University, East Lansing, 1982, 335.
42. **Simmonds, N. W.**, *Farming Systems Research: A Review*, Tech. Paper 43, The World Bank, Washington, D.C., 1985, 87.
43. **Prinz, D.**, Increasing the productivity of smallholder farming systems by introduction of planted fallows, *Plant Res. Dev.*, 24, 31, 1987.
44. **Raintree, J. B. and Warner K.**, Agroforestry pathways for the intensification of shifting cultivation, *Agrofor. Syst.*, 4, 39, 1986.

The Dynamics of Traditional Agriculture: Case Studies

Monsoon Environments

19 SMALL-SCALE AGRICULTURE IN SOUTHEAST ASIA

Gerald G. Marten

I. INTRODUCTION

Southeast Asia is a region of impressive cultural, environmental, and agricultural diversity. Much of the region is mountainous, and there is a broad range of temperatures from tropical in the lowlands to temperate in the highlands. There is also a broad range of topographic conditions, from flat alluvial valleys and coastal plains in the lowlands to undulating terrain, hills, and mountains in the uplands and highlands. Most of Southeast Asia is in the humid tropics, but there is great variation in the distinctness and duration of the dry season. Most of the land is now under agriculture (Table 1).[1]

Much of Southeast Asia is blessed with fertile volcanic or alluvial soils. Such areas, particularly the river valleys, have high human populations. Many of the mountainous areas have been dominated by forests until recent decades, but they are rapidly being transformed to agriculture as a consequence of logging and the movement of expanding human populations into land available for farming. Some areas of Southeast Asia have extremely poor soils. Until recently, most of those areas were forested and had small human populations, usually practicing shifting agriculture. Commercial logging and colonization projects are now transforming the landscape in many of those areas to agriculture.

The overriding theme of Southeast Asian agriculture is change. In addition to the spread of agriculture into forest lands, recent decades have seen major changes in agricultural technologies as a consequence of international and national programs for agricultural development. In fact, throughout the centuries there has been an influx of new agricultural technologies to Southeast Asia as various world powers have asserted their influence in the region. Farmers in the region have always been receptive to new agricultural technologies that promised to improve their lot, and the present time is no exception. However, the pace of change has quickened. Many farmers in Southeast Asia now work with a mix of traditional and modern technologies.

Along with the adoption of modern technologies, there has been a rapid transformation to a cash economy. While even remote areas of Southeast Asia bartered for certain goods (e.g., salt or cooking utensils) throughout the centuries, most farmers in the region produced almost entirely for home consumption until a few decades ago. Meeting basic household food needs is still the priority of most Southeast Asian farmers. However, most also produce as much surplus as possible to meet cash needs generated by expanding public education, rural electrification, modern communications (e.g., radio and television), and modern transport that has tied farm families to major cities in their region. Some households now specialize in one or two high-value crops and purchase most of their food.

Scientists in the Southeast Asian Universities Agroecosystem Network (SUAN) have been particularly concerned with the implications of these changes [2] Can large human populations and intensive agriculture be sustained in hilly or mountainous areas that have had forests until recently? Can introduced high-yield varieties and high-yield technologies (based on high levels of energy and chemical inputs) be expected to continue to provide high yields on a long-term basis? SUAN scientists have addressed these questions concretely in terms of the specific environmental conditions and agricultural technologies in their areas.

SUAN scientists have also been concerned with interactions between the agriculture in their areas and the social systems of farmers who practice the agriculture.[3,4] The scientists want to understand the farmers' living circumstances and their bases for making agricultural decisions, in order to know what kinds of improvements in agricultural technology will be relevant and appropriate to the farmers' needs. The scientists also have come to appreciate the need to comprehend the major social forces (e.g., transformation from a subsistence to cash economy) that are driving agricultural changes, so they can anticipate needs and guide their research accordingly.

Three broad types of small-scale agriculture are prominent throughout the region: ricefields, rainfed fields, and

TABLE 1
Land Use in Southeast Asia (1980)[1]
(thousands of hectares)

Country	Total land area	Annual cropland[a]	Perennial cropland[b]	Permanent pasture[c]	Forest and woodland[d]
Burma	65,774	9,573	450	361	32,167
Indonesia	181,135	14,200	5,300	12,000	121,800
Kampuchea	17,652	2,900	146	580	13,372
Laos	23,080	860	20	800	13,000
Malaysia	32,855	1,000	3,310	27	12,300
Philippines	29,817	7,050	2,870	1,000	12,300
Thailand	51,177	16,250	1,720	308	15,790
Vietnam	32,536	5,595	460	4,870	10,330
Total	434,026	57,428	14,276	19,946	241,149

[a] Land under temporary crops (double cropped areas are counted only once), temporary meadows for mowing pasture, land under market and kitchen gardens, and land temporarily fallow or lying idle.
[b] Land cultivated with crops that occupy the land for long periods and need not be replanted after each harvest, such as cocoa, coffee, and rubber; it includes land under shrubs, fruit trees, nut trees, and vine but excludes land under trees grown for wood or timber.
[c] Land in permanent (5 years or more) herbaceous forage crops, either cultivated or growing wild (wild prairie or grazing land).
[d] Land under natural or planted stands of trees, productive or not, includes land from which forests have been cleared but will be reforested in the foreseeable future.

homegardens. Rice is the staple food of most Southeast Asians, and flooded rice paddies dominate most agricultural landscapes in Southeast Asia. Rice is particularly prominent in floodplains and alluvial valleys, but terraced rice is also common in upland areas wherever irrigation water is available.

Rainfed fields are also common, particularly in upland areas where irrigation is not available. Rainfed fields consist primarily of annual field crops, often several different kinds interplanted in the same field. Sometimes there are fruit trees or other perennial crops scattered through the field as well. Many rainfed fields are permanent; others are rotated with a forest fallow. Paddy fields can function as rainfed fields during the dry season if they have field crops at that time.

The third major form of agriculture, homegardens, though less extensive in area than ricefields and rainfed fields, is no less ubiquitous. Nearly every house in the region is surrounded by some kind of garden, usually a mixture of shade trees and fruit trees and sometimes containing a selection of vegetables or other annual crops. Homegardens are almost always rainfed, though select crops in the garden may be hand irrigated.

This chapter will describe the agriculture of three Southeast Asia locations that have been studied by SUAN scientists (Figure 1): the Cordillera highlands in the Philippines; the uplands of West Java, Indonesia; and Chiangmai Valley in northern Thailand. These three areas represent not only a progression from highlands to lowlands but also a progression from highly traditional and subsistence agriculture to agriculture that is more modern and involved in a market economy. Each of these areas has all three types of agriculture mentioned above (i.e., ricefields, rainfed fields, and homegardens), though the details of their organization are different in each area.

Each of these areas has been selected because its agriculture has functioned so well. At the same time, the agriculture in each of these areas is changing rapidly. The changes are a response to compelling needs, but as we shall see, the changes have run into ecological and social problems. The challenge presented by these problems cannot be ignored when setting an agenda for agricultural research and development.

II. PHILIPPINE CORDILLERA (BONTOK)

The Bontok are highly traditional subsistence farmers at an altitude of 600 to 2100 m in the Cordillera of the Philippines. Villages have 600 to 3000 inhabitants who occupy a territory of 10 to 30 square kilometers. The Bontoks provide an example of agriculture that has been sustained on steep mountain slopes for centuries without ecological degradation. The following description is based on studies by the Cordillera Studies Center (University of the Philippines, Baguio) and the Institute for Environmental Science and Management (University of the Philippines, Los Baños).[5-10]

There are five major land uses: rice paddies, swidden fields, grazing areas, forest, and villages (including homegardens). Most agricultural labor is devoted to the paddy fields, which are terraced and occur primarily on the lower portions of mountain slopes wherever streams are large enough to provide irrigation water. The fronts of the terraces are held in place by stone walls. The main crop is traditional rice, which is cultivated during November to July, when sunshine is at a maximum during the dry season. A second crop, rice or field crops such as sweet potatoes, is grown in

FIGURE 1. Location of the Philippine Cordillera, West Java, and Chiangmai Valley in Southeast Asia.

paddy fields during the wet season (July to December).

Swidden fields are located higher on the slopes and are primarily rainfed. The swidden fields contain interplanted crops such as millet, sweet potatoes, beans, squash, corn, bananas, and fruit trees. Planting starts with the first rains in April, and harvests continue to the end of the rainy season in December. The timing of each swidden crop is critical, so it does not overshade or otherwise interfere with other crops in the same field. Swidden fields and traditional rice compete for labor during April to July, but the rice always has priority.

The swidden fields are not terraced, but they usually have earthen or wooden fences to hold the soil and exclude wild animals. Logs and sticks are often aligned along the contour to help hold the soil. A swidden field is cultivated for several years until yields start to decline. It is then left to fallow for about 5 years before resuming cultivation.

Higher on the slopes are pine and oak forests, which are exploited for timber (for house construction in the village), medicinal plants and herbs, mushrooms, wild animals, rattan vines, and bamboo. Level portions of the hilltops often have pastureland, which is used for grazing water buffalo and cows and as a source of grass for thatched roofing. While ricefields and swidden fields are owned individually, forest and pasturelands are owned communally by the village.

The agricultural calendar and cultivation practices of these farmers are a matter of long-standing tradition, interpreted and enforced by village elders. The elders organize labor for maintaining irrigation canals and make decisions on the timing of land preparation, transplanting, and other activities for the traditional rice crop.

Pigs occupy a central role in the village economy. They are the major source of meat for religious ceremonies and the major source of animal manure for paddy fields. Pigs are kept in stone-lined pits between the houses, where they are fed kitchen garbage, cooked sweet potatoes, and raw sweet potato tops. The pits are slanted so manure falls to one end, where it is composted with dry grass, rick husks, straw, and human wastes that are thrown into the pit. Some of the sweet potatoes are grown between the houses and constitute the major component of village homegardens.

A. Terraced Rice Paddies

Paddy field preparation begins in October by using a spade to turn over weeds and rice stalks from the previous harvest, trampling them into the mud to rot. The field is then harrowed. A man stands on the spike harrowing board as a buffalo pulls it through the mud, scattering decaying rice stalks so they cannot take root. The field is then flooded so the farmers can level the mud with the palms of their hands. Communal labor groups clean out irrigation canals at this time.

Mud from the paddy fields is packed on top of the bunds surrounding the fields and against all paddy field margins to plug holes and improve the moisture seal of the field. Taro is planted along the terrace top or laced on the rims of the paddy field. Beans or sweet potatoes may be planted at the sides of the bunds. A few weeks later it is time to weed the empty paddy fields again. The stone walls of the terraces are also weeded (with a trowel), and rat holes are stuffed with weeds. Weeds are collected from slopes immediately above the field, thrown into the fields and trampled into the mud. Pig manure compost is brought from the village in baskets and mixed into the mud.

Seedbeds are established in paddy fields with a history of high fertility and a water supply that will not be interrupted by paddy field preparation and cleaning of irrigation canals. A small section of a field is blocked off for the seedbed, so it can be drained even when the rest of the field is flooded. The seedbed is strewn with rice husks, dried bean pods, and sunflower leaves and stems, which are trampled into the mud to rot. Rice pannicles with the largest quantity of grain are chosen for the seedbed during the previous harvest.

Seedbeds are planted in November to January; pannicles about a foot in length are pushed into the mud and bent so the rice grains lie flat on the ground. Water is temporarily drained from the seedbed so the pannicles do not float. The seedbed is flooded with 1 to 2 cm of water, which discourages animals such as rats and birds from eating the seedlings. If the seedlings do not appear healthy, ashes are spread around them to increase the fertility of the bed.

Seedlings are ready for transplanting in February and March. In preparation, paddy fields receive their final smoothing, kneading, and leveling. The earth is reworked with a harrow or spading fork and subjected to another puddling with the feet. The field is drained, and the top 12 to 15 cm of soil is given a final working and leveling either with the hands or with a board that is dragged around the field by a buffalo. Farmers believe that mud in shallower water is warmer and enhances growth and flowering of the rice plants.

Women do the transplanting in groups. While some plaster mud on the margins of the fields, others bring bundles of five or six pannicles (approximately 100 to 150 seedlings) from the seedbeds. (Seedlings from seedbeds with nematodes are not used.) Different rice varieties are planted in different fields. For example, one variety does best in paddy fields that were drained during the fallow while another variety does better in fields that could not be drained.

The women tear the top leaves off the seedlings as they plant them. This is said to make the rice plant mature faster, perhaps by reducing transpiration and making the seedlings more resistant to drought. Shorter seedlings are also more resistant to being blown over by wind. The seedlings are pushed into the mud 10 to 15 cm apart with only the top 10 to 12 cm above the surface. High tillering varieties and high fertility paddies are planted less densely (i.e., 15 cm spacing). Seedlings that die after transplanting are replaced by seedlings from other parts of the field or from other fields.

A continuous supply of water is essential once the rice is transplanted. There is a village system of water rotation among the fields, but each household finds it advisable to oversee personally the delivery of water to its fields, in order to ensure its fair share. Disputes may arise and tempers wear thin as the season progresses and the supply of water diminishes.

The paddy fields are weeded 1 month after transplanting. This is the only weeding after transplanting and must be completed before appearance of rice fruitheads, which are susceptible to damage by body contact. *Azolla,* which fixes nitrogen, as well as other floating aquatic plants like *Lemna* and *Spirodela,* have spread over the surface of the water by this time and are allowed to coexist with the rice plants. These floating plants suppress the growth of rooted weeds that compete with the rice crop for soil nutrients. Fresh water snails and edible plants growing in the paddy (e.g., *Monochria vaginalis)* are gathered as food.

Once the rice heads begin to form, the main activity is setting up scarecrows. However, after a few weeks the scarecrows are no longer so effective, as the birds become accustomed to them and the maturing rice grains become more attractive. Sometimes the scarecrows are strung together with long strings attached to a paddle placed in a stream to move the string and make tin cans bump each other to produce a clanging sound. Village elders may find it necessary to announce a special day for a coordinated village effort to rid the rice fields of rats and birds. Rat holes are plugged with grass. Children and elderly people stay in small huts around the fields so they can operate noisemaker lines throughout the day.

The rice is ready for harvest by late July. This is a time of particular hazard because the typhoon season is about to start. An entire season's crop can be destroyed by typhoons if the harvest is not on schedule. Moreover, once the crop is ready, it must be harvested quickly, because even a few days' delay can lead to shattering and loss of over-ripe grain. Sufficient labor for harvesting is assured by staggering activities on the different fields of different households and deploying cooperative labor groups on each field at the proper time.

Each harvester wears a ring with a blade that is used to cut the top 30 to 35 cm from the rice stalks. Bundles of the cut pannicles (3 to 5 cm in diameter) are tied with a bamboo strip pulled tight by the front teeth. The bundles are spread on the ground to dry in the sun for at least 3 days before placing them in the granery. It is important for the grain to be completely dry to prevent molding or premature germination. Properly stored grain can last for as much as 8 to 10 years.

Yields are high, typically 6 t/ha, even though many of these paddy fields have been used continuously for centuries without chemical fertilizer. Soil fertility is maintained through organic matter and mineral inputs from pig manure compost, forest soil, weeds, and rice straw, as well as detritus and dissolved minerals in the irrigation water. Inputs of 260 kg/ha nitrogen and 120 kg/ha phosphorus have been measured from pig manure compost alone.

About half the paddy fields are drained after harvest and left to fallow the rest of the year. The other half is planted to a second rice crop or a field crop such as sweet potatoes, peanuts, beans, or garlic. The second rice crop employs varieties that mature more quickly than the first. They are ready for harvest 4 months after transplanting, but yields from the second rice crop are correspondingly lower. Cultivation procedures for the second rice crop are similar to the first, but labor groups are smaller and there is no coordination of labor by the village elders. A disadvantage of a second rice crop is that it may compete with labor for swidden fields. It is also difficult to harvest the second rice crop in time to prepare fields for the first rice crop. A delay in starting the first can be particularly serious if it is not harvested before typhoons begin.

Sweet potatoes are the most common crop in drained paddy fields during the rainy season. The field is prepared in September by ditch mounding soil over beds of rice straw. Sweet potato cuttings are inserted into the mounds, which are weeded as soon as new roots have formed and runners appear. Farmers say that drying out the paddy soil with a sweet potato crop is important because it reduces stunting and mortality of rice seedlings during the first critical week after transplanting.

The sweet potatoes are harvested in December. They are stored in a corner of the house without washing off soil clinging to the surface because it is said the soil protects them from deterioration. Young runners on sweet potato vines are gathered for human consumption, and mature leaves are used to feed the pigs.

B. Contemporary Changes

Bontok farmers have tried to be selective in adopting only those outside agricultural practices they can control, thus avoiding dependence that might interfere with village affairs. Nonetheless, there has been an increasing impact from the outside world due to greater cash needs, outmigration of villagers, their involvement in nontraditional employment, and a growing presence of the national government. Villagers are now selling some of their surplus of swidden and drained paddy-field crops such as beans, eggplant, and fruit. They have started to plant garlic in their drained paddy fields. Garlic commands a high price in the urban market but is sold locally to middlemen who pay a much lower price. Villagers at higher altitudes are growing vegetable cash crops such as potatoes, cabbages, tomatoes, stringbeans, and carrots in any paddy fields that lack sufficient irrigation water for a rice crop.

As cash has become more essential to pay for college educations, hospital bills, bad business deals, lawyers' fees, and construction of "modern" houses, villagers are working away from the village more (e.g., in nearby copper and gold mines). The result has been an agricultural labor shortage in the villages, especially during harvest time. Farmers employed away from the village are increasingly lending out their paddy fields on share-cropping arrangements or selling them for cash, and reciprocal labor groups are beginning to be paid in cash when the owner of a paddy field is employed outside the village and cannot reciprocate the labor exchange. A paddy field that is left unflooded and uncultivated for more than a year develops an overgrowth of grass and bushes whose roots may break up the stone walls holding up the terraces. The terraces may collapse if the mud at the edge of the field dries out and starts to crumble. Some of the swidden gardens have been neglected in recent years due to insufficient labor, and an increasing number have passed into fallow.

Commercial fertilizers were not taken up by the villagers during the 1960s and 1970s, but chemical fertilizers are now being adopted by farmers who have sufficient cash but are short on the household labor to carry bulky compost to the fields. Some farmers who have adopted commercial fertilizers in the past 5 years are beginning to complain about a change in the consistency and stability of their paddy field soil. Although they agree that yields have increased, they also say they are able to maintain higher yields only by increasing the amount of fertilizer they apply each year.

Farmers in the nearby Loo Valley now devote all of their land to cabbages and potatoes for the urban market. Wide swings in the market prices of these two crops (as much as fivefold in a few months) offer an opportunity for enormous profits if a large harvest is realized when the price is high, but the experience of most farmers in the Valley is much more modest.

The costs of inputs for Loo Valley have become oppressive, as it has become necessary to apply massive quantities of chicken manure (averaging 7 t/ha/year) and NPK fertilizer (averaging 1 t/ha/year). The fertilizers have caused severe soil acidity, and the large areas in continuous monoculture of these vegetables have led to severe pest and disease problems. Pest control, which has changed from traditional removal by hand and use of extracts of avocado, tobacco, and chili pepper leaves to modern pesticides, is far from satisfactory. The diamondback moth, for example, has already evolved resistance to malathion, methylparathion, DDT, diazmon, mevinphos, and carbaryl, forcing the farmers to employ restricted insecticides that are hazardous to their health and to consumers.

Cabbage yields in the Loo Valley are double what they were 20 years ago, but input costs have increased 20-fold; input costs for potatoes are ten times as great with no increase in yields. Because the farmers must borrow money to cover the costs of these inputs, they have been forced into a cycle of high interest payments and cumulative debt that recently has led some of them to turn to lower-value crops (e.g., sweet potatoes) that do not demand such costly inputs. Severe erosion, due to the lack of ground cover under continuous vegetable production and the fact that farmers cannot spare their labor for terracing or other erosion control measures, combined with diminishing water supplies from mountain streams that serve as irrigation sources, have placed in doubt the entire future of commercial farming in the Valley.

TABLE 2
Land Use in West Java (1980)[12]

Land use	Area (ha)
Ricefields	1,168,691
Plantations	670,979
Forest	968,166
Upland fields and homegardens	1,609,567
Total	4,417,403

III. UPLANDS OF WEST JAVA

West Java is an area of fertile volcanic soils and exceptionally high population density, typically 600 to 1200 inhabitants per square kilometer. The terrain is hilly in the uplands (500 to 1200 m in altitude), where gently terraced rice fields occupy the lower and flatter areas. Villages are usually located on slightly elevated land at the edges of the rice fields. Land above the village is dominated by rainfed upland fields that contain a variety of interplanted annual and perennial crops (Table 2). The highest parts of the hills usually have forest or woodlots. West Java provides examples of highly intensive agriculture that has been maintained for centuries on an ecologically sustainable basis, despite severe land scarcity. The following description is based on studies by the Institute of Ecology, Padjadjaran University, Bandung.[11-15]

Virtually every family in the uplands of West Java has a garden around its house, and most families have at least some rice paddies or upland fields (Table 3). The total area of land available to each household is small, however. Few have access to more than a hectare of land and for most it is less than half a hectare. Each household's land is usually scattered in parcels one tenth hectare or less in area. The rice paddy zone has a rather uniform appearance, but the landscape above the ricefields has more of a patchwork appearance because each of the upland fields may contain different crops. Most villages have several hundred different crop species in their homegardens and upland fields (Table 4).

The single most important crop is rice. Most paddy fields are irrigated and planted to two rice crops each year, one during the rainy season (September to February) and one during the dry season. Some irrigated paddies are planted continuously to rice, five crops every two years. Rainfed paddies can be planted to rice only once each year, during the rainy season. They may be planted to other crops such as tobacco, beans, corn, or vegetables after the rice.

Most farmers put the bulk of their labor into their ricefields. They work their homegardens in their spare time. Labor demands in ricefields, upland fields, and homegardens compete at some times of the year and complement each other at other times (Figure 2).

Production is first and foremost for home consumption, but the surplus production of most households is sufficient for them to sell about half of their total production on average. Rice is the main source of calories, protein, and

TABLE 3
Landholdings, Crop Diversity, Income Characteristics of Households in the Jatigede Area, West Java[13]

	Poor		Well-off	
	Near market	Far from market	Near market	Far from market
Percentage of households with homegardens	97	100	99	97
Percentage of households with upland fields	57	89	82	97
Percentage of households with rice fields	73	100	91	97
Average size of homegardens (m^2)[a]	223	297	417	394
Average size of upland fields (m^2)[a]	1,120	2,604	3,995	7,174
Average size of rice fields (m^2)[a]	2,050	2,426	6,853	5,240
Total area farmed (m^2)	2,254	5,037	9,962	12,366
Average number of plant species in homegardens	14.3	20.4	20.0	21.9
Average number of plant species in upland fields	4.1	10.8	10.7	11.4
Average annual cost of farm inputs (×1,000 rupiah)[b]	48	78	156	223
Average annual gross farm income (×1,000 rupiah)[b]	114	143	654	508
Average annual off-farm income (×1,000 rupiah)[b]	108	103	550	236
Average family size	4.6	4.8	4.9	3.9

[a] Average based only upon households having that agricultural system.
[b] One U.S. dollar = approximately 670 rupiah.

certain vitamins (e.g., niacin and thiamine), but upland fields and homegardens produce the fruits, vegetables, and pulses to compensate for amino acid deficiencies in rice and provide vitamins and minerals (e.g., vitamin A, vitamin C, riboflavin, calcium, and iron) for which rice is insufficient (Figure 3). Many households earn an annual net income of about $500 from their farming, but many other households earn considerably less (Table 3). A typical household purchases about one quarter of the food it consumes.

A. Homegardens

Homegardens are a mixture of trees, field crops, ornamental plants, medicinal plants, and animals on the land surrounding a house (Figure 4). Ornamentals are more frequently planted in front of the house, while valuable crops (e.g., clove, oranges, and mangoes) are also planted in the front yard where the owners can keep an eye on them. Coffee may be planted along the side and back as a hedge. Vegetables are usually grown in the front and side yards, where there is sufficient light because tall trees rarely are planted in those areas. Spices, taro, and vegetables (e.g., chayote and eggplant) are planted close to the latrine and fish pond where soil moisture is high. Crops with high nutrient requirements (e.g., banana, mango, jackfruit, and other fruit trees) are

TABLE 4
Crops Observed in Homegardens and Upland Fields of Jatigede Area, West Java[13]

Latin name	Local name	English name
Grains, starchy roots, and sugar		
*Amorphophallus campanulatus	Suweg	Elephant yam
*Amorphophallus variabillis	Iles-iles	Elephant yam
*Arenga pinnata	Aren	Sugar palm
A. saccharifera	Kawung	
*Canna edulis	Ganyong	—
*+Colocasia esculenta	Taleus, Keladi	Taro
*Colocasia nigrum	Talas	Taro
*+Dioscorea alata	Ubi manis	Yam
	Ubi tiang	
*+Dioscorea bulbifera	Ubi atas	Yam
*+Dioscorea esculenta	Gembili	Yam
	Ubi kamayung	
*+Dioscorea hispida	Gadung	Yam
+*Ipomoea batatas	Ketela rambat,	Sweet potato
	Ubi jalar	
+*Manihot glaziovii	Singkong karet	Manicoba
+*Manihot utilissima	Singkong,	Cassava
	Ubi kayu	
Maranta arundinacea	Sagu patat,	Arrowroot
	Arut	
+Oryza sativa	Padi	Rice
Oryza sativa var. glutinosa	Beras ketan	Sticky rice
Saccharum officinarum	Tebu	Sugar cane
Sorghum vulgare	Gandrung,	Sorghum
	Cantel	
+Zea mays	Jagung	Maize, corn
Fruits		
Achras zapota	Sawo	Sapodilla,
		Naseberry
*Anacardium occidentale	Jambu monyet	Cashew nut
*Ananas comosus	Nenas	Pineapple
*+Annona muricata	Nangka belanda,	Soursop
	Sirsak	
Annona reticulata	Buah nona	Sweetsop
Annona squamosa	Sarikaya	Custard apple
Artocarpus communis	Keluwih	Breadfruit
Artocrapus champeden	Sukun	Breadfruit
*+Artocarpus integra	Nangka	Jack fruit
*Averrhoa carambola	Blimbing	Star fruit
*+Carcia papaya	Pepaya	Papaya
Citrus aurantifolia	Jeruk nipis	Lime
*Citrus aurantium	Jeruk manis	Tangerine
*Citrus maxima	Jeruk bali	Pomelo
*+Cocos nucifera	Kelapa	Coconut palm
Durio zibethinus	Durian	Durian
Erioglossum rubiginosum	Kilalayu	—
*Eugenia aquea	Jambu air	Rose apple
Eugenia cumini	Juwet	Java plum
	—	Teste foure
Eugenia cymosa	Kopo	—
Eugenia malaccensis	Jambu bol	Malay rose apple
Syzgium malaccensis		
Fragaria xananasa	Arben	Strawberry
Garcinia mangostana	Manggis	Mangosteen
Lansium domesticum	Duku	Lanseh
Lansium sp.	Pisitan	—
	Langsat	
Mangifera foetida	Limus	Horse mango
	Bacang	
*+Mangifera indica	Mangga	Mango
Mangifera odorata	Kuweni	Mango

Latin name	Local name	English name
Morus alba	Murbai	Mulberry (white)
*+Musa paradisiaca,	Pisang	Banana
M. sapientum		
*Nephelium lappaceum	Rambutan	Rambutan
Persea americana	Apokat	Avocado
P. gratissima		
Phyllanthus acidus	Cereme	Gooseberry
Zalacca edulis	Salak	Salacia
Spondias dulcis	Kedongdong	Otaheite-apple
S. eytherea		
Tamarindus indica	Asam	Tamarind
Vegetables (including legumes)		
+Amaranthus hybridus	Bayam	Spinach
Apium graveolens	Seledri	Celery
+*Arachis hypogaea	Kacang tanah,	Groundnut
	Suuk	Peanut
Benincasa hispida	Beligo	Wax gourd
B. cerifera		
Cajanus indicus	Kacang kayu	Pigeon pea
C. cajan	Hiris	
Cucumis sativus	Bonteng,	Cucumber
	Ketimun	
Cucurbita moschata	Waluh	Squash
Cucurbita pepo	Waluh,	Sweet gourd
	Labu besar	Pumpkin
Dolichos lablab	Kara	Hyacinth bean,
		Bonavista bean
Glycine max	Kedele	Soybean
Gnetum gnemon	Belinjo	—
Ipomoea reptans	Kangkung	Unchoy
Kaempferia rotunda	Temu kunci	—
Lagenaria leucantha	Kukuk,	Bottle gourd
	Labu air putih	
Limnocharis flava	Genjer	Hermit's water
Luffa acutangula	Emes, Oyong	Raggourd
	Gambas	Wild petola
Momordica charantia	Paria,	Bitter melon
	Pare	
Moringa oleifera	Kelor	Horseradish
*Nothopanax scutellarium,	Mangkokan	False panax
N. fruticosum		
Pangium edule	Picung	Pangi
	Keluwak	
*+Parkia speciosa	Pete,	Locust bean
	Petai besar	
Phaseolus lunatus	Kacang uci	Rice bean
Phaseolus radiatus	Kacang hijau	Mung bean
Planchonia valida	Putat	Planchonia
Polyscias fruticosa	Kedongdong	Hedge panax
	pagar	
Psophocarpus tetragonolobus	Kecipir,	Wing bean
	Jaat	Goa bean
Raphanus sativus	Lobak	Chinese radish
Sauropus androgynus	Katuk	Sweet shoot
Sechium edule	Labu siam	Chayote
*Solanum lycopersicum	Tomat	Tomato
*Solanum melongena	Terong	Eggplant
Solanum nigrum	Leunca	Black nightshade
Solanum torvum	Takokak,	Eggplant
	Cepoka	
+Vigna cylindrica	Kacang panjang	Cowpea
Vigna sinensis	Kacang panjang	Cowpea

TABLE 4 (continued)
Crops Observed in Homegardens and Upland Fields of Jatigede Area, West Java[13]

Latin name	Local name	English name
Spices		
Aleurites molluccana	Kemiri	Candle nut, Indian walnut
Allium cepa	Brangbang, Bawang merah	Onion
+Allium fistulosum	Bawang daun	Green onion
Allium sativum	Bawang putih	Garlic
*Allium schoenoprasum	Bawang kucai	Chive
Amomum kepulaga	Kapulaga	Cardamom
Capsicum annum	Cabe	Chili
	Lombok	Red pepper
*+Capsicum frutescens	Cengek, Lombok rawit	Little pepper
Chysopogon nardus	Sereh	Lemon grass
Cymbogon citratus		
*+Curcuma domestica	Koneng, Kunir	Turmeric
*Eugenia aromatica	Cengkeh	Clove tree
Ocimum canum	Surawung, Kemangi	Hoary basil
Phaeomeria speciosa	Honje	—
Piper nigrum	Lada, Pedes, Merica	Black pepper
Sesamum indicum	Wijen	Sesame
Syzygium polyanthum	Daun salam	Bay leaf
Eugenia polyantha		
*Zingiber officinale	Jahe	Ginger
Medicinals		
Areca catechu	Jambe, Pinang	Betel palm
Blumea balsamifera	Sembung	Camphor plant
Cassia fistulosa	Ketepeng	Golden shower
C. sappans	Trengguli	Indian laburnum
Citrus hystrix	Jeruk purut	Bitter orange
Coffea arabica	Kopi	Coffee
C. robusta		
Costus speciosius	Pacing	Ginger lily
Erythrina lithosperma	Bintinu, Dadap	—
+*Kaempferia galanga	Kencur, Cikur	—
+*Languas galanga	Lengkuas, Laja	Galangal
Morinda citrifolia	Mengkudu	Fish-eye wood, Noni
Mucuna pruriens	Coas	Florida velvet bean
Nicotiana tabacum	Tembakau	Tobacco
Orthosiphon stamineus	Kumis kucing	Cat's whiskers
Piper betle	Sirih	Betel vine
+Pithecellobium lobatum	Jengkol, Jiring	—
Ricinus communis	Jarak	Castor oil plant
Strobilanthes crispus	Keji beling	—
Thea sinensis	Teh	Tea
Tinospora tuberculata	Bratawali	—
Zingiber aromaticum	Panglay	—
Zingiber odoriferum	Lempuyang	—
Trees (building material)		
*Albizia falcata	Sengon laut, Albasia	—
+Albizia sp.	Jeungjing, Sengon tarisi	—
Canangium odoratum	Kenanga	—
Dalbergia latifolia	Sonokeling	—
Dixolylon caulostachyum	Kedoya	—
+Gigantochloa apus	Bambu apus	Bamboo
+Gigantochloa verticillata	Bambu gombong	Bamboo
*Swietenia mahogani	Mahoni	Mahogany
Tectona grandis	Jati	Teak wood
Trees (fuel wood)		
Acacia auriculiformis	Akasia	—
Albizia lebbeck	Tarisi	—
Albizia procera	Kihiang	—
Bridellia monoica	Kanyere	—
Cassia bicapsularis	Bungur	—
Cassia siamea	Johar	Senna
Casuarina equisetifolia	Cemara laut	Ironwood
*Ceiba pentandra	Kapok, Randu	Kapok
Ficus virens	Bisoro	—
Hibiscus macrophyllus	Tisuk	Hibiscus
Hibiscus tiliaceus	Waru	Hau tree
H. semilis		Hibiscus
Kleinhovia hospita	Bintinu	—
Lanea grandis	Kedongdong, Jaran	—
Lantana camara	Sahari, Temblekan	Hedge flower
*Leucaena glauca	Lamtoro, Petai cina	Leucaena
Melia azedarach	Mindi	Persian lilac
Piper aduncum	Kiseureuh	—
Sansevieria trifasciata	Lidah mertua	Bowstring hemp
Semecarpus sp.	Renghas	—
Sesbania grandiflora	Turi	—
Toona sureni	Suren	—
Cendrella toona		

Note: * Common in homegardens. + Common in upland fields. *+ More common in homegardens. +* More common in upland fields.

planted near the livestock pen, fish pond, or compost heap. Plants used for daily cooking (e.g., chili peppers, languas, lemon grass, and tomatoes) are planted close to the kitchen for convenience. Coconuts and tall trees for building materials and firewood are usually grown in the back, away from the house, so the house will not be damaged if the tree falls during a storm. A portion of the front yard shaded by fruit trees is kept clear as a play area for children and for socializing with neighbors.

Homegardens have a seasonal rhythm. Annual plants are grown throughout the year, but it is usually necessary to water by hand in the dry season. Some perennials (e.g., coconuts, bananas, and jackfruit) bear fruit throughout the

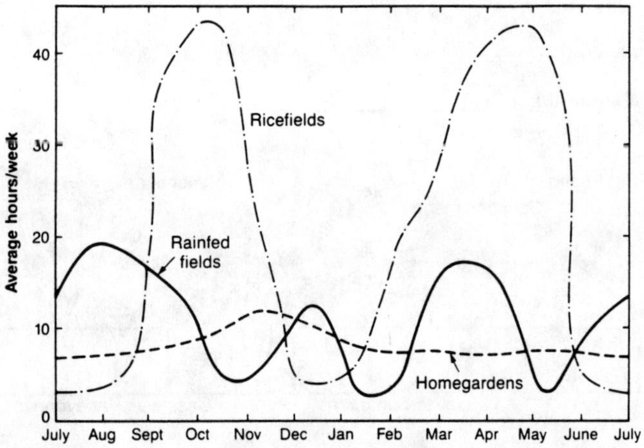

FIGURE 2. Seasonal labor patterns in ricefields, upland fields, and homegardens of the uplands of West Java. (From Iskandar, J. and Abdoellah, O. S., in *Agroecosystem Research for Rural Development*, Rerkasem, K. and Rambo, A. T., Eds., Chiang Mai University, Thailand, 1988, 237. With permission.)

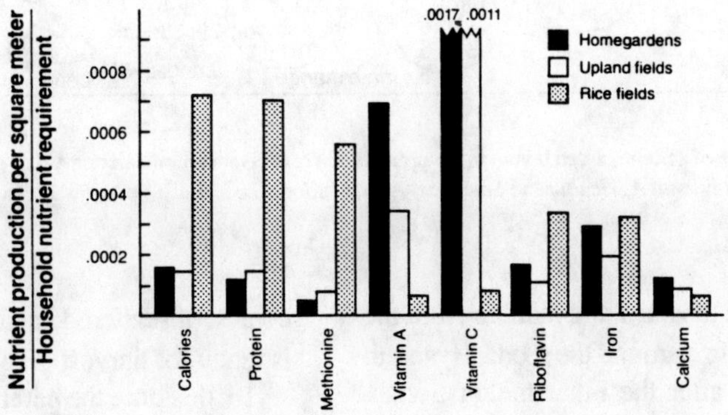

FIGURE 3. Comparison of the average production of human nutrients from a square meter of ricefield, upland field, or homegarden in the Jatigede area of West Java. (From Marten, G. G. and Abdoellah, O. S., *Ecol. Food Nutr.*, 21, 17, 1988. With permission.)

year while other perennials have restricted fruiting seasons. For example, *duku (Lansium domesticum)* fruits in December to January, *jambu semarang (Syzygium javanicum)* in April to June, mango in September to November, and durian in October to February. This pattern of harvesting provides a continuous supply of food for daily subsistence needs, minimizes risks of failure, and increases a household's financial stability by spreading cash income over the year.

B. Rainfed Upland Fields

Upland fields are typically a mixture of interplanted crops of various heights. The lowest layer consists of creeping plants such as peanuts, soybeans, cucumbers, and melon. Above them are taller vegetables such as chili peppers and eggplant. The top layer is occupied by maize, tobacco, cassava, or leguminous vines (e.g., wingbean or longbean) supported by bamboo poles. The field may also contain scattered fruit or other trees (e.g., *Albizia*).

Field preparation is in August. Farmers first weed the field and till the soil with a hoe. They cut down any unwanted perennial vegetation and leave the slash to dry. Litter, leaves, and slash are burned in small piles at the end of August. Different fields may contain completely different groups of crops, but each consists of crops which the farmers have found by experience to be compatible with one another. For example, bananas, sweet potatoes, peanuts, corn, and beans can be grown together, but cassava cannot be included because it will crowd out the other crops.

The following is an example of how an upland field consisting of bitter melon, hyacinth beans, cucumbers, bitter solanum, chili peppers, and cassava is organized. Bitter melon seeds are planted at the beginning of September, just

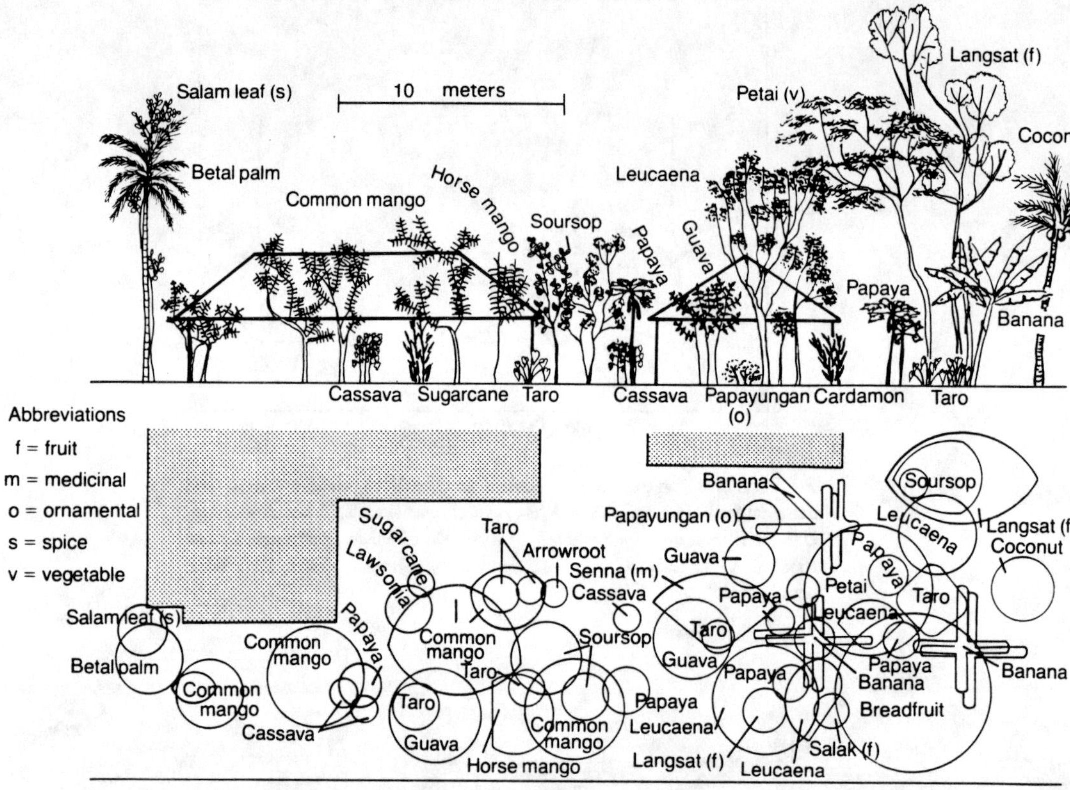

FIGURE 4. Example of a homegarden layout in the uplands of West Java. (From Christanty, L., Abdoellah, O. S., Marten, G. G., and Iskandar, J., in *Traditional Agriculture in Southeast Asia*, Marten, G. G., Ed., Westview Press, Boulder, CO, 1986, 132. With permission.)

after the first rain. Ash from the burn and manure from the homegarden are used to fertilize around the seeds. Hyacinth beans are planted 2 weeks after the bitter melons, when cassava may be planted along the edges of the field as a hedge. Cucumbers are then planted in shallow furrows between the rows of bitter melon and hyacinth beans. Finally, by the end of September, bitter solanum and chili peppers are planted around the cucumbers once they have sprouted. All crops are then fertilized with urea and compost. The crops are treated with pesticides only if a serious infestation occurs. Numerous species of insects and other animals (some of them pests) are present, but seldom is any one species sufficiently abundant to warrant pesticides.

The first harvest starts with cucumber at the end of October and continues every 5 days for 2 months. There is another weeding at the end of December to clear paths between the rows. Starting at the beginning of January, bitter melon is harvested continuously for about 3 weeks, bitter solanum is harvested weekly, and chili peppers are harvested every 2 weeks for 4 months. Hyacinth bean is harvested in mid-April. If there is enough rain after harvesting the hyacinth beans, the land can be tilled again and used to grow green beans, which are harvested at the same time as cassava in July.

Cassava is usually the main crop in the second year and is planted at the beginning of September. A higher layer may consist of scattered bananas, bamboo, or trees. The cassava is ready for harvest 9 or 10 months after planting.

By this time the perennial foliage may have expanded to where it is intercepting most of the light and limiting the possibility for growing further crops unless the perennials are cleared. It is common practice at this time to allow the perennials to take over the field for 3 or 4 years as a managed "fallow". The perennials can be fruit trees, trees for firewood and building materials, or bamboo. Fruit trees require considerable care (e.g., pruning), while fuelwood trees and bamboo can be left to take care of themselves. If trees are of different ages, harvesting can be by selective cutting, and field crops may be interplanted in the more open spaces during the rainy season. If the trees are homogeneous in age, it is common to harvest by clear-cutting in August, as preparation for returning the field to annual crops.

While the tree or bamboo fallow provides a harvest of construction materials, it also reconditions the soil for annual field crops. After 2 years of field crops, weeds have increased and organic matter and mineral nutrients have declined. (Cassava has removed particularly large quantities of phosphorus and potassium from the soil.) Further harvests can only be maintained by much heavier weeding and much heavier chemical fertilizer applications than were necessary during the previous 2 years. After several years of tree or bamboo fallow, weed seeds have diminished and litter fall has returned organic matter to the surface soil at its previous level. Ash from the burn provides mineral nutrients and a liming effect. Perhaps most important are the numerous fine roots of trees or bamboo that have permeated the soil and decompose to release mineral nutrients during the first 2 years of field crops. The mat of tree or bamboo roots at the soil surface, along with slash left around the field, helps to protect the soil from erosion.

C. Contemporary Changes

Government programs to modernize rice production have brought major changes in recent years. Traditional wet-rice varieties have disappeared from all but the most remote areas, and large quantities of pesticides and chemical fertilizers are essential to realize expected yields. Insect pests (e.g., the brown plant hopper), which were of no consequence several decades ago when traditional varieties were in use, can now cause total crop failure unless intensive control is maintained.

As wealthier people in the village (particularly those who own larger amounts of paddy land) have become more commercially oriented, their sense of obligation to other villagers has deteriorated. Traditionally, villagers who participated in the harvest received a share of the crop, but the larger landowners are now evading sharing their harvest with other villagers by selling their crop to middlemen who bring in their own crews to do the harvest.

Many homegardens and upland fields are still largely traditional. Most still feature an interplanting of local crop varieties and use chemical pesticides only when absolutely necessary. However, in areas with readily available transport to urban markets, many upland fields are now a monoculture of high-value vegetables grown from introduced seed, and large quantities of chemical fertilizers and pesticides are essential. An increasing number of households in areas connected to outside markets are also devoting a major part of their homegardens to tree crops of high commercial value, such as hybrid (dwarf) coconut, cloves, oranges, or guava. While the rest of the homegarden may still function without the benefit of pesticides and chemical fertilizer, the high-value tree crops generally require substantial quantities of these chemical inputs. Many orange plantations have been wiped out by the CVPD virus, though the virus has not been a problem for orange trees interplanted with other crops in homegardens. Clove plantations have had a serious problem with leaf disease.

IV. CHIANGMAI VALLEY

The Chiangmai Valley is a floodplain at 300 m altitude in northern Thailand. The rural population of 900,000 people farms 150,000 ha, one half of which is irrigated by a combination of local (traditional) and government irrigation systems. Agriculture in Chiangmai Valley is representative of that in many floodplains and river basins in the lowlands of Southeast Asia. The following description is based on studies by the Multiple Cropping Centre at Chiangmai University.[16-20]

Double cropping is the norm (Figure 5). The first crop is traditional glutinous rice for home consumption, grown during the wet season (July to October). The glutinous rice is followed by a cash crop during the cool season (November to March). The major cool-season crops are improved varieties of nonglutinous rice, soybeans, garlic, tobacco, shallots, chili peppers, peanuts, or vegetables.

Most households farm a bit less than a hectare of land spread over a number of small plots. About one third are tenant farmers. Yields and farm incomes on the fertile alluvial soils of Chiangmai Valley are relatively high. The average annual net income is about $500 per household. A typical household grows nearly all of its own rice but purchases most of its meat and vegetables. Homegardens provide most of the fruit as well as some local vegetables and herbs. There are usually some pigs and chickens as well.

A. Rice Cultivation

A traditional rice variety that matures quickly (in about 3 months) is used before early cool-season crops such as chili peppers or garlic (Figure 5). Late-maturing traditional rice varieties are often cultivated in sequence with soybeans. Most people in northern Thailand prefer to eat glutinous rice, but rice purchases by a government marketing agency have stimulated the cultivation of improved varieties of nonglutinous rice as a cash crop during the cool season. R.D.7 rice is also grown in the hot season (April to June), when most fields are in fallow because of insufficient water. Hot-season crops (e.g., R.D.7 rice or vegetables like Chinese cabbage) must be irrigated or planted in waterlogged areas where there is sufficient soil moisture.

About half the farmers apply chemical fertilizer to traditional rice 20 to 25 days after transplanting, and some make a second application. Most apply chemical fertilizers to improved varieties. The farmers consider animal manure and lime to be the most effective treatments for soil problems (Table 5), but they often use chemical fertilizer instead because animal manure is not available.

Heavy rain during the wet season can lead to rice diseases (e.g., blast, stem rot, or collar rot) that is followed by an attack of cutworms. A dry spell during the hot season can lead to an outbreak of thrips or armyworms on the rice. About half the farmers use chemical insecticides, and all use poisons for birds, rats, and crabs. The average annual expense for agricultural chemicals and other cash inputs is about $200 per household.

Most households have a buffalo for plowing and harrowing. Nearly all farmers use the traditional method of weeding; few use herbicides. The traditional method is to pull the weeds by hand, laying the larger ones on the paddy bund and trampling the small weeds into the soil. The bulk of the labor for all farming activities comes from household members, since every effort is made to minimize cash outlays for production.

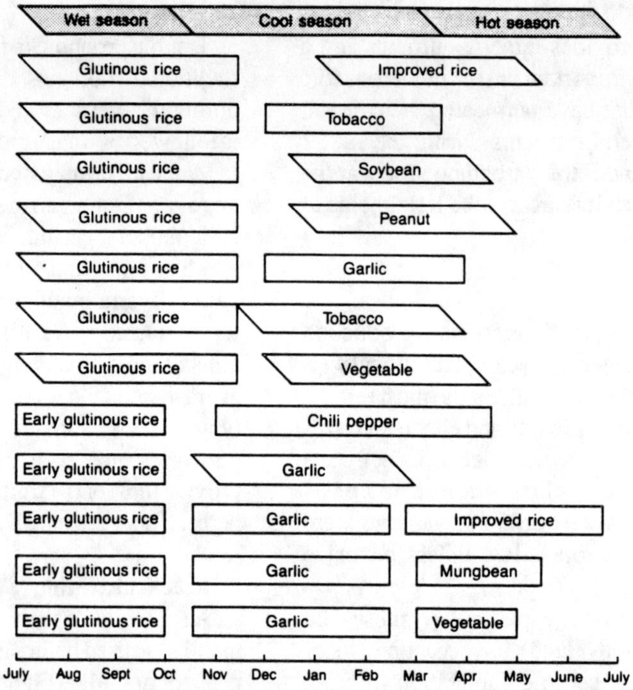

FIGURE 5. Common double cropping and triple cropping sequences in Chiangmai Valley. (Figures courtesy of Multiple Cropping Project experiment station at Chiang Mai University, Thailand.)

Table 5
Farmers' Perceptions of the Values of Various Soil Treatments for Dealing with Different Soil Problems at Chiangmai Valley, Thailand[18]

	Problems			
Treatments	**Difficult to plow**	**Low soil fertility**	**Sandy soil**	**Soil pests**
Apply animal manure	H	H	H	O
Apply lime	L	L	O	L
Irrigate before plowing	L	O	O	O
Apply chemical fertilizer	O	L	O	O
Apply green manure	O	O	L	O
Plow and sun dry	O	O	O	L

Note: H = Higher value. L = Lower value. O = No value.

B. Field Crops (Cool Season)

The crops selected for cultivation in the cool season depend to a large extent on the market price and availability of water. Planting decisions generally are based on prices from the previous year. A key question is also whether the irrigation canal will provide enough water in the dry season to sustain the crop. This depends upon the location of a particular field in the irrigation network and can be judged only from previous experience with different crops.

Chili peppers, garlic, tomatoes, vegetables, watermelon, and tobacco demand large amounts of water, labor, fertilizer, and pesticide inputs. Farmers usually grow these crops only when they have enough capital and labor. Soybeans and peanuts need smaller amounts of all these inputs, are less risky, and add nitrogen to the soil. Soybeans have become particularly popular in recent years (Figure 6). Cucumbers also need fewer inputs but are risky unless a reliable supply of water is available. Low-input crops can be grown farther from the houses because they have a lower value and do not need to be watched so carefully. The prices of low-value crops like soybeans, peanuts, and nonglutinous rice are relatively stable from year to year, but prices of the high-value crops fluctuate considerably.

Each field is a monoculture, but farmers tend to have different crops on different fields. This way they spread their risks, while also spreading out their labor as much as possible. There are two basic strategies that the farmers can follow. One strategy aims at minimizing risks in the presence of constraints on water, chemical inputs, or crop security; the other strategy aims at maximizing profits when none of these constraints is a serious problem. Most farmers follow a mixed strategy. They typically cultivate a quarter hectare with high-input, high-value crops and three quarters of a hectare with safer but lower-value crops. Other things being equal, a crop that has worked successfully in the past will be the one of choice.

C. Contemporary Change

Population growth has reduced the size of landholdings in the Valley. Twenty years ago, approximately one half the farmers had more than a hectare of land. Now only one

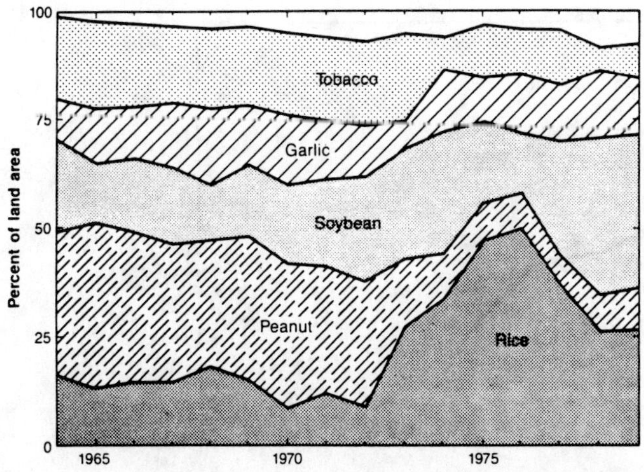

FIGURE 6. Changes in the land area planted to major cool-season crops in Chiangmai Valley. (From Rerkasem, B. and Shinawatra, B., in *Agroecosystem Research for Rural Development*, Rerkasem, K. and Rambo, A. T., Eds., Chiang Mai University, Thailand, 1988, 124. With permission.)

quarter of the farmers have this much land. The percentage of landless villagers has increased.

Cultivation practices for the traditional rice crop have not changed much. Most farmers still use few chemical inputs, and yields have been consistently around 4 t/ha over the past 10 years. However, the diversity of traditional rice varieties in the Valley is diminishing rapidly. While there were hundreds of local varieties a few decades ago, there are now only a few dozen, and many farmers no longer use them at all. Buffalos are beginning to disappear from the scene as the farmers acquire hand tractors.

The costs of inputs for cash crops have increased dramatically in recent decades. There are not enough nutrients in the soil to generate satisfactory yields unless fertilizers are applied, and pests are much more serious than before. Chemical inputs for soybeans have increased from virtually nothing 15 years ago to the present average of more than $40/ha. Soybean yields have increased 40% during the same period, but net income is lower because of higher input costs. Insect pests are more of a problem with improved varieties of rice than with traditional varieties, and birds are more of a problem for early-maturing improved varieties when their harvest is out of synchrony with the rest of the rice crop in the area.

Some of the differences between traditional and introduced cultivation practices can be illustrated with soybeans. The traditional method is to sow soybeans into the rice stubble immediately after burning the rice straw, cover them with ash, and flood the paddy. This method involves no other land preparation. Most farmers apply chemical fertilizers to the preceding rice crop on fields of low fertility and do not apply chemical fertilizer to the soybeans because they believe there is enough left over from the rice to meet the needs of the soybeans. Weeds are not a problem because weeds and weed seeds are killed when the rice straw is burned. The main problems are fungi and insects, and many farmers who otherwise follow traditional management practices now control insects with sprays.

Recently introduced practices for soybeans differ from the traditional in two major respects: preparation of raised beds and use of chemical fertilizers. The new practices have allowed soybean cultivation to expand to problem areas (e.g., infertile or waterlogged soils) where soybeans were previously not feasible. However, most farmers, particularly those who were growing soybeans already, have not adopted the new practices because on good soils the new practices provide only slightly higher yields than traditional methods, despite substantially higher costs for labor and chemical inputs.

Farmers in Chiangmai Valley are intensely interested in any new agricultural practices that will increase their income. Most farmers listen to agricultural programs on the radio for information concerning new technologies, markets, and weather. Chemical fertilizers are discussed with extension agents and neighbors more than any other aspect of soil management because farmers do not feel they know precisely which formulas and levels of application are best for each of their cropping systems on the different kinds of soils. Additional problems perceived by the farmers include a shortage of irrigation water in the dry season, rice stem borers, and weeds in their rice fields. Some problems that concern agricultural scientists are nematodes in the rice fields and the low quality of improved-variety rice seeds.

Recently triple cropping (Figure 5) has been practiced in some areas where there is a reliable supply of irrigation water throughout the year. While the third crop (usually a fast-maturing variety of nonglutinous rice) offers an opportunity for extra income, it is also somewhat marginal because risks are higher and yields generally lower than with the second crop. A serious difficulty with the third crop is the large quantity of labor necessary to squeeze three crops into a year, and the possibility that the traditional rice crop (i.e., the first crop) will be delayed if the third crop finishes late. There have also been social effects from triple cropping. Villages with triple cropping have experienced reduction in their religious festivals because people no longer have the time. Cooperative labor has also declined for the same reason, and paid labor has replaced exchange labor.

Perhaps the most serious consequence of triple cropping has been a reduction in yields (often as much as 50%) after triple cropping for 5 years (Figure 7). All crops are affected, including the first and second crops. Two of the problems are soil acidity and boron depletion, the correction of which can return some field crops to their normal yields, but so far a complete solution has not been found for some of the other crops, including rice. As a consequence, many farmers that ventured into triple cropping have now abandoned it to return to double cropping.

V. SALIENT FEATURES OF SMALL-SCALE AGRICULTURE IN SOUTHEAST ASIA

A. Farm Integration

Most small-scale farms in Southeast Asia are managed by

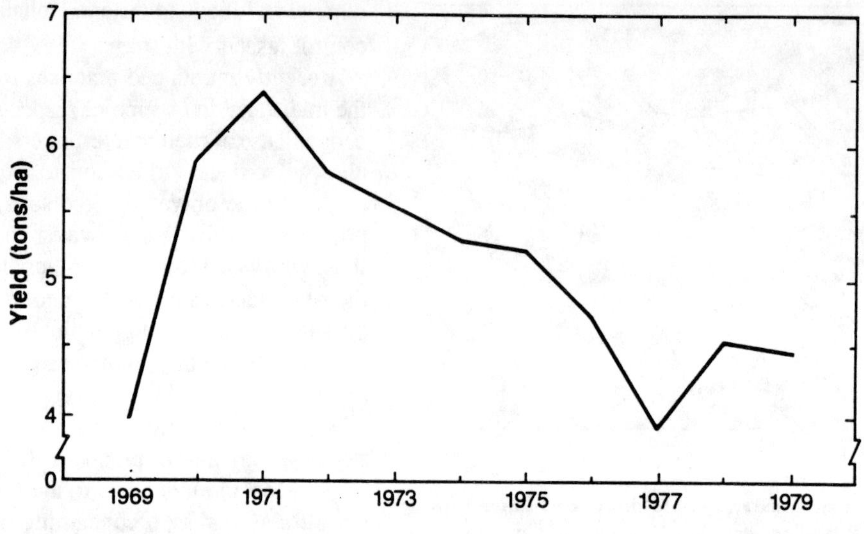

FIGURE 7. Increase in rice yields due to modern technology (1969—1971) followed by decline under triple cropping (1971—1979).[16] These figures are from the Multiple Cropping Project experiment station at Chiangmai University; the same changes occurred on farmers' fields.

extended-family households. Because a household may have several parcels of land, each with its unique capabilities, most households plant different crops on different parcels and on the same parcel at different seasons of the year. When deciding which crops to plant in which parcels, and during which seasons of the year, the first priority of most households is to satisfy the bulk of their food needs. The second priority is to produce additional crops to generate as much income as possible from available household labor.

The basic strategy is to select crops with (a) an assured market at reasonably high prices, (b) reliable yields, and (c) low cash inputs for hired labor, fertilizers, pesticides, and other costs such as rental of tractors or draft animals. Conditions for farming may be more favorable during some seasons of the year than others, but an effort is made to spread farming activities across the year in order to take full advantage of household labor throughout as much of the year as possible. At the same time, the total demand for labor from various overlapping farm activities at any one time must be kept within bounds so it does not exceed the capacity of the household to do the work.

A prominent feature of small farm organization is full utilization of all products. For example, rice is not the only product from rice paddies. They are also a major source of fish, and in many areas paddies are public domain for harvesting animals such as rice rats and aquatic invertebrates (e.g., crabs). This practice not only leads to full utilization of animal production in the paddies but also provides an incentive for keeping the numbers of edible animal pests (e.g., rats that feed on rice grains or crabs that cut the stalks of rice seedlings) below the point where they do serious damage to the rice. The same principle applies in swidden fields, where traps may be set for bush pigs that are attracted to the crops.

The various crops employed by a single household are often mutually reinforcing by providing materials or services to one another. Residues from one crop may be used as green manure for other crops that follow on the same field or for crops on different fields. Livestock are used to transfer crop residues from one field to another by grazing the livestock on one field after harvest, collecting manure in the corral, and distributing it to other fields. Chemical fertilizers applied to one crop (e.g., rice) may also provide nutrients for the crop that follows (e.g., vegetables). Irrigation for rice can provide residual soil moisture for vegetables that follow. Mineral nutrients (from chemical or organic fertilizer applications) that are lost from an irrigated field as water drains from the field are carried by the same water as inputs to other fields.

B. Diversity

Perhaps the most conspicuous feature of small-scale agriculture in Southeast Asia is its overwhelming diversity. Variation in environmental conditions from lowlands to highlands and from flat lands to hilly or mountainous lands leads to a corresponding variation in crops to fit the conditions. Even in one place the landscape is usually a patchwork of hundreds of small fields, each with its own crop or mixture of crops. Lowland areas of extensive rice cultivation may not present such a varied landscape, but on close inspection even these areas have considerable diversity. A variety of field crops or trees may be planted along the bunds, and the gardens around the houses may present a diversity of dozens or even hundreds of plant species. Once the rice season has passed, paddies may be converted to a patchwork of dozens of different field crops.

Even a continuous expanse of rice may not be as uniform as it appears. Many traditional villagers maintain dozens of local rice varieties, each adapted to perform best under particular environmental conditions. A single household may plant half a dozen different varieties during the same season, a monoculture of each variety in each paddy but different varieties in different paddies, based on explicit knowledge of the characteristics of each variety and how each fits into a total cropping strategy. A similar varietal diversity exists for most of the others of the hundreds of crops in Southeast Asia, though very little of it has been documented.

C. Flexibility

The diversity of Southeast Asian agriculture translates into flexibility that underlies (1) adaption of the agriculture to local environmental and social conditions and (2) adaptiveness of the agriculture through time.

The particular crops and crop varieties, and details of cultivation practices, are different in each region in accord with topography, soil conditions, and seasonal weather patterns. Equally important, the crops, varieties, and practices are tailored to each agricultural field and even to varying conditions within a field. This is essential for dealing with the heterogeneity of soil conditions on undulating or hilly topography, a situation in which a large percentage of Southeast Asian agriculture must be pursued. Decisions on these matters derive largely from the knowledge and experience of each farmer.

The agriculture is also adaptive through time. There is nothing unusual about this. Southeast Asian agriculture has always been in a state of evolution, even in the most remote and subsistence-oriented societies. Most farmers are constantly seeking information from other farmers (and any other available sources) on new crops and new production technologies that enable them to improve the returns from their agricultural activities. Many continuously experiment with promising new possibilities on a small portion of their land and cautiously expand the scale when results are encouraging. Keeping open as many options as possible is the key to their flexibility—assuring that they can modify their agriculture in response to changing needs or opportunities and withdraw from changes that do not work out.

D. Intensification

The most significant trend throughout Southeast Asia in recent years has been an increase in agricultural production. This increase has been driven not only by the increase in human population but also by the spread of a market economy that has created a demand for cash earnings. The area under cultivation has increased immensely. Only a few decades ago many parts of Southeast Asia had agriculture only in the flatter areas. Sloping or hilly areas were covered by forest. Now agriculture is moving progressively up the hills as new settlements and new fields displace the forest. One of the major consequences of this trend is an increasingly quantity of agriculture on land that is marginal with respect to soil quality or water supply, where yields (and returns to labor) are lower and permanent cropping may not be possible. Soil erosion on steeply sloping lands can be severe, and land degradation is exacerbated by the many farmers in hilly areas who are recent immigrants from the lowlands and lack the knowledge (and sometimes the motivation) to practice sustainable agriculture under these conditions.

The other way that agricultural production has been increased has been through higher production per hectare. A major part of the strategy has been more intensive chemical inputs—fertilizers to increase soil fertility and pesticides to reduce crop damage—along with improved crop varieties to take full advantage of the inputs. This kind of intensification has been most pronounced for rice, but it is also becoming common for other major food crops (e.g., corn, peanuts, potatoes, and various kinds of beans) and commercial vegetables (e.g., cabbage). In general, tropical fruit trees have experienced less intensification of this sort, with local varieties still prevailing in many areas, but a number of temperate fruits and a few select tropical tree crops (e.g., cloves and some citrus fruits) have experienced considerable intensification because of their high commercial value.

Multiple cropping (i.e., a sequence of crops on the same field during a year) has been another common means of increasing total agricultural production per hectare. In rainfed areas with a distinct dry season, where until recently there was only one crop a year (e.g., rice during the rainy season), many farmers are now attempting to grow a second crop (e.g., vegetables) after the rice. In many cases the second crop is risky because of unreliable water supplies.[21] In irrigated areas, where water is not a problem and where until recently there were two crops a year, some farmers are now attempting to squeeze in a third crop. Many are running into the kinds of difficulties described for Chiangmai Valley.

In upland areas where swidden cultivation prevails, a common means of intensification has been to shorten the forest fallow between cropping periods. For example, in the hills around Chiangmai Valley, the traditional fallow is 10 years in duration. However, an increasing number of farmers in that area do not have enough land at their disposal to leave their fields in fallow so long. Many are returning their fields to crops after a fallow of only 2 or 3 years, insufficient time to complete soil regenerative processes performed by the fallow.

While intensification has increased the quantity of agricultural production from available land, it often has reduced the efficiency of production with regard to labor or cash inputs. It is common for national policies to place a top priority on increasing the production of rice or other basic foodcrops, in order to attain national self sufficiency or even produce surpluses for export. However, such policies may not be in the interests of farmers whose production is limited primarily by the supply of household labor or cash for agricultural inputs.

E. Autonomy vs. Dependence

A prominent feature of most small-scale agriculture in Southeast Asia is its ecological self sufficiency. One of the keystones is mineral nutrient recycling.[22] In traditional subsistence agriculture, where nearly all of a household's agricultural production is used for home consumption, kitchen wastes (and the mineral nutrients they contain) are returned to the fields, thereby completing the cycle. A field's loss of nutrients through harvests, leaching, or erosion is balanced by nutrient inputs from kitchen wastes, animal manures, weeds, and other organic materials that are thrown onto the field. Much of the agriculture is also energetically self sufficient. Energy for land preparation, harvesting, weeding, and other agricultural activities comes from human labor and draft animals, which derive their energy from pasture and farm produce. The traditional farm is also genetically self sufficient, employing local crop varieties adapted to local environmental and social conditions.

Much of this is changing today with agricultural intensification and modernization and expansion of a cash economy. Cash incomes of highland farmers depend to a large extent on urban demands for temperate vegetables. Cash incomes can even depend upon international marketing arrangements, such as cassava in northeast Thailand, which is sold to Europe for animal feed. Farmers are adopting high yielding varieties and need to purchase the various chemical inputs necessary to realize their potential. Because pesticides have rendered many ricefields unfit for fish, many farm households now have to purchase fish to replace the ones they no longer obtain from their own fields. Petroleum energy is replacing animal energy as minitractors replace water buffalos. As buffalo disappear from the scene, their manure also disappears, increasing the need to purchase chemical fertilizers as a substitute. Buffalo wallows are disappearing, and with them a variety of ecological services they have performed, such as reservoirs for ricefield fish during the dry season.

F. Risk Management

Most farmers operate in the face of numerous risks such as short-term droughts, attacks by crop pests or diseases, typhoons, or uncertain prices for their products. They deal with these risks, first, by employing crops and cropping practices that are explicitly directed toward ameliorating the hazards. Local varieties are adapted to drought, frost, pests, etc. The planting of crops is timed to avoid periods of droughts or typhoons and to avoid seasons of the year that are particularly favorable for key pests.

Almost everyone employs a diversified cropping strategy to reduce risks. With "fallback" crops, at least some production is likely even under the worst of circumstances. Some rice varieties are more resistant to drought than others. Shorter varieties are more resistant to lodging while taller varieties are more resistant to flooding. At least one will survive whether the year is unusually dry or unusually wet. For cash returns, farmers who grow several different kinds of vegetables have a hedge against price fluctuations.

In paddy fields on elevated topography, where the irrigation supply is unreliable, vegetables may be planted to replace the rice halfway through the wet season if there has not been enough water to expect a good rice harvest. In rainfed fields, the highest yields of corn can be realized from varieties that take the longest time to mature, varieties that risk failure because they must be planted at the beginning of the growing season when rainfall is unreliable. However, if a long-term variety fails at the beginning of the growing season, it can be replaced with a faster maturing (but lower yielding) variety for the rest of the season.

G. Sustainability

Most traditional agriculture in Southeast Asia has continued for centuries on a sustainable basis. By recycling plant residues and animal manure to the fields, soil quality is maintained without depending on materials or energy from outside the farm. Pest damage to crops is maintained at tolerable levels by using pest resistant crop varieties, intercropping different crop species so pests do not find large quantities of their host plants in any one place, and relying upon natural enemies of the pests to keep them from getting out of hand.[23] Weeds are removed from the field by burning or by letting it fallow. Weeds are also used to provide soil cover for preventing soil erosion. In general, traditional agriculture has been structured as much as possible to mimic natural ecosystems, so farmers expend as little effort as possible struggling against natural ecological processes.[24,25]

The trend in modernizing and intensifying agriculture has been to make the fields less like natural ecosystems so a higher share of ecosystem production can be channeled to human use. The benefits are indisputable, but they have been achieved at a cost. As a rule, higher yields can only be sustained by means of costly chemical inputs. More intensive cropping often means that the soil is exposed to erosion and leaching more of the year, and more mineral nutrients are removed in the harvest.

A reduction in soil fertility can be compensated by chemical fertilizer applications, but the organic matter content of the soil often declines as chemical fertilizers replace organic fertilizers. With consequent reduction in the soil's cation exchange capacity and moisture storage capacity, even heavier chemical fertilizer inputs become necessary. With some of the fertilizers in common use, this leads to progressive soil acidification, which may affect the crop not only directly, but also indirectly if there is phosphorus fixation by iron or aluminum oxides, leading to a need for increasingly heavy phosphorus fertilizer applications.

The same cycle of increasing chemical applications can also occur with pesticides. Pesticides often kill natural enemies (e.g., predators and parasitoids) more effectively than the pests themselves and this can lead to a loss of control by these natural enemies. As pests evolve resistance to the pesticides, the quantity and number of different kinds of pesticides that are applied continually increases. As the quantities of chemical fertilizers and pesticides increases, cash outlays for these inputs also increase. This trend has been exacerbated because the chemical inputs are tied to increases in the price of petroleum energy. In the end, it may be necessary to abandon the crop.

There is also the risk that modernized agriculture cannot

be resilient to sudden and unexpected hazards such as the appearance of a pesticide resistant pest biotype, a drop in market prices (or loss of the market if it depends upon contractual arrangements), unusually severe weather conditions (e.g., drought, floods, or typhoons), or sudden increases in the cost of inputs. This trend reflects the general principle that increases in production, and improvements in the stability of production, often reduce resilience because of (a) dependence on outside resources that are not reliable or (b) a failure to exercise a capacity for dealing with particular hazards. As an example of the latter, if irrigation is brought to a rainfed area, farmers can realize higher and more reliable yields without resorting to their traditional drought resistant varieties or specialized cultivation techniques for coping with vagaries of rainfall. These varieties and techniques may eventually drop from the scene, even though they could be highly useful if siltation or some other problem should someday prevent the irrigation system from providing a reliable water supply.

VI. CONCLUSIONS

While the modernization of Southeast Asian agriculture is generally increasing agricultural production and farm incomes, in many instances there are serious doubts as to how reliably and for how long the gains can be sustained. There are also doubts about their costs in social and environmental terms.

Resisting change is not a realistic option; modernization is a fact of life in Southeast Asia. Population growth of the past 50 years had rendered yesterday's agriculture obsolete, and aspirations aroused by the universal spread of a cash economy mean that farmers will change their agriculture to increase their earnings in any way available to them. It is a responsibility of agricultural scientists to provide farmers with new technologies that offer not only higher yields and higher incomes but also sustainable yields and prudently low levels of environmental degradation and social disruption.

It is not reasonable to expect centrally developed agricultural packages to satisfy these criteria.[25] The environmental and social conditions in which agriculture must be applied are so variable that new technologies must be tailored not only to a region of the country but also to different villages, different households, and even different fields. While agricultural research and extension can offer farmers some of the tools they need to improve their agriculture, a major part of the research and development process will have to lie with the farmers themselves. Most farmers are well prepared for such a task by virtue of their experience experimenting with new crops and techniques to fit their farm operation.

Effective agriculture should be adaptive through time. One major way to maintain the flexibility necessary for adaptive agricultural development is to retain as much of the existing agricultural diversity in Southeast Asia as possible. This includes maintaining the wealth of crop varieties and production technologies in the region's traditional agriculture. Traditional agriculture in most areas cannot be expected to meet modern needs without modification, but its proven record of sustainability is a strength that farmers cannot afford to ignore. Agricultural scientists and farmers have a responsibility to ensure that this valuable heritage is not lost in the course of modernization. The challenge is to combine the best of traditional and modern technologies.

There is ample experience that high-yield agriculture is not sustainable when it is socially or environmentally exploitative, but there is also ample evidence that agriculture can be organized to make increased productivity and sustainability mutually reinforcing.[26] Although the use of chemical fertilizer can lead to soil degradation if the fertilizer is used to compensate for soil-degrading agriculture practices, higher yields from chemical fertilizers can in fact be used to improve and sustain soil fertility if a significant proportion of the crop biomass is returned to the soil. To take another example, multistory intercropping that makes fullest use of space, soil, and water (as described for homegardens and rainfed fields in West Java) can also provide an environment where crop pests are not a serious problem even without the use of pesticides. Identifying ways to improve production on a sustainable basis may involve retention or adaptation of agricultural design features (e.g., recycling crop residues or intercropping) that are still common in Southeast Asia. It may also involve new designs appropriate to changing conditions and changing needs, but the prospects for new designs will be most promising if they build on insights from past experience with agricultural and natural ecosystems.

REFERENCES

1. **Capistrano, A. D. and Marten, G. G.,** Agriculture in Southeast Asia, in *Traditional Agriculture in Southeast Asia,* Marten, G. G., Ed., Westview Press, Boulder, CO, 1986, 6.
2. **Rambo, A. T. and Sajise, P. E.,** Developing a regional network for interdisciplinary research on rural ecology: the Southeast Asian Universities Agroecosystem Network (SUAN) experience, *The Environmental Professional,* 7, 289, 1985.
3. **Rambo, A. T.,** Human ecology research on tropical agroecosystems in Southeast Asia, *Singapore J. Trop. Geogr.,* 3, 86, 1982.
4. **Rambo, A. T. and Sajise, P. E.,** Human ecology research on tropical agriculture in Southeast Asia, in *An Introduction to Human Ecology Research on Agricultural Systems in Southeast Asia,* Rambo, A. T. and Sajise, P. E., Eds., University of the Philippines, Los Baños, 1985, 1.
5. **Tapang, B.,** The cash flow in a typical farming household in the Loo Valley rural ecosystem, in *Agroecosystem Research for Rural Development,* Rerkasem, K. and Rambo, A. T., Eds., Chiang Mai University, Thailand, 1988, 224.
6. **Brett, J. P.,** Stone walls and waterfalls: irrigation and ritual regulation in the Central Cordillera, northern Luzon, Philippines, in *Cultural Values and Human Ecology in Southeast Asia,* Hutterer, K. L., Rambo, A. T., and Lovelace, G. W., Eds., Center for South and Southeast Asian Studies, University of Michigan, Ann Arbor,

1985, 125.

7. **Brett, J. P.,** The Bontok: traditional wet rice and swidden cultivators of the Philippines, in *Traditional Agriculture in Southeast Asia,* Marten, G. G., Ed., Westview Press, Boulder, CO, 1986, 54.
8. **Omengan, E.,** Nitrogen and Phosphorus Cycles in a Bontoc Paddy Rice System, M.S. thesis, University of the Philippines, Los Baños, 1981.
9. **DeRaedt, C.,** Relations between autonomy and solidarity in an upland commercial vegetable growing village in Benguet province, in *Agroecosystem Research for Rural Development,* Rerkasem, K. and Rambo, A. T., Eds., Chiang Mai University, Thailand, 1988, 194.
10. **Alcantara, A.,** Sustainability of commercial vegetable production, in *Agroecosystem Research for Rural Development,* Rerkasem, K. and Rambo, A. T., Eds., Chiang Mai University, Thailand, 1988, 211.
11. **Soemarwoto, O. and Soemarwoto, I.,** The Javanese rural ecosystem, in *An Introduction to Human Ecology Research on Agricultural Systems in Southeast Asia,* Rambo, A. T. and Sajise, P. E., University of the Philippines, Los Baños, 1985, 254.
12. **Christanty, L., Abdoellah, O. S., Marten, G. G., and Iskandar, J.,** Traditional Agroforestry in West Java: the *pekarangan* (homegarden) and *kebun-talun* (annual-perennial rotation) cropping systems, in *Traditional Agriculture in Southeast Asia,* Marten, G. G., Ed., Westview Press, Boulder, CO, 1986, 132.
13. **Marten, G. G. and Abdoellah, O. S.,** Crop diversity and nutrition in West Java, *Ecol. Food Nutr.,* 21, 17, 1988.
14. **Iskandar, J. and Abdoellah, O. S.,** Agroecosystem analysis: a case study in West Java, in *Agroecosystem Research for Rural Development,* Rerkasem, K. and Rambo, A. T., Eds., Chiang Mai University, Thailand, 1988, 237.
15. **Hadikusumah, H. V.,** Development of traditional agriculture in West Java: a case study in Cihampelas village and Sadu village, *Third SUAN Regional Symposium on Agroecosystem Research,* Chiang Mai, Thailand, October, 1986.
16. **Gypmantasiri, P., Wiboonpongse, A., Rerkasem, B., Craig, I., Rerkasem, K., Ganjanapan, L., Titayawan, M., Seetisarn, M., Thani, P., Jaisaard, R., Ongprasert, S., Radenachaless, T., and Conway, G.,** An interdisciplinary perspective of cropping systems in the Chiang Mai Valley, key questions for research, Faculty of Agriculture, University of Chiang Mai, Chiang Mai, Thailand, 1980.
17. **Rerkasem, B. and Rerkasem, B.,** The agroecological niche and farmer selection of rice varieties in Chiang Mai Valley, in *An Introduction to Human Ecology Research on Agricultural Systems in Southeast Asia,* Rambo, A. T. and Sajise, P. E., University of the Philippines, Los Baños, 1988, 303.
18. **Srimongkol, K. and Marten, G. G.,** Traditional agriculture in Northern Thailand, in *Traditional Agriculture in Southeast Asia,* Marten, G. G., Ed., Westview Press, Boulder, CO, 1986, 85.
19. **Rerkasem, B. and Shinawatra, B.,** On the systems properties of the Chiangmai Valley agroecosystem: implications for development, in *Agroecosystem Research for Rural Development,* Rerkasem, K. and Rambo, A. T., Eds., Chiang Mai University, Thailand, 1988, 124.
20. **Ramitanondh, S.,** Impact of development on human activity systems in a Northern Thai village, in *Impact of Development on Human Activity Systems in Southeast Asia,* Soemarwoto, O. and Rambo, A. T., Eds., Institute of Ecology, Padjadjaran University, Bandung, Indonesia, 1987, 123.
21. **Polthanee, A. and Marten, G. G.,** Rainfall cropping systems in Northeast Thailand, in *Traditional Agriculture in Southeast Asia,* Marten, G. G., Ed., Westview Press, Boulder, CO, 1986, 103.
22. **Marten, G. G. and Vityakon, P.,** Soil management in traditional agriculture, in *Traditional Agriculture in Southeast Asia,* Marten, G. G., Ed., Westview Press, Boulder, CO, 1986, 199.
23. **Brown, B. J. and Marten, G. G.,** The ecology of traditional pest management in Southeast Asia, in *Traditional Agriculture in Southeast Asia,* Marten, G. G., Ed., Westview Press, Boulder, CO, 1986, 241.
24. **Marten, G. G.,** The tropical rainforest as an ecosystem, in *An Introduction to Human Ecology Research on Agricultural Systems in Southeast Asia,* Rambo, A. T. and Sajise, P. E., Eds., University of the Philippines, Los Baños, 1985, 61.
25. **Marten, G. G.,** Traditional agriculture and agricultural research in Southeast Asia, in *Traditional Agriculture in Southeast Asia,* Marten, G. G., Ed., Westview Press, Boulder, CO, 1986, 326.
26. **Marten, G. G.,** Productivity stability sustainability equitability and autonomy as properties for agro-ecosystem assessment, *Agric. Syst.,* 26, 291, 1988.

The Dynamics of Traditional Agriculture: Case Studies

Tropical Environments

20 INDIGENOUS AGROFORESTRY SYSTEMS IN THE LATIN AMERICAN TROPICS

Janis B. Alcorn

I. INTRODUCTION

A diverse array of traditional, Latin American agroforestry systems have potential for small farm development for a variety of reasons. Over generations, these systems have been adapted to meet small farmers' needs under local environmental and socio-economic constraints. They have proven their sustainability in areas considered marginal for conventional agriculture. An incredible diversity of economically valuable species are adapted to cultivation in these agroforestry systems: indigenous Latin American agroforestry systems often incorporate 50 to several hundred plant species.

Flexible decision points in management plans give farmers control over how a structure matures; a farmer is not committed to a particular endproduct mixture of species when he begins to create an agroforestry plot. Different structures can be started with the same initial steps. In addition, farmers can opt to follow different management subroutines in different areas of a plot in order to create different patches within an agroforestry plot of a given age. This flexibility and diversity serve small farmers well, allowing farmers to respond to the farm family's changing needs and opportunities.

Indigenous agroforestry systems also share the numerous advantages of modern agroforestry systems.[1-6] Agroforestry systems, modern or traditional, serve to integrate trees into farming systems where trees serve a variety of agricultural uses and contribute to the sustainability of the agricultural enterprise. Trees make productive use of vertical space. Trees protect watersheds. Trees improve and protect soil. Fallen leaves contribute organic matter to the soil; deep roots bring subsoil nutrients and captured leached nutrients back to the upper levels of the soil; superficial root systems improve soil aeration and soil structure; mycorrhizae and root nodules enable trees to prosper in marginal conditions; and extensive root systems and protective canopies prevent soil erosion. Trees provide niches for a variety of game animals and useful predators of animals and insects that damage crops. Trees can be manipulated to create diverse micro-climates for growing a variety of companion crops within a small area. There is a tremendous range of marketable products farmers can produce from trees. By doing agroforestry, farmers can directly produce firewood, wood, livestock forage, medicines, construction materials, and other materials they need. At the same time, living capital stands in the form of timber trees farmers can harvest for sale when cash is needed quickly. Farmers also reduce their risks by growing a diverse range of products to protect themselves from unpredictable crop failures and market fluctuations.

II. CHARACTERISTICS OF INDIGENOUS AGROFORESTRY SYSTEMS

Despite shared advantages, Latin American indigenous agroforestry systems differ from most modern systems in at least seven ways:

1. **Indigenous agroforestry systems integrate native species.** Native vegetarian is managed in two basic ways: en masse or as individuals.[7] En masse management relies on natural succession to make available communities of associated species which arise predictably after a given type of disturbance — as after slashing, burning, and weeding a plot. But farmers also protect, encourage, or plant individuals of native species that arise in their fields, along trails, or around homes. These activities are integrated into all stages of an agroforestry system.
2. **Farmers using indigenous systems make use of natural environmental variation.** Farmers acknowledge different biotopes (different natural envi-

ronments, and their associated biota, related to topographic variation) and often choose to site their farms where they can use different biotopes. Each biotope offers different natural resources and different farming options. Farmers also use and create microenvironments within a field. Thus, different types of agroforestry mixtures are created in the various biotopes of a farmer's land, and different species mixtures are used in patches within a given field according to microenvironmental differences. This means there are regional and local variants of basic agroforestry systems.

3. **Natural successional processes are utilized as management tools.** In the humid areas, sequential agroforestry systems often terminate in managed, enriched successional forests initiated in earlier stages of the agroforestry plot. From a conservation and sustained production perspective, this aspect of indigenous systems is very important. Instead of viewing secondary succession as a weed problem, the successional process is used as a subsidy from nature to improve soils, provide resources, and reduce pest problems.[8]
4. **A large number of species are included in the system.** The advantages of this are risk avoidance, subsistence production, and pest control.[9] In addition to well-known cultigens and native plants, farmers also often cultivate minor crops of regional importance.
5. **Agroforestry plots are integrated into a diversified farmstead.** At any given time, a farmer has access to a variety of patches: different types of agroforestry plots, a variety of modules within a given type of agroforestry plot,[10] different stages of agroforestry plots, orchards, gardens, pasture, managed forest, and, often, intensively cultivated cash crop plots. His farming strategy spreads risks and aims at sustained production of many commodities.[11]
6. **There is variation among farms using the same system.** Different farmers in the same cultural group make different choices of crop mixtures according to their household needs and resources. Experimentation with novel mixtures or new cultivars is an integral part of the system. Different crops may be stressed in different communities of the same ethnic group.
7. **Each farm is designed to meet all the farm family's needs for independent survival within the context of the local community's land use.** Each family has access to food, firewood, medicines, construction materials, and utilitarian items either on their own land, on neighbors' land, on the uncultivated wildlands available between their fields, or in the uncultivated, forested hills or savannahs that constitute part of the community's resource base — the tree resources of which are, in a sense, agroforestry adjuncts to the more intensively managed fields.

III. AGRI-SILVICULTURAL SYSTEMS

Sequential agri-silvicultural schemes create five basic structures: fields of staples, homegardens, orchard, managed fallow thickets, and artificial forests. It is useful to classify agri-sylvicultural systems based on the dominant staple crop component, because the staple cultigen has a strong impact on the agroforestry structure, because a given staple crop is associated with a particular set of other cultigens, and because the staple choice tends to follow cultural and geographic divisions.

A. Maize-Initiated Systems

The milpa system of Middle America, popularly synonymous with fields of maize, beans and squash, is actually a complex sequential agroforestry system. The milpa cycle integrates maize fields, homegardens, and managed fallow thickets. Based on ethnohistorical, ethnographic, and ecological analyses, tree culture within the milpa system was important prior to the Spanish conquest.[12-18]

The products of the milpa system include staples, fruit, vegetables, marketable tree and vine crops, firewood, construction materials, livestock feed, and wild game. There are many local variations on the milpa system, depending on local soil, topography, vegetation, and climate as well as on the socio-economic, historical, and ethnic relations of farmers who operate the system.

Planting of maize entails removal of trees from a site. Some useful trees are commonly spared during this process depending on a farmer's needs, resources, and knowledge. The process of slashing down standing trees to clear a spot for milpa is also the process of harvesting firewood. The details of cutting and burning are affected by firewood concerns and vary from area to area. Mexican Yucatec Maya, for example, leave individuals of good quality firewood standing and cut them down after burning the slash so that less wood is destroyed during the burn.[19] Alternatively, they leave a zone of good firewood trees next to the new milpa, and the fire kills those trees when the slash on the adjacent milpa site is burned. After the firewood is harvested, the zone is then incorporated into the milpa. Farmers use their knowledge of how the woods of different species burn to harvest firewood and produce ash for the maize on the same site. Mexican Huastec Maya cut the young trees on fallows into appropriate lengths for firewood as they slash them down.[20] Some firewood is removed from the milpa site before it is burnt, but a light burning is said to preserve firewood better. Most firewood is usually picked up in the milpa while the maize is growing.

The first planting on a site is usually only maize or maize and beans although in some cases squash, sesame, or other minor crops are included. These cultigens develop in the rootmat of the previous stage, the fallow. Depending upon the region, different species common to secondary vegetation communities share the milpa with the maize almost as soon as it germinates. The number of years that a site is replanted after harvest of the first planting of maize varies considerably. The Huastec Maya of northeastern Mexico, for example, usually fallow the plot after one harvest, except in sections occasionally developed into milpa gardens.[21] Their milpa operations are designed primarily for the production of firewood by secondary vegetation. The Totonac, also of northeastern Mexico, are more likely to plant two crops a year for 2 years.[22-23] The number of maize crops harvested depends to a large part on the amount of land available for the milpa system. Milpas made on older fallows produce more crops of maize.

Polycultural homegardens are part of most milpa systems. These milpa gardens contain a wide variety of annual and perennial crops, often 40 or more species. After the initial maize harvest, bananas, papayas, guavas, and other fruit trees are gradually interplanted with root crops, tomatoes, beans, chiles, spices, other cultigens, and spared native plants in part of the original plot. These areas are maintained for 5 or more years after the rest of the milpa has been fallowed. The exact species composition, species zone patterning, and management varies from region to region, and from farmer to farmer. While the Huastec milpa garden (k'alumlaab), for example, contains concentrations of certain species such as manioc, sweet potato and bananas, individuals of other species are scattered through the garden.[24] Trees are not an important component of the k'alumlaab unless the plot is destined to become a managed forest. The Kekchi Mayan milpa gardens of Guatemala (sak'ecuaj), however, contain many trees.[25] Lancandon Maya of Chiapas plant tree crops after maize has been produced on the site for several years.[26] For commercial production of allspice, Totonac farmers in northern Veracruz transplant native *Pimienta dioica* into their milpas,[27] and, in addition to planting and sparing fruit trees, they plant cuttings of *Bursera simaruba* and *Gliricidia sepium* around the borders of the milpa for extra, fast firewood production. These milpa gardens typically achieve the sort of multistoried diversity also found in dooryard gardens. But, instead of domestic livestock, these gardens and adjacent fallows include wild game harvested by milpa farmers.

The fallow stage which follows the milpa or milpa-garden is also productive and is managed in various ways. I am referring to this stage as a thicket, a term used by Gordon[28] to distinguish it from older forests. It ranges from 2 to 3 meter tall bush to taller secondary forest.

Beans are cultivated in thickets in southern Yucatan, Mexico.[29] The area is prepared by removing herbaceous undergrowth and small saplings, usually at the same time trees on a new, adjacent milpa site are being cut down. Larger trees are then felled by cutting their trunk on one side about 1 m from the ground; these leaning, dead trees serve as support for common bean and lima bean vines. After the bean harvest, firewood is harvested, and the site is either prepared to enter the milpa cycle with the adjacent plot begun when the beans were planted, or it is fallowed.

Vanilla is commercially produced in the fallow milpa thickets of the Totonac of northern Veracruz, Mexico.[30,31] In some cases, when a site is selected for a new milpa (usually in an old fallow), a band of secondary vegetation is left uncut to form a thicket border where vanilla will be planted. In most cases, however, the vanilla is planted on a plot after four to six harvests of maize (about 2 years after the site was cleared). When a new milpa site is initially slashed, if it is destined to be used for vanilla cultivation, many individuals of useful trees are spared. Then, during each weeding or cleaning operation after maize harvest, the native trees that will support vanilla vines are spared. These support trees ("tutors") have a rough bark for the vine's sturdy attachment, exhibit rapid growth, and have a thin canopy which lets filtered light through onto the vanilla vines.[32] The support species used include *Eugenia capuli, Nectandra aff. sanguinea, Adelia triloba, Ardisia escallonoides,* and *Trichilia havanensis,* among others. After the last harvest of maize, a light clearing of undergrowth is done, and the vanilla vines are planted. Three years later the vanilla begins to produce significant yields. Maintenance involves limited pollarding of support trees, but the dense undergrowth is not weeded, nor is firewood or other wood removed from the vanilla thicket. Spared fruit trees are harvested during this time, however. When vanilla thickets are 10 to 12 years old, they are fallowed or prepared for maize planting again.

After the initial stages of the milpa agroforestry system, crop species diversity drops, but total species diversity increases as native species develop on the site. In many areas, fallows are enriched with useful species because of selective weeding and sparing that occurs during initial clearing and during maize cultivation. For example, Huaves on the Pacific coast of Oaxaca, Mexico favor certain species in secondary vegetation according to individual farmer's goals.[33] Species serving for construction may be protected or transplanted into a plot. Nine species are planted or protected from animals for food production. Many species are favored by protection or dispersal of seed for medicinal or other utilitarian uses. These same practices are also carried out in wooded areas unsuitable for cultivation.

Yucatecan farmers establish bees in clearings in old fallows, and then selectively remove trees not of value to bees from the areas surrounding the gap.[34] These patches are used for 8 to 10 years.

In some cases, farmers encourage native species known to improve soil fertility. For example, Tzeltal Maya in Chiapas, Mexico, protect *Leucaena brachycarpa* and several other species with the consequence that some fallows contain almost pure stands of *L. brachycarpa.*[35]

In Costa Rica, *Lippia torresii*, a fast growing, stump-sprouting firewood producer, is protected in milpas, and other species are selectively removed in order to create a nearly pure stand of this species.[36] In other cases, the management regime itself favors a particular subset of firewood species. For example, the Mexican Huastec Maya short fallow system favors fast growing trees that stump sprout — good firewood producing species, some of which are leguminous and therefore probably fix nitrogen.[37-38] Firewood is an important product on small farms. In Central America, three quarters of the households use firewood for cooking fuel.[39] Yet, firewood production is not economically profitable enough for plantations to be established to meet this need. Management of native species in this manner may be more profitable than weeding out native successional species.[40] Efforts to experiment with native practices of enriching and directing secondary succession using native species, for this and other purposes, are rare, however.[41]

Although there is minimal intervention during the fallow thicket stage, fallows are not unused. Farmers harvest fruits, firewood, and construction materials, especially from native trees spared earlier during milpa clearing and weeding. Cattle are often tied in fallows to forage on native trees. But, in some cases, the fallow stage is managed to create another stage, the managed or artificial forest. This last stage is discussed in a later section.

In upland areas where plow agriculture is used, maize fields are generally under permanent cultivation. In these systems, (1) native trees or cacti are spared or planted in maize fields, (2) cultivated fruit trees or maguey are interplanted with maize and associated crops, (3) trees or maguey are spared or planted on field or terrace borders, or (4) trees are maintained adjacent to fields for litter production to improve field fertility.

In the arid Zapotec-occupied highland areas around Mitla, Oaxaca, Mexico, two types of maguey (for production of the mescal and pulque beverages) are interplanted in rows between maize.[42] In addition to interplanted maguey, in Tlaxcala, Mexico, native trees are left scattered through maize fields and maintained along field borders for many reasons including their positive effect on soil nutrients.[43] The practice of sparing leguminous trees (especially *Prosopis* spp., *Leucaena esculenta,* and *Pithecellobium* spp.) in fields is common from Puebla and Tehuacan south to Oaxaca, but conversion from animal-drawn plow to tractor is correlated with removal of these trees from the fields.[44] In Ostuncalco, Guatemala, leaves and branches of native *Sambucus mexicanus* spared in maize fields are annually harvested, chopped, and mixed with the soil to improve the fields' fertility.[45]

In Tlaxcala, Mexico, on the slopes of Popocatepetl and Ixtaccihuatl, and elsewhere in highland Middle America, apples, pears, peaches, plums, apricots, walnuts, and spared native fruit trees are intercropped with maize and beans.[46-47] In the Zapotec highlands of Oaxaca, Mexico, maize, vegetables, beans and alfalfa are planted under terraced orchards of mamey, avocado, and sapodilla,[48] and native mesquite is spared in fields and on field borders.[49] On the Urubamba River in Yucay, Peru, fruit trees are planted on bunds between irrigated fields of maize, beans, and vegetables.[50] In the Vilcanota Valley of Peru, peach, pear and plum trees are scattered in maize fields at altitudes of 2400 to 3300 m.[51] The Aymara near Chulumani, Bolivia interplant some fruit and coffee trees on their coca terraces.[52]

Alnus jorullensis is planted in contour lines in Guatemalan mountain maize fields to prevent soil erosion.[53] Its leaves add nutrients to the soil, and the trunk is used for firewood or construction material.

In the huamil system of arid-land agriculture practiced in Guanajuato, Mexico, nopal cacti *(Opuntia* spp.) are intercropped with maize, beans and squash.[54] These tall cacti are used for livestock feed, medicines and fuel, and they produce two marketable products: pads and delicious fruit. Huamil fields are created on dry, rocky slopes by cutting back most of the existing scrub. Useful species, including individuals of the native *Opuntia,* are spared as the field is prepared for plowing. If commercial production is planned, the farmer will also plant domesticated varieties of *Opuntia* and other cacti. Propagules (seed and cuttings) are carefully selected by farmers. The cacti are weeded and protected from damage by the plow; they are also pruned and supported when necessary. Domesticated types require more care, fertilizer application and weeding. Fields range from Opuntia orchards for commercial production to polycultural maize fields containing a community of cacti species for a variety of uses.

Annually or biannually cropped maize fields are sometimes slowly converted into orchards or woodlots in the Latin American highlands and lowlands. For example, riverdwellers of Yanallpa on the Ucayali River, Peru plant maize under spared palms and tropical cedar *(Cedrela odorata)* in annually flooded areas.[55] As the number of spared trees increases, bananas are substituted for maize until the plot eventually becomes a concentration of timber trees. In the foothills of the Sierra Madre Oriental in Mexico, the Huastec Maya interplant maize in new orange groves while the permanent crop is getting established.[56] Chilean campesinos farming in a Mediterranean climate interplant maize, potatoes, and beans in young orchards of apricot, quince, cherry, figs, and olives in order to reap yields from the land before the trees become productive.[57]

B. Manioc-Initiated Systems

The manioc producing systems of Central and South America and the Caribbean have several regional names (chacra, conuco, roça) as well as indigenous names. The term chacra will be used here. Chacras are polycultural systems, but manioc is generally the single most important species (although the term chacra is also applied to some systems at higher elevations that lack manioc altogether). Bananas, plantains or maize are emphasized in some areas, however. The chacra system is used from Costa Rica southward across Amazonia and the lower slopes of the Andes.

The traditional chacra is an ideal candidate for modernization, because it is sustainable and flexible, contains numerous crop combinations adapted to local conditions, and is used by settlers as well as by Indians (and has therefore proven acceptable to settlers).[58] A few efforts have been made to improve the traditional Amazonian chacra system[59]

The initial stages of the chacra are more homegarden-like than the initial stages of a milpa. The succeeding garden-orchard thicket stage created in a subsection of the plot is more orchard-like in comparison with the garden-like extended-use subsection of milpa systems. A chacra plot (usually half a hectare or less) is smaller than a milpa plot (usually one hectare). Generally speaking, the chacra is a longer cycle system, so that manioc-producing plots are a small percentage of the area operated by a community, and old fallow forest is a larger percentage than in milpa operating regions. Chacra systems are operated in a wide range of forest associations; thus the native species that are part of the system vary from area to area. Over a hundred cultigens and fruit trees are included. The cultigen list is relatively uniform across South America, although a few groups cultivate few tree species in the chacras.[60-63] Because so many species are cultivated, there are many variations in crop associations between regions, between farmers, and between a given farmer's set of fields. Unfortunately, despite the chacra's wide range, published chacra accounts are primarily limited to lists of cultivated species. The integration of cultivated trees into the system is particularly under-reported. Maps of chacras often do not include the trees listed as cultivated. Details on selection, planting, and care of particular tree species are rarely published.

Chacras are usually established by clearing undergrowth and then cutting down trees in old fallows or forest. A few individuals of useful or rare trees are spared, but the list of spared species is long, and the actual individuals spared vary from farmer to farmer and region to region. Trees are usually felled so that they fall away from the center of the clearing or downslope. Crowns fall into surrounding forest, knocking down adjacent trees and reducing the shadow cast by adjacent forest. This practice also ensures that successional tree species, which eventually must seed into the cultivated zone, will get established near the cultivated area quickly. The slash is usually then burned once, and branches which remain intact are then stacked and reburned in piles where minor root crops are to be planted. Large trunks, however, are left criss-crossing the field to the extent that a passerby can, by walking on trunks, quickly cross a new manioc field without touching his/her foot to the ground. But most of these same trunks will have rotted away by the third or fourth year. Some of the slightly charred trunks are harvested for construction purposes, and others serve for firewood. In some cases, the slash is not burned at all; instead a slash and mulch system is followed.[64-66]

Planting sequences vary, as do planting patterns. Usually planting takes place over a period of several weeks, and all crops (tree and nontree) are planted at that time. There are two basic planting zone patterns: rings and clumps. Regardless of the pattern, manioc (or sometimes introduced plantains) usually occupies the largest segment of the initial garden. Most families plant much more manioc than they need as insurance against failed crops or accidents. Rice and maize are sometimes interplanted with manioc, or manioc is not planted until after the grain is harvested. Other zones are dominated by peanuts, barbasco fish poison, banana/plantains, cacao, pineapples, sweet potatoes, tobacco, fruit trees, sugarcane, or arrow cane. Trees are either planted on the garden perimeter or in scattered patches inside the garden space. Different manioc varieties or mixtures of varieties may be planted in separate zones as well.[67] Minor crops and individuals of spared native species are scattered throughout the different zones. In some cases, one small polycultural plot is maintained in a larger manioc plot.[68] Bora farmers of the Peruvian Upper Amazon restrict their tree crops to patches because they feel the roots interfere with manioc production and harvest.[69] They site tree patches along ridges where trails pass so that harvesting fruits from fallow fields is easy (often done for refreshment on the way home from harvesting manioc, working in a new chacra, or hunting). Within a given region, there are many different agroforestry combinations. Interviews with 55 Panamanian Kuna farmers who maintained agroforestry plots revealed 72 agroforestry associations developed by these 55 farmers.[70] The reasons for these variations in planting pattern and species combinations are related to topography, soil differences, and harvest access as well as to species interactions and ethnic patterns.

As manioc is harvested and replanted daily, piles of manioc plants, weeds, and partially burned branches from the original forest are burned in piles, a practice which improves soil fertility and controls pest and pathogen populations. Runa of Ecuador scatter beans among manioc solely for the purpose of soil improvement.[71] Kayapó, of central Brazil, mulch their crops with palm leaves high in phosphorus.[72]

When a garden is 3 years old, fruit trees often begin production, although the Cocamilla of the Upper Amazon do not plant their tree crops until this time.[73] Longer-lived trees do not come into production until after manioc harvest has ceased. Ethnohistorical comments about the wonderful fruit gardens of the Caribbean coast of South America[74] indicate that the orchard stage was at one time a more important part of chacras in northern South America and Panama. Pejibaye *(Bactris gasipaes)*, a staple during part of the year in some areas, was probably once more important in the Upper Amazon where study of the species indicates it was most thoroughly domesticated and modified by selection.[75] Increased nomadism as Indians attempted to hide from Spaniards may have led to decreased planting of tree crops. Today the most popular trees are relatively short, quick maturing trees (papaya, guava, cashew, caimito, annatto and others), but taller trees (such as pejibaye, macambo, and brazil nut) are also included. Campesinos and caboclos (peasants who speak Spanish, or Portugese, and identify with national culture), as well as native Indian groups, include trees in their chacras.[76-79] Maintenance of the orchard area or individual long-lived trees such as pejibaye only requires occasional removal of vines and minimal weeding.

The fruit stage also attracts game. While most groups see game harvest as a side-benefit of fruit trees, Ecuadorian Runa establish chacras high in fruits distant from their homes specifically for hunting trips[80] and plan to increase fruit planting to feed game near settlements.[81]

The fallowed fruit stage of the chacra is sometimes slashed back for planting manioc 10 to 12 years after the initial clearing, but 20 years is a more common cycle with the main cultigen-harvesting stages lasting 6 to 10 of those years. Older gardens are sometimes carefully managed, however. Crop-tree associations are well developed in old Kayapó gardens in central Brazil where established sets of companion crops are cultivated around plantains, papaya, genipapa *(Genipa americana)*, and annatto.[82] The Aguaruna of Peru sometimes interrupt the fallowing process by clearing older gardens to plant plantains.[83]

Native trees are an important part of the chacra system. The rate at which they move into a cleared area varies widely. Usually, however, native vegetation moves in from the edges as the cultivated core shrinks. While some native tree seedlings are spared during weeding, the repeated uprooting of seedlings associated with manioc cultivation (as opposed to maize cultivation in milpas) requires that forest regeneration begin from the edges or under fruit trees if the chacra cycle is to continue.[84] Some groups plant cultigens under the developing fallow vegetation. For example, Brazilian Kayapó plant root crops under the canopy of secondary trees,[85] and Ecuadorian Runa clear the underbrush, mulch with compost, and plant maize, naranjilla *(Solanum quitoense)*, and minor crops.[86] Many root and tree crops persist in the chacra thicket after weeding has ceased, but people don't regularly harvest them when there are safer and easier places to find the same crops.[87]

The species composition of old fallows is influenced by activities that occur during the earlier chacra stages. Runa of Ecuador, for example, view old fallows as the product of one's work.[88] Management during the chacra stage includes systematic elimination of unwanted stump-sprouts as well as sparing and planting of desirable native species. The resulting fallow thickets are concentrations of many desirable resources and wild game.

More permanent orchards are created in some areas. For example, river-dwellers near Iquitos, Peru manage their chacra agroforestry system for market production[89-92] and emphasize marketable fruits. Umarí *(Pouraqueiba sericea)* and brazil nuts are particularly marketable. Umarí comes into production 5 or 6 years after the chacra is begun and continues to produce for 20 years. The plot is weeded several times a year, and secondary forest never becomes established. When umarí yields begin to drop, umarí trees are cut, and their wood is made into charcoal. Only then are native vegetation and weeds allowed to grow on the plot for 5 to 10 years, after which time the cycle is repeated. These farmers also maintain chacras without perennials for producing their staples.

Guarana *(Paullinia cupana)*, sold for use in beverages and medicine, is interplanted with manioc in chacras managed by the Saterê-Mawé near Manaus, Brazil.[93] Wild seedlings of the shrubby plant are collected from old fallows and forest near the chacra. The established guarana plantation contains scattered fruit trees. Efforts to improve guarana production in commercial plantations have revealed the value of Saterê-Mawé practices.[94]

In the Upper Xingu Basin of Brazil, piqui *(Caryocar brasiliensis)* orchards are established in fields where manioc is already being grown at the edges of the cerrado (thickly wooded scrub plains).[95,96] Large groves of piqui are common around villages of Kalapalo and other Indian groups in this area.[97-99] The trees do not produce their nutritious, oily fruit[100] until many years after they are planted. Once established, the groves (owned by the family whose parents and grandparents planted the trees) are maintained by periodically burning off the undergrowth.

Groves of pejibaye *(Bactris gasipaes)*, a species of great agroforestry potential,[101-105] are found in old Central American fallows created by the very polycultural chacra system of the Caribbean coast. The Guaymi of Panama, for example, broadcast maize seed and plant a variety of root crops in addition to fruit trees, pejibaye, and useful native trees.[106] A large number of native trees are protected in these chacras which eventually contain tree gardens. Pejibaye is one of the top four food sources produced in Guaymi chacras. At the time of the Spanish conquest, large pejibaye groves were widespread in Costa Rica.[107-108] Today, Costa Rican Chirripó[109] and farmers in the Bocas region of Panama[110,111] interplant pejibaye with bananas and plantains as well as maintain almost pure stands of pejibaye.

Polycultural semi-permanent orchards are created in the Orinoco delta of Venezuela.[112] After the initial harvests of maize and root crops, one section of the chacra is fallowed and another section is developed into a "hacienda" stage. The composition of the carefully maintained hacienda orchard depends on individual farmers' decisions based on market conditions. Pineapple and tobacco may also be cultivated here although most species are trees. A careful planting plan is followed, and certain crops and trees are grown in associations. In the fallowed section, farmers encourage development of secondary trees in banana chacras. They plant *Erythrina poepigiana*, cut back herbs, and protect native trees that volunteer.

Some chacra systems are initiated with bananas and plantains. For example, the Chirripó of Costa Rica[113] and the Shipibo of Peru[114] create polycultural fields containing root crops, plantains, and cultivated trees.

C. Wheat-Initiated System

In Totonicapán, Guatemala, pine trees are spared in terraced wheat fields.[115] Over time, the field contains more and more pine until it is no longer cultivated. When the pine stand is mature, it is harvested and wheat cultivation begins again.

D. Sugarcane System

Trees are integrated into smallholders' sugarcane plots. For example, the Huastec of Mexico maintain native trees and palms destined for construction materials in their sugarcane.[116] Pejibaye is scattered in the sugarcane of smallholders in Panama.[117] Several Panamanian Kuna agroforestry combinations include sugarcane and tree crops.[118]

IV. ARTIFICIAL FORESTS

The artificial forest ranges from a managed forest high in native species, to polyspecific orchards with a few native species, to plantations (fincas) of a single tree crop species mixed with a few native trees. This type of agroforestry component may be initiated in the early stages of the agrisilvicultural systems as described above, but it may also be initiated in a forest, an old fallow, or even in a savannah. Once created, the artificial forest is a semi-permanent structure maintained in conjunction with other staple-generating sources — either in close association with a market where tree products are sold in order to purchase staples, or in association with agricultural structures that produce staples.

As an adjunct to intensively managed cash crops, it has great potential for modernized small farms. As a forest seed and gene bank, it serves the longterm interests of small farmers in the tropical, humid lowlands.[119]

The artificial forest is poorly documented, because it has been overlooked by fieldworkers who tended to classify trees growing outside neatly maintained fields as elements of wild forest. In its most common form, the managed forest is a heterogeneous, apparently helter-skelter, association of minimally cultivated woodland enriched with useful species. It is different from a thicket in that it reaches a much older age, it may be cycled through the staple-producing field, orchard-grown, and fallow stages occasionally, but, relative to other traditional field types, it is a long-lived structure designed primarily for production from trees. Creation and maintenance rely upon selective removal of unwanted native species, removal of unproductive or mature individuals of protected species, minimal weeding of undergrowth, and planting of native and introduced species.

The Huastec Maya of northeastern Mexico maintain managed forests (te'lom) on steep slopes and ridges or as greenbelts along creeks where they serve to protect water and prevent erosion.[120] One quarter of a community's land is typically maintained in such forest plots. Over 300 species are found in the te'lom.[121] Thirty three species produce construction materials; 81 species produce edible fruits, leaves or other parts; 65 have utilitarian uses, and over 200 are medicinal plants. Leaves and fruit produced in the te'lom also serve as livestock feed, and livestock are tied in certain areas for protection from the sun during midday. Different sections of a given te'lom may be managed for different products (e.g., avocados, firewood, construction materials, oranges, coffee), but in all sections native species also abound. Thus, within the larger managed forest plot, farmers create different subpatches concentrating different species in a background matrix of native vegetation. In addition, in small clearings and around the open edges of the te'lom, small gardens of manioc, chiles, bananas, nopales, and other cultigens are established.

A similar system has been reported for the Totonac of Veracruz.[122] Among both the Huastec and Totonac, there are communally owned managed forest plots. Totonac communities in Veracruz maintain such artificial forests near population centers expressly for their children's recreation.

Coffee fincas based on this management pattern are found throughout Middle America. The degree to which native trees, cultivated fruit trees, and timber trees are integrated into the upper, middle, and lower stories varies from area to area. Thirty one different coffee finca structures have been documented in Mexico,[123,124] many of which incorporate native trees with multiple uses and introduced fruit trees. On the slopes of the Sierra Madre Oriental, Mexico, both mestizos and Indians have created coffee fincas under spared native trees.[125-127] In many of these areas, the dense overstory functions to protect the coffee from frost damage.[128] In the wet mountains of Vegaguas and Cocle provinces in Panama, campesinos manage a similar system utilizing native species.[129] Coffee is also grown under native trees by the Yukpa in Venezuela.[130] Smallholders intercrop coffee with timber and fruit trees in the premontane forest regions of Venezuela.[131]

In Costa Rica, 65% of coffee fincas have fruit trees for shade.[132] Many small farmers include numerous species, some of them native, in their diversified coffee fincas, because they have to produce firewood, lumber, food, and other materials on a limited area. Other farmers limit the numbers of non-coffee species to a few.[133]

Cacao has been integrated into managed forests in a similar way among the Kekchi Maya[134] and in the Orinoco Delta of Venezuela.[135]

Walled artificial forests, "pet kotoob," were maintained until recently by Yucatec Maya.[136] They were established at the end of a milpa cycle or in a forested site. Cultivation and other intervention was minimal. Native seeds and seedlings were casually introduced from nearby, useful individuals were spared, and domesticated trees were planted in order to create an area where a person could find everything he/she needed in one place. These forested areas also served as shady resting spots with fresh fruits for farmers taking a break from work in nearby milpas. In the past, pet kotoob may have been used to confine semi-domesticated animals (turkey, peccary, deer) to be fed on fallen fruits. Although they have abandoned the pet kotoob, Yucatec Maya continue to develop artificial forest patches on the steep sides and bottoms of sinkholes where avocados and other domesticated tree species flourish among spared native trees. Yucatec Maya thereby exploit naturally humid settings available in their area of karst topography.[137]

Savannah forest islands (apêté) are created in central Brazil by the Kayapó.[138-140] Compost, prepared in nearby forest, is piled into depressions in a selected savannah spot and mixed with soil from ant and termite nests. Seeds and seedlings of native species are introduced, the island of vegetation slowly expands, and an artificial forest grows. Different zones are established within the apêté according to the microhabitat preferences of different species. Some apêté are created to have a shady center with more sunny edges; in others farmers create open gaps in the interior. Apêté are protected by firebreaks when the savannah is regularly burned. Some apêté, however, are created of species adapted to fire; in which case fire is used as a management tool within the apêté. In a representative mature apêté, 85% of the plants present were planted.[141] Medicinal plants, plants with various utilitarian uses, game-attracting plants, shade trees, and food plants are among the over 140 species concentrated here.[142] Kayapó think of apêté as similar to old fallows except that apêté are surrounded by savannah instead of forest.[143]

Enriched areas are also created within natural forests. Kayapó create concentrations of useful trees in forested areas through which they trek,[144] a practice common (though rarely reported) among tropical forest farmer-hunters.[145-147] Casually managed greenbelts are found along trails and between fields throughout the lowlands of Latin America.

In some cases, forests are used to improve field fertility. For example, in western Guatemala, farmers take litter from area pine-oak stands, put it into animal pens to soak up urine, and then spread it onto their fields.[148] A typical .2 ha vegetable plot is supported by 2 ha of forest. Stands of mature forest near the farmed lands of many indigenous people are harvested for numerous products, but their management has not been carefully studied.

Artificial forests have recently been reported among varzean peasant families near Belém, Brazil.[149] These families rely on dooryard gardens, and on managed and unmanaged floodplain forest plots for income to purchase staples in the nearby city. The managed forest plots are selectively weeded and enriched with useful species to enhance production of marketable products.

Long-lived orchards with relatively few native trees are established in highland and lowland areas to complement production from other agricultural plots. River dwellers of Yanallpa on the Ucayali River in Peru maintain mixed orchards of 24 species that survive occasional inundation.[150] Along the Balsas River in arid areas of Guerrero, Mexico, Nahuatl farmers maintain orchards to complement production of annuals in riverbed fields.[151] In the Mediterranean climates of Chile, campesino farmers include orchard-vineyards in their diversified farmsteads that also contain pastures, woodlots of *Pinus radiata,* and annually cropped fields.[152] Farmers manage the orchard-vineyards differently depending on the priority given the tree crops and the grapes. If grapes are of secondary interest, they are removed when the fruit tree canopies begin to close. If grapes are of primary interest, fruit trees are scattered in vineyards. If the two have equal value, trees are more numerous but spaced to allow understory grapes adequate sunlight, either by planting trees and grapes in alternate rows or by scattering grapes between trees. For honey production, farmers place bee hives in these orchard-vineyards.

Commercial coconut orchards of the Panamanian Kuna are established without an annual crop stage and contain other trees including pejibaye, avocado, and mango.[153]

V. DOORYARD GARDENS

Homegardens are maintained in dooryards, fields, and managed forests. The latter two types have been discussed above. The dooryard garden usually produces few staples; rather it is polycultural, often multi-storied, mixture of minor garden crops, fruits, ornamentals, medicinal plants and plants with utilitarian uses. It often contains several different subpatches managed for different products. Useful native species are frequently included, as are experimental plots of new cultigens or cultivars. Children learn about useful plants and their maintenance here.[154]

The dooryard garden deserves special attention for three reasons. Because the dooryard garden is often more permanent than the shifting fields of many traditional farms, and because its location makes it easy for family members to spend spare time working in it, the dooryard garden is a more intensively managed polycultural mix of trees, shrubs, vines, and herbaceous plants than other garden-orchards of the farmstead. Second, common household livestock (chickens, turkeys, ducks, and pigs) are integrated into the system. Finally, it is, of all the traditional agroforestry systems, the most adaptive to modules designed for colonists living on farms or in towns. Planned colonial settlements should allow sufficient space around homes for such gardens, and settlement site choice should include on-ground investigation of the site's suitability for gardening.[155] The economic contributions of these gardens have been little studied, but indications are that they make significant contributions to income as well as to nutrition and subsistence.[156-158] Development planners have recognized the attractiveness of this productive agroforestry structure and increased attention is now being paid to dooryard gardens.[159-161]

Most indigenous groups, whether they live in highlands or lowlands, or arid or moist environments, maintain some useful plants near the house. Yet dooryard complexes have been poorly documented in Latin America; students of Latin American gardens lag far behind students of Indonesian gardens.[162] Existing descriptions of dooryard gardens are dispersed in the literature of many different disciplines and a complete review of those descriptions has never been done. General articles on dooryard gardens have tended to cite the same few examples[163-165] as Latin American models. These are good descriptions of certain gardens, but their overuse has created a stereotypic dooryard garden that does not adequately portray the tremendous diversity of gardens found around Latin American homes. I will focus on a few, less well-known examples in order to demonstrate the variations of this traditional agroforestry system.

The intensively managed Yucatec Mayan dooryard, surrounded by a stone wall, is interesting because the dooryard is a miniature farm. It contains a patch of secondary vegetation, where livestock fodder is harvested, and a small patch of maize in addition to various subtypes of the typical orchard-garden.[166,167] Of the 92 species commonly found in Yucatec dooryards, 61% are native plants.[168] Only half of the production is for household use.

Firewood trees *(Bursera simaruba* and *Gliricidia sepium)* are cultivated in Totonac dooryards in Veracruz, Mexico.[169] A typical dooryard is fenced by densely planted *Opuntia ficus-indica* (a nopal cactus producing edible pads) on three sides and the firewood producer *Gliricidia sepium* on the fourth. Ninety one species are found in Totonac dooryards; 46% are medicinal; 39% are edible, 8% serve for construction, and 6% serve for firewood.

The Huastec Maya of Mexico generally have dooryards high in secondary vegetation.[170] They are unfenced and have ambiguous boundaries. Most dooryards contain fruit and shade trees and a small patch of ornamentals, but the ground surrounding the open area around the house is seldom weeded, and numerous species of medicinal plants many of them native, are scattered in this secondary thicket. Some households, however, do maintain a small, intensively weeded patch containing 80 or more medicinal and food species within the larger, less intensively managed dooryards. One hundred and eighty two species were recorded in Huastec dooryards. Each family designs a dooryard the species composition of which reflects that family's ideas about what species should be maintained nearby.

Large dooryard gardens were once intensively managed by black families in the Cauca Valley of Colombia.[171] Families supported themselves on 2 ha plots from the labor of a couple working 60 days a year. They maintained pigs, poultry, and over 70 species of plants mixed with plantains, cacao, and coffee under native shade trees. This system was disrupted in the 1970s by government programs encouraging peasants to uproot perennials and plant seed crops. The destruction of the perennials resulted in reduced income per unit of land, and 2.3 times the labor days were spent per unit of land for that reduced income.

Trees and other perennials tend to be more important in the dooryard gardens of South American chacra farmers than in their chacras.[172-179] Tree crops, particularly pejibaye, are also important in the communal gardens maintained by South American Indians who share a communal house.[180-183] The basic chacra management plan, described above, is followed in establishing the chacra farmers' dooryard gardens, but more trees and individuals of minor importance are included. Transplanted individuals of wild species are a common part of these gardens,[184,185] and seedbeds for cultivated trees are sometimes found here as well.[186] Many indigenous chacra operators move their homes every 10 years or so (although they sometimes re-use house sites after a fallow period), and this affects the garden's life cycle.

Luzo-Brazilian families who manage varzea floodplain forest near Belém eliminate trees with minor uses from dooryards in order to open the canopy to cultivate other useful species in more sunlight.[187] Of the 68 species in a representative dooryard, 66% were not native to the area. This in contrast to the rest of their holdings where native trees are more important.

In cities and towns throughout Latin American, the Luso-Latin style garden,[188] a small urban garden of fruit trees, ornamentals and herbs, is found in the courtyards and patios of the rich and poor.[189-191]

A dooryard garden-like complex is often found around temporary dwellings and worksites throughout Latin America. These are areas where more delicate plants can be easily tended. And they are places where people need the refreshments and medicines that a garden contains.

Not every home in Latin America has an elaborate garden complex. Dooryard gardens are not found among many Central Brazilian groups living in the gallery forests of the cerrado.[192,193] And dooryard gardens of the arid zones in northeastern Brazil are reported to be impoverished and uncommon.[194] Compared to the diverse dooryard gardens containing a hundred or more species maintained by indigenous families, Costa Rican smallholders' gardens, maintained on 70% of small farms,[195] only contain an average of 16 species (8 of them trees) in their dooryard gardens.[196]

Dooryard gardens have a long historical tradition in Latin America,[197-200] but gardening-promotion programs have emphasized temperate-zone style row gardens planted in carrots and peas.[201-203] Neither the native cultigens nor the complex native gardens have been accorded the respect or the promotion they deserve.

VI. AGRO-SILVO-PASTORAL SYSTEMS

Silvo-pastoral systems are best developed in northeast Brazil where livestock graze under perennial crops (cashew, coconut, carnauba wax palm, African oil palm, babassu palm and others) on large and small holdings.[204] These pasture systems also support an unusual form of shifting cultivation. Laborers on the large plantations fence small plots under these plantations to cultivate rice for rent and subsistence crops for themselves.[205,206] The babassu palm pastures are especially interesting. The babassu has multiple uses,[207] and densities of 100 trees/ha do not affect the productivity of pasture grasses.[208] Babassu forests are also used for shifting agriculture; the leaves are burned for ash, the nut husks are burned as firewood, and cattle graze the tender leaves of young secondary vegetation in fallows.[209] Unproductive trees are removed. The babassu forests in Cebara were probably established by indigenous cultivators.[210]

Palm-pasture combinations are also common in seasonally inundated areas near Quibdo, Colombia.[211] Pineapples are cultivated among the palms in the early stages of pasture establishment.

On the Gulf Coast slopes of the Sierra Madre Oriental in Mexico, Huastec,[212] Totonac,[213] and mestizo farmers spare native timber trees (especially *Cedrela odorata)* as shade trees in small pasture plots. Secondary trees on which cattle forage are also maintained, as are palms for thatch. These pastures are occasionally burned and planted with maize.

In the arid dry forest and scrub desert areas of Venezuela, silvo-pastoral systems were maintained in which native trees served for shade and forage for goats, but mechanized pasture maintenance has led to the removal of most trees.[214]

In the Mediterranean climate of central Chile, farmers integrate animal and tree production in a variety of ways. Trees are spared in pastures for shade and for wood production. Farmers' cows, sheep, pigs, geese, chickens, turkeys, and rabbits forage beneath orchards while leaving manure to enrich the soil. And, during the spring and early summer, shepherds from the Talca region of Chile bring their sheep down to graze on the fresh cover crop of weeds growing under orchards at lower elevations.[215]

VII. WETLAND SYSTEMS

Groves of useful palms are often found in agriculturally-useless swampy areas of South and Central America. Such habitats are frequently found interspersed with cultivable ground, and hence are found on, or bordering, small farms. Some indigenous groups rely heavily on palms during part of the year,[216] and most groups utilize many palm products.[217,218] For example, the Warau of Venezuelan mangrove swamps harvest many food products, including sago from *Mauritia flexuosa* and *Manicaria saccifera,* from the palmetums near their homes.[219] Their "traditional form of arboriculture"[220] is not well described, however. River dwellers near lquitos, Peru harvest fruit from native stands of *Mauritia flexuosa* for sale.[221,222] Generally, useful palms in swampy areas do not seem to be actively managed, although it is reported that the Guato of Paraguay cultivated the acuri palm on raised fields in swamps.[223] In most cases, useful swamp palms seem to be conserved in agroecosystems because of the nature of their habitat. It has been suggested that modern efforts to cultivate some of the native oil palms should center on a similar strategy: develop palm stands in their natural swampy habitats.[224]

In the ancient chinampa system of Central Mexico, willows are planted to secure the soil on the sides of raised islands created in swamps.[225] In the chinampa fields themselves, plantains and bananas are grown in addition to maize, beans, manioc and vegetables. Efforts have been made to adapt raised-field systems to small farms in the Mexican lowlands, and there small farmers spontaneously added fruit trees to the crop mix.[226-228] In the Tlaxcala region of Mexico, farmers harvest fruits, basketry materials, and firewood from native trees and shrubs encouraged along river and drainage canal banks in an area where staples are produced in raised fields.[229]

VIII. LIVING FENCES

Living fences often function to integrate trees among the other crops of a farm. In addition to demarcating property, living fences are pruned for firewood and livestock fodder and in some cases produce fruit or nuts. In Costa Rica, 85% of small farms have living fences.[230,231] In Nicaragua, 60 species are used in living fences[232] that cover 50,000 ha.[233] On the northeastern Gulf Coast slope of the Sierra Madre Oriental of Mexico, Totonac farmers,[234] Huastec farmers,[235] and mestizo ranchers rely primarily on two species, *Bursera simaruba* and *Gliricidia sepium,* for living fences. In most places, wire is strung between the individual trees of these living fences, but in some places the trees alone make the fence. For example, the Huaves, who live on the Pacific coast of Oaxaca, Mexico, have very complex living fences — a dense network of fruit trees, cacti, shrubs and palms which supply a variety of useful products to the farmers in addition to serving as windbreaks and protecting fields from livestock.[236] Forty different species (native and cultigen) are used in four different agroforestry combinations. In northern Mexico, living fences are used to capture soil that would be washed from fields on seasonal floodplains.[237] In arid Central Mexico, pulque magueys *(Agave* spp.) are planted as living fences called "metapancles" to delimit land, control erosion, and serve as windbreaks.[238] In lowland and highland areas of Mexico, cacti with edible parts are frequently used as living fences around dooryards. At high altitudes in the Peruvian Andes, several multiple-use, native tree species are used for living fences which serve as windbreaks as well as producers of firewood, construction material and leaves important for soil improvement.[239] Living fences and windbreaks are a common feature of the Chilean landscape where multiple use tree species are combined with grapes, berry vines, and pita along field and property borderlines.[240]

IX. CONCLUSION

These agroforestry systems are operated on farms containing several production units (pastures, agroforestry plots, intensively cropped fields, gardens, fishing streams, forest corridors between fields, etc.). The integration of these units is an area of great importance for small farmers and to the sustainability of the farm,[241-245] but a discussion of that integration is beyond the scope of this paper.

I have focused on agroforestry structures. Lists of the numerous economically valuable species found in these structures can be located in the references cited and elsewhere.[246-252] Many of these agroforestry systems maintain valuable crop germplasm in the form of traditional cultivars. The gene bank functions of these systems should not be discarded during their modernization.[253,254]

Despite their abundance, few indigenous Latin American agroforestry systems have been carefully documented.[255] Study of indigenous agroforestry systems is urgently needed if the knowledge of their operation is to contribute to small farm development.[256-259]

REFERENCES

1. **CATIE/OTS,** *Sistemas Agroforestales: Principios y Aplicaciones en los Tropicos,* Organización para Estudios Tropicales (OTS) and Centro Agronomico Tropical de Investigación y Enseñanza (CATIE), San José, Costa Rica, 1986, 818.
2. **Raintree, J. B.,** Agroforestry pathways: land tenure, shifting cultivation and sustainable agriculture, *Unasylva,* 38, 3, 1986.
3. **Ewel, J. J.,** Designing agricultural ecosystems for the humid tropics, *Annu. Rev. Ecol. Syst.,* 17, 245, 1986.
4. **Hecht, S. B.,** Agroforestry in the Amazon Basin: Practice, theory and limits of a promising land use, in *Amazonia: Agriculture and Land Use Research,* Hecht, S. B., Ed., CIAT, Cali, Colombia, 1982, 331.
5. **Bainbridge, D. A.,** Agroforestry: Promise and reality, The need for institutional reform, presented at the Society of Economic Anthropology meeting, Riverside, 1987.
6. **Budowski, G.,** Applicability of agro-forestry systems, in *Agro-Forestry in the African Humid Tropics,* MacDonald, L. H., Ed., United Nations University, Tokyo, Japan, 1982, 13.
7. **Alcorn, J. B.,** *Huastec Mayan Ethnobotany,* University of Texas Press, Austin, 1984, 982.
8. **Alcorn, J. B.,** Process as resource: The traditional agricultural ideology of Bora and Huastec resource management and its implications for research, in *Resource Management in Amazonia: Indigenous and Folk Strategies,* Posey, D. A. and W. Balee, Eds., Advances in Economic Botany Series, New York Botanical Gardens, Bronx, 1989, 63.
9. **Glass, E. H. and Thurston, H. D.,** Traditional and modern crop protection in perspective, *BioScience,* 28, 109, 1978.
10. **Oldeman, R. A. A.,** The design of ecologically sound agroforests, *Plant Research and Agroforestry,* Huxley, P. A., Ed., ICRAF, Nairobi, 1983.
11. **Peck, R. B.,** Traditional forestation strategies of local farmers in the tropics, in *Strategies and Designs for Afforestation, Reforestation, and Tree Planting,* Wiersum, K. F., Ed., Pudoc, Wageningen, Netherlands, 1984, 205.
12. **Rico-Gray, V. A., Gómez-Pompa, and C. Chan,** Las selvas manejadas por los Mayas de Yohaltun, Campeche, México, *Biótica,* 10, 321, 1985.
13. **Gómez-Pompa, A.,** Tropical deforestation and Maya silviculture: An ecological paradox, *Tulane Studies in Biology,* 1987, in press.
14. **Gómez-Pompa, A.,** On Maya silviculture, *Mexican Studies,* 3, 1, 1987.
15. **Turner, B. L., II and Miksicek, C. H.,** Economic plant species associated with prehistoric agriculture in the Maya lowlands, *Econ. Bot.,* 38, 179, 1984.
16. **Flannery, K. V., Ed.,** *Maya Subsistence,* Academic Press, New York, 1982, 368.
17. **Harrison, P. D. and Turner, B. L., II, Eds.,** *Pre-Hispanic Maya Agriculture,* University of New Mexico Press, Albuquerque, 1978, 414.
18. **Netting, R. McC.,** Maya subsistence: mythologies, analogies, possibilities, in *The Origins of Maya Civilization,* Adams, R. E. W., Ed., University of New Mexico Press, Albuquerque, 1977, 299.
19. **Sanabria, O. L.,** *El Uso y Manejo Forestal en la Comunidad de Xul. en el Sur de Yucatán,* Etnoflora Yucatanense Fasciculo 2, INIREB, Xalapa, Veracruz, Mexico, 1986, 192.
20. **Alcorn, J. B.,** *Huastec Mayan Ethnobotany,* University of Texas Press, Austin, 1984, 982.
21. **Alcorn, J. B.,** *Huastec Mayan Ethnobotany,* University of Texas Press, Austin, 1984, 982.
22. **Kelly, I. and A. Palerm,** *The Tajín Totonac,* Smithsonian Institution Institute of Social Anthropology Publication No. 13, U.S. Government Printing Office, Washington, D.C., 1952, 369.
23. **Medellín Morales, S. G.,** *Uso y Manejo de las Especies Vegetales Comestibles, Medicinales, Para Construcción y Combustibles en una Comunidad Totonaca de la Costa (Plan de Hidalgo, Papantla, Veracruz, México),* Programa Formación de Recursos Humanos, INIREB, Xalapa, Veracruz, Mexico, 1986, 79.
24. **Alcorn, J. B.,** *Huastec Mayan Ethnobotany,* University of Texas Press, Austin, 1984, 982.
25. **Wilk, R. R.,** Agriculture, Ecology, and Domestic Organization Among the Kekchi Maya, Ph.D. thesis, University of Arizona, Tucson, 1981.
26. **Nations, J. D. and R. B. Nigh,** The evolutionary potential of Lacandon Maya sustained-yield tropical forest agriculture, *J. Anthropolog. Res.,* 36, 1, 1980.
27. **Medellín Morales, S. G.,** *Uso y Manejo de las Especies Vegetales Comestibles, Medicinales, Para Construcción y Combustibles en una Comunidad Totonaca de la Costa (Plan de Hidalgo, Papantla, Veracruz, México),* Programa Formación de Recursos Humanos, INIREB, Xalapa, Veracruz, Mexico, 1986, 79.
28. **Gordon, B. L.,** *A Panama Forest and Shore: Natural History and Amerindian Culture in Bocas del Toro,* Boxwood Press, Pacific Grove, CA, 1982, 178.
29. **Sanabria, O. L.,** *El Uso y Manejo Forestal en la Comunidad de Xul, en el Sur de Yucatan,* Etnoflora Yucatanense Fasciculo 2, INIREB, Xalapa, Veracruz, Mexico, 1986, 192.
30. **Medellín Morales, S. G.,** *Uso y Manejo de las Especies Vegetales Comestibles, Medicinales, Para Construcción y Combustibles en una Comunidad Totonaca de la Costa (Plan de Hidalgo, Papantla, Veracruz, México),* Programa Formación de Recursos Humanos, INIREB, Xalapa, Veracruz, Mexico, 1986, 79.
31. **Kelly, I. and A. Palerm,** *The Tajín Totonac,* Smithsonian Institution Institute of Social Anthropology Publication No. 13, U.S. Government Printing Office, Washington, D.C., 1952, 369.
32. **Medellín Morales, S. G.,** *Uso y Manejo de las Especies Vegetales Comestibles, Medicinales, Para Construcción y Combustibles en una Comunidad Totonaca de la Costa (Plan de Hidalgo, Papantla, Veracruz, México),* Programa Formación de Recursos Humanos, INIREB, Xalapa, Veracruz, Mexico, 1986, 79.
33. **Zizumbo V. D., and P. Colunga García-Marin,** *Los Huaves: La Apropiación de los Recursos Naturales,* Universidad Autonoma Chapingo, Chapingo, Mexico, 1982, 277.
34. **Sanabria, O. L.,** *El Uso y Manejo Forestal en la Comunidad de Xul, en el Sur de Yucatan,* Etnoflora Yucatanense Fasciculo 2, INIREB, Xalapa, Veracruz, Mexico, 1986, 192.
35. **Nigh, R. B.,** Evolutionary Ecology of Maya Agriculture in Highland Chiapas, Mexico, Ph.D. thesis, Stanford University, Stanford, CA, 1976.
36. **Beer, J.,** *Research and Development Work by Costa Rican Farmers: Lessons for Agroforesters,* CATIE, Turrialba, Costa Rica, 1982,6.
37. **Alcorn, J. B.,** *Huastec Mayan Ethnobotany,* University of Texas Press, Austin, 1984, 982.
38. **Alcorn, J. B.,** Process as resource: The traditional agricultural ideology of Bora and Huastec resource management and its implications for research, in *Resource Management in Amazonia: Indigenous and Folk Strategies,* Posey, D. A. and W. Balee, Eds., Advances in Economic Botany Series, New York Botanical Gardens, Bronx, 1989, 63.
39. **Jones, J. R.,** The Central American energy problem: Anthropological perspectives on fuelwood supply and production, *Culture Agric.,* 22, 6, 1984.
40. **Beer, J.,** *Research and Development Work by Costa Rican Farmers: Lessons for Agroforesters,* CATIE, Turrialba, Costa Rica, 1982,6.

41. **del Amo R., Sylvia,** Management of secondary vegetation for artificially creating useful rain forest in Uxpanapa, Veracruz, Mexico, in *International Workshop on Rainforest Regeneration and Management*, Gómez-Pompa, A. and T. C. Whitmore, Eds., UNESCO, Geneva, in press.
42. **Messer, E.** Zapotec Plant Knowledge: Classification, Uses, and Communication and Plants in Mitla, Ph.D. thesis, University of Michigan, Ann Arbor, 1975.
43. **Altieri, M. A., J. Trujillo, and J. G. Farrell,** The agroecology of corn production in Tlaxcala, Mexico, unpublished manuscript, 1987.
44. **Wilken, G. C.,** Integrating forest and small-scale farm systems in Middle America, *Agro-Ecosystems*, 3, 291, 1977.
45. **Wilken, G. C.,** Integrating forest and small-scale farm systems in Middle America, *Agro-Ecosystems*, 3, 291, 1977.
46. **Altieri, M. A., J. Trujillo, and J. G. Farrell,** The agroecology of corn production in Tlaxcala, Mexico, unpublished manuscript, 1987.
47. **Wilken, G. C.,** Management of productive space in traditional farming, *Actes du XLII Congres International des Americanistes, Congres du Centenaire*, 2, 409, 1979.
48. **Donkin, R. A.,** *Agricultural Terracing in the Aboriginal New World*, Viking Fund Publications in Anthropology No. 56, University of Arizona Press, Tucson, 1979, 196.
49. **Messer, E.** Zapotec Plant Knowledge: Classification, Uses, and Communication about Plants in Mitla, Ph.D. thesis, University of Michigan, Ann Arbor, 1975.
50. **Donkin, R. A.,** *Agricultural Terracing in the Aboriginal New World*, Viking Fund Publications in Anthropology No. 56, University of Arizona Press, Tucson, 1979, 196.
51. **Gade, D. W.,** *Plants, Man and the Land in the Vilcanota Valley of Peru*, Dr. W. Junk, B. V., The Hague, 1975, 240.
52. **Donkin, R. A.,** *Agricultural Terracing in the Aboriginal New World*, Viking Fund Publications in Anthropology No. 56, University of Arizona Press, Tucson, 1979, 196.
53. **Pretell Chichlote, J., D. Ocaña V., R. Jon Jap, and E. Baranhona C.,** *Apuntes Sobre Algunas Especies Forestales Nativas de la Sierra Peruana*, FAO, Lima, Peru, 1985, 120.
54. **Colunga García-Marin, P., E. Hernández X., and A. Castillo Morales,** Variación morfológica, manejo agrícola tradicional y grado de domesticación de *Opuntia* spp. en el Bajío Guanajuatense, *Agrociencia*, 65, 7, 1986.
55. **Padoch, C. and W. de Jong,** Traditional agroforestry practices of native and ribereno farmers in the lowland Peruvian Amazon, in *Agroforestry: Realities, Possibilities, and Potentials*, Gholz, H., Ed., Martinus Nijhoff, Dordrecht, Netherlands, 1987, in press.
56. **Alcorn, J. B.,** *Huastec Mayan Ethnobotany*, University of Texas Press, Austin, 1984, 982.
57. **Altieri, M. A. and J. Farrell,** Traditional farming systems of south-central Chile, with special emphasis on agroforestry, *Agroforestry Systems*, 2, 3, 1984.
58. **Moran, E. F.,** *Developing the Amazon*, Indiana University Press, Bloomington, 1981, 320.
59. **Flores Paitan, S.,** *Potencial agroforestal de bosques secundarios bajo sistema Bora de agricultura migratoria en la Amazonía Peruana*, UNAP-MAB-IIAP, Iquitos, Peru, 1983.
60. **Beckerman, S.,** Swidden in Amazonia and the Amazon Rim, in *Comparative Farming Systems*, Turner, B. L., II, and S. B. Brush, Eds., Guilford Press, New York, 1987, 55.
61. **Hames, R.,** personal communication, 1987.
62. **Lizot, J.,** La agricultura Yanomami, *Antropológica*, 53, 3, 1980.
63. **Smole, W. J.,** *The Yanoama Indians*, University of Texas Press, Austin, 1976.
64. **West, R. C.,** *The Pacific Lowlands of Colombia*, Louisiana State University Press, Baton Rouge, 1975, 278.
65. **Whitten, N. E.,** *Sacha Runa: Ethnicity and Adaptation of Ecuadorian Jungle Quichua*, University of Illinois Press, Chicago, 1976, 348.
66. **Bort, J.,** Ecology and subsistence on opposite sides of the Talamancan range, in *Adaptive Radiations in Prehistoric Panama*, Peabody Museum of Archaeology and Ethnology, Linares, O. F. and A. J. Ranere, Eds., Harvard University, Cambridge, 1980, 499.
67. **Boster, J.,** A comparison of the diversity of Jivaroan gardens with that of the tropical forest, *Hum. Ecol.*, 11, 47, 1983.
68. **Eden, M. J.,** Ecological aspects of development among Piaroa and Guahibo Indians of the Upper Orinoco Basin, *Antropológica*, 39, 25, 1974.
69. **Denevan, W. M., Treacy, J. M., Alcorn, J. B., Padoch, C., Denslow, J., and S. Flores P.,** Indigenous agroforestry in the Peruvian Amazon: Bora Indian management of swidden fallows, *Interciencia*, 9, 346, 1984.
70. **Castillo, G. and J. W. Beer,** *Utilización del Bosque y de Sistemas Agroforestales in la Región de Gardi, Kuna Yala (San Blas. Panama)*, CATIE, Turrialba, Costa Rica, 1983, 93.
71. **Whitten, N. E.,** *Sacha Runa: Ethnicity and Adaptation of Ecuadorian Jungle Quichua*, University of Illinois Press, Chicago, 1976, 348.
72. **Hecht, S. B. and D. A. Posey,** Preliminary results on soil management techniques of the Kayapo Indians, in *Resource Management in Amazonia: Indigenous and Folk Strategies*, Posee, D. A. and W. Balee, Eds., Advances in Economic Botany Series, New York Botanical Gardens, Bronx, 1989, 174.
73. **Stocks, A.,** personal communication, 1987.
74. **Castellanos, Juan De,** *Elegias de Varones Ilustres de Indias*, Vol. 4, Biblioteca de Autores Españoles, Madrid, 1857.
75. **Clement, C. R.,** *Domestication of the Pejibaye Palm (Bactris gasipaes): Past and Present*, Advances in Economic Botany, New York Botanical Garden, Bronx, in press.
76. **Clark, K. E. and C. Uhl,** Deterioro de la vida de subsistencia tradicional en San Carlos de Río Negro, *Interciencia*, 9, 358, 1984.
77. **Padoch, C., J. Chota Inuma, W. de Jong, and J. Unruh,** Amazonian agroforestry: A market-oriented system in Peru, *Agrofor. Syst.*, 3, 47, 1985.
78. **Hiraoka, M.,** Changing floodplain livelihood patterns in the Peruvian Amazon, *Tsukuba Studies in Human Geography*, 9, 243, 1985.
79. **Moran, E. F.,** *Developing the Amazon*, Indiana University Press, Bloomington, 1981, 320.
80. **Whitten, N. E.,** *Sacha Runa: Ethnicity and Adaptation of Ecuadorian Jungle Quichua*, University of Illinois Press, Chicago, 1976, 348.
81. **Irvine, D.,** Resource Management by the Runa Indians of the Ecuadorian Amazon, Ph.D. thesis, Stanford University, Stanford, CA, 1987.
82. **Posey, D. A.,** A preliminary report on diversified management of tropical forest by the Kayapo Indians of the Brazilian Amazon, *Advances in Economic Botany*, 1, 112, 1984.
83. **Brown, M. F.,** *Tsewa's Gift: Magic and Meaning in an Amazonian Society*, Smithsonian Institution Press, Washington, D.C., 217.
84. **Uhl, C.,** Factors controlling succession following slash-and-burn agriculture in Amazonia, *J. Ecol.*, 75, 377, 1987.
85. **Posey, D. A.,** A preliminary report on diversified management of tropical forest by the Kayapo Indians of the Brazilian Amazon, *Advances in Economic Botany*, 1, 112, 184.
86. **Whitten, N. E.,** *Sacha Runa: Ethnicity and Adaptation of Ecuadorian Jungle Quichua*, University of Illinois Press, Chicago, 1976, 348.
87. **Hames, R.,** personal communication, 1987.
88. **Irvine, D.,** Resource Management by the Runa Indians of the Ecuadorian Amazon, Ph.D. thesis, Stanford University, Stanford, CA, 1987.

89. **Padoch, C., J. Chota Inuma, W. de Jong, and J. Unruh,** Amazonian agroforestry: A market-oriented system in Peru, *Agrofor. Syst.*, 3, 47, 1985.
90. **Padoch, C. and W. de Jong,** Traditional agroforestry practices of native and ribereño farmers in the lowland Peruvian Amazon, in *Agroforestry: Realities, Possibilities, and Potentials,* Gholz, H., Ed., Martinus Nijhoff, Dordrecht, Netherlands, 1987, in press.
91. **Hiraoka, M.,** Mestizo subsistence in riparian Amazonia, *Natl. Geogr. Res.*, 1, 236, 1985.
92. **Hiraoka, M.,** Changing floodplain livelihood patterns in the Peruvian Amazon, *Tsukuba Studies in Human Geography*, 9, 243, 1985.
93. **Henman, A. R.,** Guarana *(Paullinia cupana* var. *sorbilis):* Ecological and social perspectives on an economic plant of the Central Amazon Basin, *J. Ethnopharmacol.*, 6, 311, 1982.
94. **Henman, A. R.,** Guarana *(Paullinia cupana* var. *sorbilis):* Ecological and social perspectives on an economic plant of the Central Amazon Basin, *J. Ethnopharmacol.*, 6, 311, 1982.
95. **Basso, E. B.,** *The Kalapalo Indians of Central Brazil*, Holt, Rinehart and Winston, New York, 1973.
96. **Carneiro, R. L.,** The cultivation of manioc among the Kuikuru of the Upper Xingu, in *Adaptive Responses of Native Amazonians,* Hames, R. B. and W. T. Vickers, Eds., Academic Press, New York, 1983, 65.
97. **Basso, E. B.,** *The Kalapalo Indians of Central Brazil*, Holt, Rinehart and Winston, New York, 1973.
98. **Schmidt, M.,** Comments on cultivated plants and agricultural methods of South American Indians, in *Native South Americans,* Lyon, P. J., Ed., Little, Brown, Boston, 1974, 60.
99. **Levi-Strauss, C.,** The use of wild plants in tropical South America, in *Handbook of South American Indians*, Volume 6, Steward, J. H., Ed., Smithsonian Institution, Washington, D.C., 1950, 465.
100. **NAS,** *Underexploited Tropical Plants with Promising Economic Value,* National Academy of Sciences, Washington, D.C., 1975, 186.
101. **Clement, C. R. and J. E. Mora Urpi,** Pejibaye palm *(Bactris gasipaes,* Arecaceae): Multi-use potential for the lowland humid tropics, *Economic Botany,* 41, 302, 1987.
102. **Clement C. R.,** El pejibaye: Resultados y necesidades de investigación, *Useful Palms of Tropical Americas Newsletter*, 2, 2, 1986.
103. **Clement, C. R.,** *Domestication of the Pejibaye Palm (Bactris gasipaes): Past and Present,* Advances in Economic Botany, New York Botanical Garden, Bronx, in press.
104. **Hunter, J. R.,** The lack of acceptance of the pejibaye palm and a relative comparison of its productivity to that of maize, *Econ. Bot.*, 23, 237, 1969.
105. **Clement, C. R.,** The pejibaye palm *(Bactris gasipaes* H.B.K.) as an agroforestry component, *Agrofor. Syst.*, 4, 205, 1986.
106. **Gordon, B. L.,** *A Panama Forest and Shore: Natural History and Amerindian Culture in Bocas del Toro,* Boxwood Press, Pacific Grove, CA, 1982, 178.
107. **Stone, D.,** La definición de dos culturas distintas vistas en la antropología de América Central, in *Homenaja al Dr. Alfonso Caso,* Imprenta Mundo, Mexico, 1951, 353.
108. **Prance, G.,** The pejibaye, or peach palm, *(Guilielma gasipaes* HBK) Bailey, and the papaya, *Carica papaya* L., in *Pre-Colombian Plant Migration,* Stone, D. Ed., Papers of the Peabody Museum of Archaeology and Ethnology Volume 76, Harvard University Press, Cambridge, MA, 1984, 85.
109. **Bergman, R. W.,** Shifting Cultivation in the High Rainforest: The Chirripó Indians, Costa Rica, M.S. thesis, University of Wisconsin, Madison, 1969.
110. **Bort, J.,** Ecology and subsistence on opposite sides of the Talamancan range, in *Adaptive Radiations in Prehistoric Panama,* Linares, O. F. and A. J. Ranere, Eds., Peabody Museum of Archaeology and Ethnology, Harvard University, Cambridge, 1980, 499.
111. **Young, P. D.,** Notes on Guaymi traditional culture, *Adaptive Radiations in Prehistoric Panama,* Linares, O. F. and A. J. Ranere, Eds., Peabody Museum of Archaeology and Ethnology, Harvard University, Cambridge, 1980, 224.
112. **Ruddle, K. and R. Chesterfield,** *Education for Traditional Food Procurement in the Orinoco Delta,* Ibero-Americana 53, University of California, Berkeley, 1977,172.
113. **Bergman, R. W.,** Shifting Cultivation in the High Rainforest: The Chirripó Indians, Costa Rica, M.S. thesis, University of Wisconsin, Madison, 1969.
114. **Bergman, R. W.,** *Amazon Economics: The Simplicity of Shipibo Indian Wealth,* University Microfilms International, Ann Arbor, 1980, 249.
115. **Wilken, G. C.,** Integrating forest and small-scale farm systems in Middle America, *Agro-Ecosystems,* 3, 291, 1977.
116. **Alcorn, J. B.,** *Huastec Mayan Ethnobotany,* University of Texas Press, Austin, 1984, 982.
117. **Guzman, L. E.,** *Farming and Farmlands in Panama,* Department of Geography Research Paper No. 44, University of Chicago, Chicago, 1956, 137.
118. **Castillo, G. and J. W. Beer,** *Utilización del Bosque y de Sistemas Agroforestales in la Región de Gardi, Kuna Yala (San Blas. Panama),* CATIE, Turrialba, Costa Rica, 1983, 93.
119. **Alcorn, J. B.,** Development policy, forests, and peasant farms: Reflections on Huastec managed forests' contributions to commercial production and resource conservation, *Econ. Bot.*, 38, 389, 1984.
120. **Alcorn, J. B.,** Development policy, forests, and peasant farms: Reflections on Huastec managed forests' contributions to commercial production and resource conservation, *Econ. Bot.*, 38, 389, 1984.
121. **Alcorn, J. B.,** El te'lom huasteco: Presente, pasado, y futuro de un sistema de silvicultura indigena, *Biotica,* 8, 315, 1983.
122. **Medellín Morales, S. G.,** *Uso y Manejo de las Especies Vegetales Comestibles, Medicinales, Para Construcción y Combustibles en una Comunidad Totonaca de la Costa (Plan de Hidalgo, Papantla, Veracruz, México),* Programa Formación de Recursos Humanos, INIREB, Xalapa, Veracruz, Mexico, 1986, 79.
123. **Toledo, V., J. Carabias, C. Mapes, and C. Toledo,** *Ecología Autosuficiencia Alimentaria,* Siglo Veintiuno, Mexico City, 1985, 118.
124. **Fuentes Flores, R.,** Coffee production farming systems in Mexico, in *Agroforestry Systems in Latin America,* de las Salas, G. Ed., CATIE, Turrialba, Costa Rica, 1979, 60.
125. **Alcorn, J. B.,** *Huastec Mayan Ethnobotany,* University of Texas Press, Austin, 1984, 982.
126. **Toledo, V., J. Carabias, C. Mapes, and C. Toledo,** *Ecología Autosuficiencia Alimentaria,* Siglo Veintiuno, Mexico City, 1985, 118.
127. **Medellín Morales, S. G.,** *Uso y Manejo de las Especies Vegetales Comestibles, Medicinales, Para Construcción y Combustibles en una Comunidad Totonaca de la Costa (Plan de Hidalgo, Papantla, Veracruz, México),* Programa Formación de Recursos Humanos, INIREB, Xalapa, Veracruz, Mexico, 1986, 79.
128. **Barradas, V. L. and L. Fanjul,** La importancia de la cobertura arborea en la temperatura del agroecosistema cafetalero, *Biótica,* 9, 415, 1984.
129. **Guzman, L. E.,** *Farming and Farmlands in Panama,* Department of Geography Research Paper No. 44, University of Chicago, Chicago, 1956, 137.
130. **Ruddle, K.,** *The Yukpa Cultivation System,* Ibero-Americana 52, University of California, Berkeley, 1974, 197.
131. **Escalante, E. E.,** Promising agroforestry systems in Venezuela, *Agrofor. Syst.*, 3, 209, 1985.
132. **Jones, J. R. and N. Price,** Agroforestry: An application of the farming systems approach to forestry, *Hum. Organ.*, 44, 322, 1985.

133. **CATIE (Centro Agronomico Tropical de Investigacion y Ensenanza),** *Informe de Progreso 1983,* Serie Institutional Informe Anual No. 5, CATIE, Turrialba, Costa Rica, 1984, 100.
134. **Wilk, R. R.,** Agriculture, Ecology, and Domestic Organization Among the Kekchi Maya, Ph.D. thesis, University of Arizona, Tucson, 1981.
135. **Ruddle, K. and R. Chesterfield,** *Education for Traditional Food Procurement in the Orinoco Delta,* Ibero-Americana 53, University of California, Berkeley, 1977,172.
136. **Gómez-Pompa, A., J. Salvador Flores, and V. Sosa,** The 'pet kot': a man-made tropical forest of the Maya, *Interciencia,* 12, 10, 1987.
137. **Gómez-Pompa, A.,** On Maya silviculture, *Mex. Stud.,* 3, 1, 1987.
138. **Posey, D. A.,** Indigenous management of tropical forest ecosystems: the case of the Kayapó Indians of the Brazilian Amazon, *Agrofor. Syst.,* 3, 139, 1985.
139. **Anderson, A. B. and D. A. Posey,** Manejo de cerrado pelos índios Kayapó, *Boletim do Museu Paraense Emilio Goeldi Botânica,* 2, 77, 1985.
140. **Posey, D. A.,** The keepers of the campo, *Garden,* 8, 8, 1984.
141. **Posey, D. A.,** Indigenous management of tropical forest ecosystems: The case of the Kayapó Indians of the Brazilian Amazon, *Agrofor. Syst.,* 3, 139, 1985.
142. **Posey, D. A.,** The keepers of the campo, *Garden,* 8, 8, 1984.
143. **Posey, D. A.,** Indigenous management of tropical forest ecosystems: The case of the Kayapó Indians of the Brazilian Amazon, *Agrofor. Syst.,* 3, 139, 1985.
144. **Posey, D. A.,** A preliminary report on diversified management of tropical forest by the Kayapó Indians of the Brazilian Amazon, *Adv. Econ. Bot.,* 1, 112, 1984.
145. **Torres, H.,** personal communication, 1986.
146. **Colchester, M.,** Rethinking Stone Age economics: some speculations concerning the Pre-Colombian Yanoama economy, *Hum. Ecol.,* 12, 291, 1984.
147. **Holmberg, A. R.,** *Nomads of the Long Bow,* The Natural History Press, Garden City, NY, 1969, 294.
148. **Wilken, G. C.,** Integrating forest and small-scale farm systems in Middle America, *Agro-Ecosystems,* 3, 291, 1977.
149. **Anderson, A., A. Gely, J. Strudwick, G. L. Sobel, and M. das Cracas C. Pinto,** Un sistema agroforestal na várzea do estúario amazônico (ilha das Onças, município de Barcarena, estado do Pará), *Acta Amazônica,* Suppl., 15, 195, 1985.
150. **Padoch, C. and W. de Jong,** Traditional agroforestry practices of native and ribereno farmers in the lowland Peruvian Amazon, in *Agroforestry: Realities, Possibilities, and Potentials,* Gholz, H., Ed., Martinus Nijhoff, Dordrecht, Netherlands, 1987, in press.
151. **del Amo R., S., R. Aguilar Lojero, and M. A. Delgado,** The tecallis: a traditional water management system, unpublished manuscript, 1987.
152. **Altieri, M. A. and J. Farrell,** Traditional farming systems of south-central Chile, with special emphasis on agroforestry, *Agroforestry Systems,* 2, 3, 1984.
153. **Castillo, G. and J. W. Beer,** *Utilización del Bosque y de Sistemas Agroforestales in la Región de Gardi, Kuna Yala (San Blas. Panama),* CATIE, Turrialba, Costa Rica, 1983, 93.
154. **Ruddle, K. and R. Chesterfield,** *Education for Traditional Food Procurement in the Orinoco Delta,* Ibero-Americana 53, University of California, Berkeley, 1977, 172.
155. **Ewell, P. T. and T. T. Poleman,** *Uxpanapa: Agricultural Development in the Mexican Tropics,* Pergamon Press, New York, 1980, 206.
156. **Gonzalez-Jacome, A.,** Home gardens in Central Mexico, *Prehistoric Intensive Agriculture in the Tropics,* Farrington, I. S., Ed., BAR International Series 232, BAR, Oxford, England, 1985, 521.
157. **Ninez, V. K.,** *Household Gardens: Theoretical Considerations on an Old Survival Strategy,* International Potato Center (CIP), Lima, Peru, 1984, 44.
158. **Truman, K.,** personal communication, 1987.
159. **Budowski, G.,** *Homegardens in Tropical America: A Review,* CATIE, Turrialba, Costa Rica, 1985, 16.
160. **Brownrigg, L.,** *Home Gardening in International Development: What the Literature Shows,* League for International Food Education, Washington, D.C., 1985, 332.
161. **Price, N.,** *The Tropical Mixed Garden: An Agroforestry Component of the Small Farm,* CATIE, Turrialba, Costa Rica, 1983, 38.
162. **Budowski, G.,** *Homegardens in Tropical America: A Review,* CATIE, Turrialba, Costa Rica, 1985, 16.
163. **Anderson, E.,** *Plants, Man and Life,* University of California, Berkeley, California, 1952, 251.
164. **Covich, A. P. and N. H. Nickerson,** Studies of cultivated plants in Choco dwelling clearings, Darien, Panama, *Econ. Bot.,* 20, 285, 1966.
165. **Lathrap, D. W.,** Our father the cayman, our mother the gourd, *Origins of Agriculture,* Reed, C. A., Ed., Mouton, The Hague, Netherlands, 1977, 713.
166. **Sanabria, O. L.,** *El Uso y Manejo Forestal en la Comunidad de Xul, en el Sur de Yucatán,* Etnoflora Yucatanense Fasciculo 2, INIREB, Xalapa, Veracruz, Mexico, 1986, 192.
167. **Vara Moran, A.,** La dinámica de la milpa en Yucatán: El solar, in *Seminario Sobre Producción Agricola en Yucatán,* Hernández, X. E. and Rafael Padilla y Ortega, Eds., Colegio de Postgraduados, Chapingo, Mexico, 1980, 305.
168. **Barrera, A.,** Sobre la unidad de habitación tradicional campesina y el manejo de recursos bióticos en el área Maya Yucatanense I. Arboles y arbustos de las huertas familiares, *Biótica,* 5, 115, 1980.
169. **Medellín Morales, S. G.,** *Uso y Manejo de las Especies Vegetales Comestibles, Medicinales, Para Construcción y Combustibles en una Comunidad Totonaca de la Costa (Plan de Hidalgo, Papantla, Veracruz, México),* Programa Formación de Recursos Humanos, INIREB, Xalapa, Veracruz, Mexico, 1986, 79.
170. **Alcorn, J. B.,** *Huastec Mayan Ethnobotany,* University of Texas Press, Austin, 1984, 982.
171. **Taussig, M.,** Peasant economics and the development of capitalist agriculture in the Cauca Valley, Colombia, *Latin American Perspectives,* 5, 62, 1978.
172. **Boster, J.,** personal communication, 1987.
173. **Vickers, W. T.,** Tropical forest mimicry in swiddens: a reassessment of Geertz's model with Amazonian data, *Hum. Ecol.,* 11, 35, 1983.
174. **Ruddle, K.,** *The Yukpa Cultivation System,* Ibero-Americana 52, University of California, Berkeley, 1974, 197.
175. **Boom, B. M.,** *Ethnobotany of the Chacobo Indians,* Advances in Economic Botany, Vol. 4, New York Botanical Garden, Bronx, 1987, 68.
176. **Eden, M. J.,** Ecological aspects of development among Piaroa and Guahibo Indians of the Upper Orinoco Basin, *Antropologica,* 39, 25, 1974.
177. **Lathrap, D. W.,** Our father the cayman, our mother the gourd, *Origins of Agriculture,* Reed, C. A., Ed., Mouton, The Hague, Netherlands, 1977, 713.
178. **Boster, J.,** personal communication, 1987.
179. **Denevan, W. M., and C. Padoch, Eds.,** *Swidden-Fallow Agroforestry in the Peruvian Amazon,* Advances in Economic Botany, New York Botanical Garden, Bronx, 1988, 107.
180. **Hames, R.,** personal communication, 1987.
181. **von Hildebrand, P.,** Observaciones preliminares sobre utilización de tierras y fauna por los indígenas del rio Miriti-Parana, *Revista Colombiana de Antropolgía,* 18, 183, 1975.
182. **Casanova, J.,** El sistema de cultivo secoya, in *Ecología y Etnicidad,* Chiribi, A., Ed., Lima, Peru, 1978, 41.
183. **Denevan, W. M., Treacy, J. M., Alcorn, J. B., Padoch, C., Denslow, J. and S. Fores P.,** Indigenous agroforestry in the Peruvian Amazon: Bora Indian management of swidden fallows, *Interciencia,* 9, 346, 1984.

184. **Johnson, A.,** Machiguenga gardens, in *Adaptive Responses of Native Amazonians,* Hames, R. B. and W. T. Vickers, Eds., Academic Press, New York, 1983, 29.

185. **Vickers, W. T. and T. Plowman,** *Useful Plants of the Siona and Secoya Indians of Eastern Ecuador,* Fieldiana Botany New Series No. 15, Field Museum of Natural History, Chicago, 1974, 63.

186. **Vickers, W. T.,** Tropical forest mimicry in swiddens: a reassessment of Geertz's model with Amazonian data, *Hum. Ecol.,* 11, 35, 1983.

187. **Anderson, A., A. Gely, J. Strudwick, G. L. Sobel, and M. das Cracas C. Pinto,** Un sistema agroforestal na várzea do estúario amazônico (ilha das Onças, município de Barcarena, estado do Pará), *Acta Amazônica,* Suppl., 15, 195, 1985.

188. **Cleveland, D. A.,** Culture and horticulture in Mexico, *Culture Agric.,* 29, 1, 1986.

189. **Brownrigg, L.,** *Home Gardening in International Development: What the Literature Shows,* League for International Food Education, Washington, D.C., 1985, 332.

190. **Moran, E. F.,** personal communication, 1987.

191. **Cleveland, D. A.,** Culture and horticulture in Mexico, *Culture Agric.,* 29, 1, 1986.

192. **Flowers, N. M., D. R. Gross, M. L. Ritter, and D. W. Werner,** Variation in swidden practices in four Central Brazilian Indian societies, *Hum. Ecol.,* 10, 203, 1982.

193. **Flowers, N. M.,** Forager-Farmers: The Xavante Indians of Central Brazil, Ph.D. thesis, City University of New York, 1983.

194. **Anderson, A. B.,** personal communication, 1987.

195. **Jones, J. R. and N. Price,** Agroforestry: An application of the farming systems approach to forestry, *Hum. Organ.,* 44, 322, 1985.

196. **Price, N.,** *The Tropical Mixed Garden: An Agroforestry Component of the Small Farm,* CATIE, Turrialba, Costa Rica, 1983, 38.

197. **Folan, W. J., L. A. Fletcher, and E. R. King,** Fruit, fiber, bark, and resin: Social organization of a Maya urban center, *Science,* 204, 697, 1979.

198. **Wilken, G. C.,** Food-producing systems available to the ancient Maya, *American Antiquity,* 36, 432, 1971.

199. **Matheny, R. T. and D. L. Gurr,** Variation in prehistoric agricultural systems of the New World, *Annual Review of Anthropology,* 12, 79, 1983.

200. **Brownrigg, L.,** *Home Gardening in International Development: What the Literature Shows,* League for International Food Education, Washington, D.C., 1985, 332.

201. **Cleveland, D. A.,** Culture and horticulture in Mexico, *Culture Agric.,* 29, 1, 1986.

202. **Ewell, P. T. and T. T. Poleman,** *Uxpanapa: Agricultural Development in the Mexican Tropics,* Pergamon Press, New York, 1980, 206.

203. **Alcorn, J. B.,** *Huastec Mayan Ethnobotany,* University of Texas Press, Austin, 1984, 982.

204. **Johnson, D. V. and P. K. R. Nair,** Perennial crop-based agroforestry systems in Northeast Brazil, *Agrofor. Syst.,* 2, 281, 1985.

205. **May, P. H., A. B. Anderson, J. M. F. Frazao, and M. J. Balick,** Babassu palm in the agroforestry systems in Brazil's Mid-North region, *Agrofor. Syst.,* 3, 275, 1985.

206. **Johnson, D. V. and P. K. R. Nair,** Perennial crop-based agroforestry systems in Northeast Brazil, *Agrofor. Syst.,* 2, 281, 1985.

207. **NAS,** *Underexploited Tropical Plants with Promising Economic Value,* National Academy of Sciences, Washington, D.C., 1975, 186.

208. **May, P. H., A. B. Anderson, J. M. F. Frazao, and M. J. Balick,** Babassu palm in the agroforestry systems in Brazil's Mid-North region, *Agrofor. Syst.,* 3, 275, 1985.

209. **May, P. H., A. B. Anderson, J. M. F. Frazao, and M. J. Balick,** Babassu palm in the agroforestry systems in Brazil's Mid-North region, *Agrofor. Syst.,* 3, 275, 1985.

210. **Anderson, A. and S. Anderson,** A "tree of life" grows in Brazil, *Nat. Hist.,* 12/85, 40, 1985.

211. **Balick, M.,** personal communication, 1987.

212. **Alcorn, J. B.,** *Huastec Mayan Ethnobotany,* University of Texas Press, Austin, 1984, 982.

213. **Medellín Morales, S. G.,** *Uso y Manejo de las Especies Vegetales Comestibles, Medicinales, Para Construcción y Combustibles en una Comunidad Totonaca de la Costa (Plan de Hidalgo, Papantla, Veracruz, México),* Programa Formación de Recursos Humanos, INIREB, Xalapa, Veracruz, Mexico, 1986, 79.

214. **Escalante, E. E.,** Promising agroforestry systems in Venezuela, *Agrofor. Syst.,* 3, 209, 1985.

215. **Altieri, M. A. and J. Farrell,** Traditional farming systems of south-central Chile, with special emphasis on agroforestry, *Agroforestry Systems,* 2, 3, 1984.

216. **Balick, M. J.,** Economic botany of the Guahibo I. Palmae, *Econ. Bot.,* 33, 361, 1979.

217. **Pilgrim, T.,** *Sourcelist of Uses of Native American Palms,* Dept. of Anthropology, University of California, Berkeley, 1987, 138.

218. **Levi-Strauss, C.,** The use of wild plants in tropical South America, in *Handbook of South American Indians,* Volume 6, Steward, J. H., Ed., Smithsonian Institution, Washington, D.C., 1950, 465.

219. **Wilbert, J.,** *Manicaria saccifera* and its cultural significance among the Warao Indians of Venezuela, *Botanical Museum Leaflets,* 24, 275, 1976.

220. **Wilbert, J.,** *Manicaria saccifera* and its cultural significance among the Warao Indians of Venezuela, *Botanical Museum Leaflets,* 24, 275, 1976.

221. **Hiraoka, M.,** Mestizo subsistence in riparian Amazonia, *Natl. Geogr. Res.,* 1, 236, 1985.

222. **Padoch, C. and W. de Jong,** Traditional agroforestry practices of native and ribereno farmers in the lowland Peruvian Amazon, in *Agroforestry: Realities, Possibilities, and Potentials,* Gholz, H., Ed., Martinus Nijhoff, Dordrecht, Netherlands, 1987, in press.

223. **Schmidt, M.,** Comments on cultivated plants and agricultural methods of South American Indians, in *Native South Americans,* Lyon, P. J., Ed., Little, Brown, Boston, 1974, 60.

224. **Balick, M. J.,** Amazonian oil palms of promise: A survey, *Econ. Bot.,* 33, 11, 1979.

225. **Siemens, A. H.,** Wetland agriculture in pre-historic Mesoamerica, *Geogr. Rev.,* 73, 166, 1983.

226. **Denevan, W. M.,** Hydraulic agriculture in the American tropics: Forms, measures, and recent research, in *Maya Subsistence,* Flannery, Kent V., Ed., Academic Press, New York, 1982, 181.

227. **Gómez-Pompa, A., H. L. Morales, E. Jímenez A., And J. Jímenez A.,** Experiences in traditional hydraulic agriculture, in *Maya Subsistence,* Flannery, Kent V., Ed., Academic Press, New York, 1982, 327.

228. **Morales, H. L.,** Rural development and the management of tropical integrated production units, *Tropical Ecology and Development,* 2, 427, 1980.

229. **Wilken, G. C.,** The ecology of gathering in a Mexican farming region, *Econ. Bot.,* 24, 286, 1970.

230. **Jones, J. R. and N. Price,** Agroforestry: An application of the farming systems approach to forestry, *Hum. Organ.,* 44, 322, 1985.

231. **Sauer, J. D.,** *Living Fences in Costa Rican Agriculture,* Department of Geography, University of California, Los Angeles, 1977, 19.

232. **Jones, J. R. and N. Price,** Agroforestry: An application of the farming systems approach to forestry, *Hum. Organ.,* 44, 322, 1985.

233. **Jones, J. R.,** The Central American energy problem: Anthropological perspectives on fuelwood supply and production, *Culture Agric.,* 22, 6, 1984.

234. **Medellín Morales, S. G.,** *Uso y Manejo de las Especies Vegetales Comestibles, Medicinales, Para Construcción y Combustibles en una Comunidad Totonaca de la Costa (Plan de Hidalgo, Papantla, Veracruz, México),* Programa Formación de Recursos Humanos, INIREB, Xalapa, Veracruz, Mexico, 1986, 79.

235. **Alcorn, J. B.,** *Huastec Mayan Ethnobotany,* University of Texas Press, Austin, 1984, 982.
236. **Zizumbo, V. D. and P. Colunga García-Marin,** *Los Huaves: La Apropiación de los Recursos Naturales,* Universidad Autonoma Chapingo, Chapingo, Mexico, 1982, 277.
237. **Nabhan, G. H. and T. E. Sheridan,** Living fencerows of the Rio San Miguel, Sonora, Mexico, *Hum. Ecol.,* 5, 96, 1977.
238. **Gómez-Pompa, A.,** *Los Recursos Bióticos de Mexico,* INIREB, Xalapa, Veracruz, 1985, 122.
239. **Pretell Chichlote, J., D. Ocaña V., R. Jon Jap, and E. Baranhona C.,** *Apuntes Sobre Algunas Especies Forestales Nativas de la Sierra Peruana,* FAO, Lima, Peru, 1985, 120.
240. **Altieri, M. A. and J. Farrell,** Traditional farming systems of south-central Chile, with special emphasis on agroforestry, *Agrofor. Syst.,* 2, 3, 1984.
241. **Gliessman, S. R., R. García E., and M. Amador A.,** The ecological basis for the application of traditional agricultural technology in the management of tropical agro-ecosystems, *Agro-Ecosystems,* 7, 173, 1981.
242. **Toledo, V. M., A. Argueta, P. Rojas, C. Mapes, and J. Caballero,** Uso múliple del ecosistema, estrategias del ecodesarrollo, *Ciencia y desarrollo,* 11, 33, 1976.
243. **Toledo, V., J. Carabias, C. Mapes, and C. Toledo,** *Ecologia y Autosuficiencia Allmentaria,* Siglo Veintiuno, Mexico City, 1985, 118.
244. **Ewel, J. J.,** Designing agricultural ecosystems for the humid tropics, *Annu. Rev. Ecol. Syst.,* 17, 245, 1986.
245. **Hiraoka, M.,** Zonation of mestizo riverine farming systems in Northeast Peru, *Natl. Geogr. Res.,* 2, 354, 1986.
246. **Johnson, D. V.,** Multi-purpose palms in agroforestry: A classification and assessment, *Int. Tree Crops J.,* 2, 217, 1983.
247. **Johnson, D. V., R. W. Reed, and M. J. Balick,** *Economic Botany and Threatened Species of the Palm Family in Latin America and the Caribbean,* Final Report WWF 3322, World Wildlife Fund, Washington, D.C., 1986, 150.
248. **Pilgrim, T.,** *Sourcelist of Uses of Native American Palms,* Dept. of Anthropology, University of California, Berkeley, 1987, 138.
249. **Cavalcante, P. B.,** *Frutas Comestíveis da Amazônia,* Instituto Nacional de Pesquisas da Amazônia (INPA), Manaus, Amazônas, Brazil, 1976, 157.
250. **Correa, M. P.,** *Dicionario das Plantas Uteis do Brasil e das Exoticas Cultivadas,* Servico de Informaçao Agricola, Ministerio de Agricultura, Rio de Janeiro, Brazil, 1952.
251. **Williams, L. O.,** *Useful Plants of Central America,* Ceiba, Vol. 24, Tegucigalpa, Honduras, 342.
252. **Patino, V. M.,** *Plantas Cultivadas y Animales Domésticos en América Equinoccial,* Imp. Dept., Cali, Colombia, 1963.
253. **Oldfield, M. L. and J. B. Alcorn,** Conservation of traditional agroecosystems, *BioScience,* 37, 199, 1987.
254. **Altieri, M. A. and L. C. Merrick,** In situ conservation of crop genetic resources through maintenance of traditional farming systems, *Econ. Bot.,* 41, 86, 1987.
255. **Nair, P. K. R.,** personal communication, 1987.
256. **NRC (National Research Council),** *Ecological Aspects of Development in the Humid Tropics,* National Academy Press, Washington, D.C., 1982, 297.
257. **Gliessman, S. R.,** The use of traditional knowledge in development and implementation of resource use systems that sustain tropical forest and woodland resources, paper prepared for the Office of Technology Assessment, U.S. Congress, 1982.
258. **Olofson, H.,** Indigenous agroforestry systems, *Philipp. Q. Culture Soc.,* 11, 149, 1983.
259. **Posey, D. A., J. Frechione, J. Eddins, and L. Francello da Silva,** Ethnoecology as applied anthropology in Amazonian development, *Hum. Organ.,* 4, 95, 1984.

Research Methodologies

21 RAPID RURAL APPRAISAL AND AGROECOSYSTEM ANALYSIS

Gordon R. Conway and Jennifer A. McCracken

I. INTRODUCTION

Those concerned with research and development for small farmers face a number of difficult challenges. Small farms, as earlier chapters in this book make abundantly clear, are complex systems. Since their complexity arises primarily from the interaction of ecological and socio-economic processes, effective research and development must rely on the combined skills and insights of a wide range of disciplines, spanning the natural and social sciences. Small farms are also under increasing pressures from growing rural populations and declining resource availability, and while, in the past, farms on marginal lands were isolated from modern industrialized society, this is no longer the case. Expectations are rising and technological innovations are reaching the remotest corners of the globe. Deciding which innovations or interventions are appropriate, in both the long and short term, requires careful analysis and a genuine dialogue between development specialists, policy makers and, neither last nor least, the farmers themselves.

Moreover, the challenge is not solely one of ensuring that analyses are careful, insightful and multidisciplinary in nature. Poor farmers cannot wait for the results of time-consuming academic studies. The need is for methods that are powerful and quick and, given the current economic climate of development aid, are also inexpensive.

In this chapter we briefly review a number of such methods which fall under the general headings of Rapid Rural Appraisal (RRA) and Agroecosystem Analysis (AA). We begin by describing a number of basic organizing concepts that provide a theoretical framework for these methods. The philosophy, history and techniques of RRA are then introduced, followed by descriptions of the methods of Agroecosystem Analysis and Topical, Participatory and Monitoring RRA. We conclude with suggestions as to how these various methods fit in the design and management of rural development projects.

II. BASIC CONCEPTS

If multidisciplinary analysis is to be effective and useful it has to produce insights that significantly transcend those of the individual participating disciplines. This can be aided by creating a research or development team that works well together and is sensitive to the requirements of good communication, both between disciplines and between development professionals, and policy makers and farmers. However, although this is a necessary condition, it is not by itself sufficient. Experience suggests that the generation of good interdisciplinary insights also requires organizing concepts and relatively formal, i.e., semi-structured, working procedures.[1]

Such concepts have to be fairly simple and involve a minimal set of assumptions that are acceptable to all disciplines. They also have to be understandable, at least in essence, by those with whom the development professionals are working, that is both policy makers and the farmers. Three basic concepts that appear to meet these requirements are the agroecosystem, the agroecosystem hierarchy and the concept of agroecosystem properties.

A. Agroecosystems

Agroecosystems may be defined as ecological systems modified by human beings to produce food, fiber and other agricultural products.[2] In the transformation, the great diversity of wildlife in the original natural system is reduced to a restricted assemblage of crops, pests and weeds (Figure 1). The basic ecological processes — competition, herbivory and predation — still remain, but these are now overlain and regulated by the agricultural processes of cultivation, subsidy, control, harvesting and marketing. Recognizable system goals become apparent that are sought through human social and economic cooperation and competition. One consequence is that the system boundary acquires a socio-economic dimension. It is this new complex agro-socio-

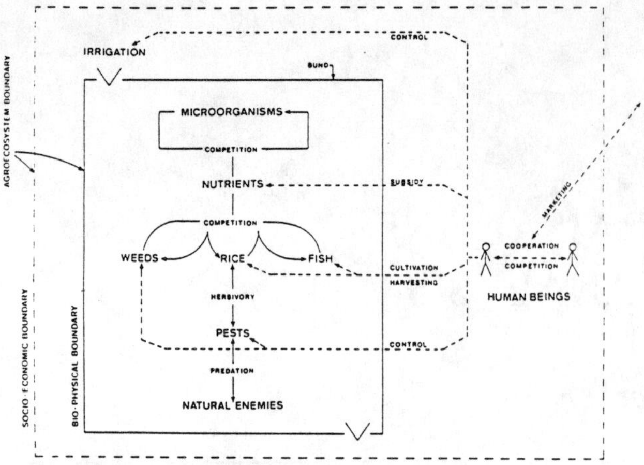

FIGURE 1. A ricefield agroecosystem.[2]

economic-ecological system, bounded in several dimensions, that we refer to as an agroecosystem.

B. The Hierarchy of Agroecosystems

The most widely recognized agroecosystem is the crop field conceptualized in Figure 1, or the livestock paddock. But if agroecosystems are defined so as to include both ecological and socio-economic components, then we can envisage a classical hierarchy of such systems (Figure 2). At the bottom of the hierarchy is the agroecosystem comprising the individual plant or animal, its immediate micro-environment, and the people who tend and harvest it. The next level is the field or paddock and the hierarchy continues upwards in this way, each agroecosystem forming a component of the agroecosystem at the next level. Near the top is the national agroecosystem, composed of regional agroecosystems linked by national markets, and above that the world agroecosystem, consisting of national agroecosystems linked by international trade.

C. Agroecosystem Properties

Systems theory holds that the behavior of higher systems in such a hierarchy is not readily discovered simply from a study of lower systems, and vice versa. This has important consequences not only for analysis but also for rural development.[2] It implies that such development cannot be based solely, or largely, on genetic engineering, or macro-environment policy, or even on farming systems research. Each level in the agroecosystem hierarchy has to be analyzed and developed both in its own right and in relation to the other levels above and below. To achieve this, with rigor and speed, is a difficult task but is greatly helped by a common approach to analysis and, in particular, a set of well defined common properties for each level in the hierarchy that can be related to each other, within and between levels. Conway[2] has suggested that four such key properties exist (Figure 3).

1. Productivity

Productivity is defined as the output of valued product per unit of resource input. Common measures of productivity are yield or income per hectare, or total production of agricultural goods and services per household or nation. But a large number of different measures are possible: Yield may be in terms of kilograms of grain or meat or any other consumable or marketable product; alternatively it may be converted to value in calories, proteins or vitamins or to its monetary value at the market. Frequently the valued product may not be yield in conventional agricultural terms. It may be employment generation, or an item of amenity of aesthetic value or one of a wide range of products that contribute, in ways that are difficult to measure, to social, psychological and spiritual well being.[3]

2. Stability

Stability may be defined as the constancy of productivity in the face of small disturbing forces arising from the normal fluctuations and cycles in the surrounding environment. Fluctuations, for example, may be in the climate or in the market demand for agricultural products. Since productivity may be level, rising or falling, stability will refer to the variability about a trend.

3. Sustainability

Sustainability is defined as the ability of an agroecosys-

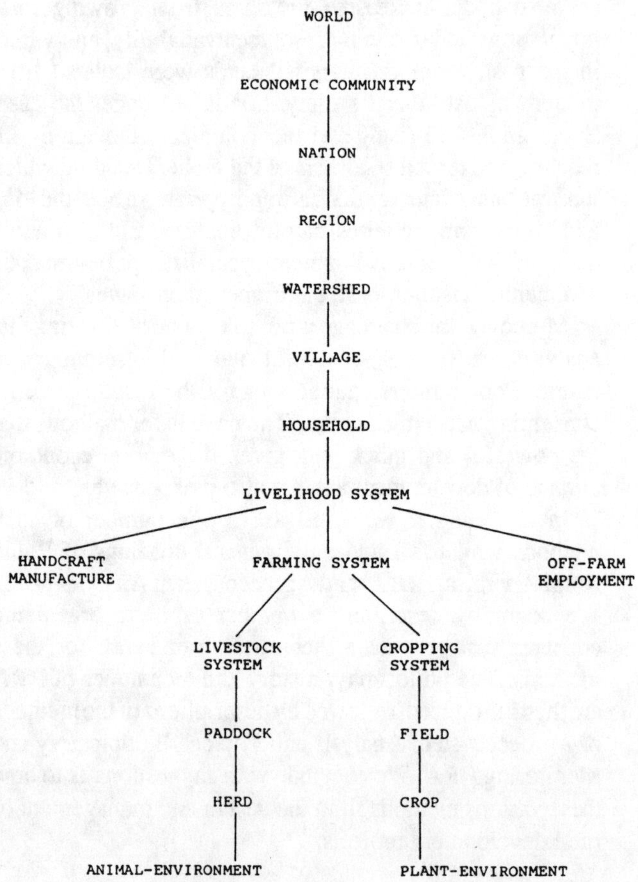

FIGURE 2. The hierarchy of agroecosystems.[1]

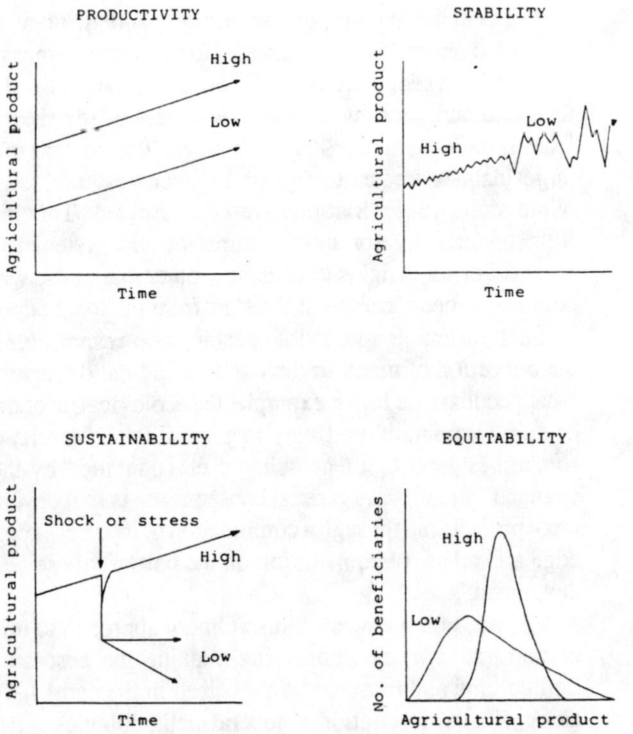

FIGURE 3. The properties of agroecosystems.

tem to maintain productivity when subject to a major disturbing force. The actual or potential disturbance may be caused by an intensive stress, where stress is defined as a frequent, sometimes continuous, relatively small and predictable disturbing force that has a large cumulative effect. Salinity, toxicity, erosion, indebtedness or declining market demand are examples of such forces. Alternatively the disturbance may be caused by a shock, defined here as an infrequent, relatively large and unpredictable disturbing force which has the potential of creating an immediate, large disturbance or perturbation. Examples of shocks include a rare drought or flood, or a new pest or the sudden rise in an input price, such as that of oil in the mid 1970s.

Following a stress or shock the productivity of the agroecosystem may be unaffected, or may fall and then return to the previous level or trend, or settle to a new lower, or sometimes higher, level or may disappear altogether. Sustainability thus determines the persistence or durability of an agroecosystem's productivity under known or possible conditions. It is a function of the intrinsic characteristics of the agroecosystem, of the nature and strength of the stresses and shocks to which it is subject, and of the human inputs that may be introduced to counter these stresses and shocks.

4. Equitability

Equitability is defined as the evenness of distribution of the productivity of the agroecosystem among the human beneficiaries. Commonly, it refers to the distribution of the total production of goods and services of the agroecosystem under consideration, i.e., the field, farm, village or nation. The human beneficiaries may be the farm household or the members of a village or a national population.

Equitability may be measured by a Lorenz curve, Gini coefficient or some other related index. In practice, though, it is difficult to define equitability in a purely positive sense, the measures available reflecting different value judgements. Equitability is thus often the evenness of distribution of productivity among the human beneficiaries according to need.

Although the four properties, described above, are the key properties in determining the social value of an agroecosystem there are a large number of other ways that can be used to characterize an agroecosystem. Examples include energy and materials conservation, diversity, autonomy, market penetration and some measure of cultural acceptability. Each of these, however, can be shown to contribute to social value through one or more of the four primary properties. Thus materials conservation can contribute to productivity but also has a major effect on stability and sustainability; diversity contributes to all four primary properties, although in rather complex ways; and cultural acceptability is an important component of sustainability and equitability.

D. Development Goals

These four key properties are essentially descriptive in nature, summarizing the status of the agroecosystem. But they can also be employed normatively as indicators of performance and hence can be used in the design and evaluation of development projects and of individual innovations and interventions.

However, agricultural development, especially as it applies to small farm agroecosystems, almost inevitably involves some degree of trade-off between the different system properties. The introduction of new technologies frequently increases productivity, but often at the expense of lowered values of one or more of the other properties. Short term increases in production may result in reduced sustainability, and hence a greater risk of collapse in the future, or may benefit some members of the community at the expense of others. The challenge facing development specialists is to analyze agroecosystems in such a way as to make these actual and potential trade-offs explicit and hence enable identification of innovations and interventions that lead to smaller trade-offs. The "trick" is to find ways of increasing productivity while simultaneously, at the very least, not reducing the stability or sustainability of production or the equitability of its distribution.

III. RAPID RURAL APPRAISAL

A. Philosophy of RRA

Rapid Rural Appraisal (RRA) may be defined as a systematic, but semi-structured, activity carried out in the field by a multidisciplinary team and designed to acquire quickly new information on, and new hypotheses about, rural life.

Two themes are central to the philosophy of RRA. The first is the pursuit of 'optimal ignorance' (a term first used by

Ilchman[4]). This implies that both the amount and the detail of information required to formulate useful hypotheses in a limited period of time are regarded as expenses to be kept to a minimum. In terms of the concepts presented earlier, the aim of the multidisciplinary team is to arrive at an agreed sufficiency of knowledge of the key agroecosystem processes and properties relevant to the objectives of the RRA and not to exceed this by investigating irrelevant aspects or being concerned with unnecessary detail.

The second theme is diversity of analysis. This is pursued through the process of 'triangulation' — that is the use of several different sources of, and means of gathering, information. Notwithstanding the self-imposed limits of time and resources, the accuracy and completeness of an RRA study is maximized by investigating each aspect of the situation in a variety of ways. "Truth" is approached through the rapid build up of diverse information rather than via statistical replication. Secondary data, direct observation in the field, semi-structured interviews, the preparation of diagrams all contribute to a progressively accurate analysis of the situation under investigation.

These themes in turn lead to five key features of good RRAs, namely that they are

- **Iterative** — the process and goals of the study are not immutably fixed beforehand, but modified as the team realizes what is or is not relevant.
- **Innovative** — there is no simple, standardized methodology. Techniques are developed for particular situations depending on the skills and knowledge available.
- **Interactive** — all team members and disciplines combine together in a way that fosters serendipity and interdisciplinary insights.
- **Informal** — the emphasis is, in contrast to the formality of other approaches, on partly structured and informal interviews and discussions.
- **In the field** — the aim is not just to gather data for later analysis. Learning takes place largely in the field "as you go", or immediately after, in short intensive workshops.

B. History of RRA

Studies of local rural situations in developing countries have often concentrated on only one set of conditions, investigating for instance the economic, social, ecological *or* agricultural aspects. Where several sectors are included, as in project designs, they are often still considered in isolation from each other, at best being collected together in a single voluminous report. Extensive data collections, involving many researchers, over a long period of time and costing large sums of money, are often regarded as integral to the process. These are usually followed by equally extensive statistical analyses, although often remaining narrow in their focus and assumptions.

The obvious logistical problems of such an approach are frequently accompanied by other, more serious, shortcomings. Local inhabitants are seldom consulted, or at best through fixed and formal channels, for instance by means of a written questionnaire with the questions determined beforehand and unchanged from day to day of the study, or from farm to farm or village to village. The context of the target data is frequently ignored; "averages" are sought, while significant variations are often missed. This gives little opportunity for new features of the system to be revealed or for insights to be gained other than those which could have been learned at the start from the local people.

Such inflexible methodologies are also responsible for the collection of many irrelevant data and the disregard of local peculiarities in, for example, the ecological, economic or cultural conditions. Delay in providing the results can sometimes lead to them being useless in the "by-then-changed" situation. A general consequence is that development projects fail through a combination of incorrect knowledge and a lack of cooperation on the part of "those being developed".

The need for a new multidisciplinary approach to information gathering and analyzing, making the best use of limited time and resources, was voiced in the early 1970s, along with the description of general methodologies of RRA to fulfil such a need. Those involved in this initial development of RRA included Robert Chambers, Peter Hildebrand and Michael Collinson.

The work of the early practitioners of RRA was brought together in conferences at the Institute of Development Studies, University of Sussex in October 1978 and December 1979.[5] A more recent conference was held in September 1985 at Khon Kaen University, Thailand.[6-20] The papers from these conferences still account for the main body of formal RRA literature.

C. Suite of Techniques

The experiences of the many RRA practitioners over the past ten years have resulted in a basic suite of techniques that reflect the themes and features of efficient and useful rapid appraisal. The five described here are not an exhaustive list and may be supplemented by others appropriate to particular situations — e.g., simple field experiments or opinion polls.

1. Secondary Data Review

Secondary data are published or unpublished data acquired by other people at an earlier time that are relevant to the topic or agroecosystem which is the target of the RRA. Such data are to be found in government agencies, universities, research centers, marketing bodies and other similar institutions. They take a great variety of forms, including project documents, research papers, annual reports, computer data files, survey results, maps, photographs and satellite images. Even travel books, newspaper articles and other ephemera should be consulted.

Time spent on quickly reviewing and summarizing secondary data in the form of simple tables, diagrams or brief written notes can be of considerable value in setting the RRA task in the context of previous work. In particular, it can help avoid time wasted in repeating studies and, by revealing gaps or biases in the existing data, it can also stimulate ideas and suggest paths of investigation for the researchers. Nevertheless, it is important not to spend time on this that could be better spent in the field.

2. Semi-Structured Interviews

Probably the most powerful of the RRA techniques are semi-structured interviews (SSIs). These are normally associated with farm, field or village visits and the interviewees may be individual farmers or groups of farmers, or key informants such as village leaders, school teachers or local development officials.

The essential features of an SSI are: it is informal in nature; it occurs *in situ*, i.e., in the normal surroundings of the interviewee; apart from exceptional cases it does not involve a written questionnaire; some of the questions are predetermined, but new questions or lines of questioning are determined by the interviewee's responses and the characteristics of his/her land, farm, household, and livelihood. The key to success is the interaction that occurs between the knowledge and experience of the interviewer and the interviewee.

SSIs normally do not exceed an hour, but may be repeated after the analysis of information from several different visits to obtain further information or verify previous answers. The RRA team usually splits up into one or two individual members for the interviews and interpreters are sometimes required, though not recommended. Notes are taken during the SSIs and rewritten during the subsequent analyses. Grandstaff and Grandstaff[12] provide practical advice on conducting SSIs.

3. Direct Observation

This is relatively straightforward and encompasses any direct observation of field objects, events, processes, relationships or people that is recorded by the team in note or diagrammatic form.

Some of the more innovative forms of direct observation rely on skillfully chosen indicators. These are events, processes or relationships which are easily observed or measured but can be used as a indicator of some other variable that is more difficult or impossible to observe.

4. Conceptual Models

Besides SSI the other key RRA technique is the use of conceptual and, in particular, diagrammatic models. They are employed throughout the discussion and analysis stages of RRA, as part of SSIs, as a record of direct observation, in the team's daily field analysis of the information obtained, and in the workshops that follow field visits, as described below. Such tools are real, tangible aids to, and initiators of, the knowledge, and insight of those involved in the RRA — both outsiders and locals.

Useful diagrams include sketch maps, transects, seasonal calendars, bar diagrams, time trends, venn diagrams, decision trees and various forms of flow diagram (see Section D.1). Experience suggests that such diagrams, providing they can be kept fairly simple are powerful aids to communication between different disciplines and serve to focus attention on the best opportunities for development action.

5. Workshops

The fifth technique we describe is that of the workshop. This is typically an intensive session involving the field team, and often other outsiders introduced for their skills and experience, that analyzes the information gathered in order to arrive at a consensus of opinion. The numbers involved may be quite small and the session last for 1 or 2 hours, or several tens of people may be involved over several days.

In general some degree of structure is involved (see the Agroecosystem Analysis workshop described below) if the use of time is to be efficient and productive. Interaction is greatly facilitated by the use of diagrams and an essential requirement for workshops is the provision of overhead projectors, white boards, etc. Table 1 compares and contrasts the use of different techniques in RRA and classical formal approaches.

D. Classes of RRA

The various techniques described above will be used in various combinations depending on the objective of the RRA. Very broadly there are four principal classes of RRA, which ideally follow one another in the sequence of development activity:

- **Exploratory RRAs** — to obtain initial information about a new topic or agroecosystem. The output is usually a set of preliminary key questions and hypotheses (see Section D.1).
- **Topical RRA** — to investigate a specific topic, often in the form of a key question and hypothesis generated by the exploratory RRA. The output is usually a detailed and extended hypothesis that can be used as a strong basis for research or development (see Section D.2)
- **Participatory RRA** — to involve villagers and local officials in decisions about further action based on the hypotheses produced by the exploratory or topical RRAs. The output is farmer-managed trials or a development activity in which the villagers are closely involved.
- **Monitoring RRA** — to monitor progress in the trials and experiments and in the implementation of the development activity. The output is usually a revised hypothesis together with consequent changes in the trials or development intervention which will hopefully bring about improved benefits.

1. Agroecosystem Analysis

A number of different exploratory RRAs have been developed by various workers. They include the "sondeo"

TABLE 1
Classical and RRA Approaches Compared

Techniques employed	Classical	RRA
Statistical analysis	Often a major part	Little or none, uses triangulation
Formal questionnaires	Often included	Avoided
Interviews with local farmers and key informants	Through *formal* questionnaire if at all	A major component using semi-structured interviewing
Qualitative discriptions and diagrams	Not as important as the 'hard data'	Considered at least equally as important
Sampling	Statistically acceptale sample sizes regarded as necessary. Often random sampling	Often small sample size, selecting "key' areas or farms, households etc. "Statistical" requirements not always adhered to
Consulting secondary data sources	Yes	Yes
Measurements	Detailed, accurate	Qualitative or indicators used
Group discussion	Informal sessions	Via workshops and brainstorming

approach by Hildebrand,[21] the Informal Agricultural Survey of Rhoades,[13,22] the Rapid Reconnaissance of Honadle,[23] the Reconnaissance Survey of Shaner et al.,[24] the Exploratory Survey of Collinson,[25] and the Rapid Assessment Technique of Alton and Craig.[26]

Here we will confine ourselves to the method known as Agroecosystem Analysis as developed by Conway and co-workers.[1,27] Development of the method began in 1978 and has been subsequently used in various situations in Thailand,[28] Indonesia[29-31] and Pakistan.[7,32,33]

This method is based explicitly on the concepts described earlier in this paper. It can take as its focus any level in the agroecosystem hierarchy, although villages have so far been the common target. Following a secondary data review the analysis begins with a field visit whose aim is to produce a series of diagrams of the agroecosystem based on direct observation and semi-structured interview. These diagrams are designed to relate to four basic system patterns — in space and time, and of flows and decision making. Maps and transects are used to describe the spatial patterns. The transect in Figure 4, for example, was used to identify the specific location of particular problems and of opportunities for development. Seasonal calendars and graphs summarize patterns in time. The calendar in Figure 5 shows the interrelationships between a wide variety of activities and indicates when particular constraints arise and 'slots' in time when new activities can be undertaken. Sources and flows of income are summarized in bar diagrams based on semi-structured interviews. Flow diagrams are used to summarize production and marketing cycles and the actual or potential impact of major innovations or interventions (Figure 6). Finally, decision trees (Figure 7) describe the choices of different farming strategies and the factors affecting these choices, while venn diagrams are used to analyze institutional interactions in decision making. In each case the diagrams serve to pinpoint where problems occur and where there are opportunities for innovation and intervention.

In the village Agroecosystem Analyses in Pakistan[7,32,33] the field visits take about 2 days per village. The diagrams are then converted to overhead transparencies and used as the focus of a half to one day workshop. This involves the field team and other members of the development group and is structured around the procedure shown in Figure 8. Following a brief discussion of objectives and system boundaries the team concentrates on the analysis of the diagrams. This has two outcomes: first, as an intermediate step, a table of the most important factors affecting the system properties of productivity, stability, sustainability and equitability; and, second, a set of key questions and related hypotheses that the team as a whole agree are of primary importance.

The final stage is then for the team to assess the innovations and interventions implicit in the key questions in terms of their effects on system properties and to rank them for priority attention (Table 2). Some of the questions are key research questions with related research hypotheses; others are key development questions with related working hypotheses. The first category lead to further research, either involving formal analysis and experimentation or a form of topical RRA. The second category result in development action. An example of a development key question might be: "How can soil development be speeded up while at the same time providing a high return?" and the related working hypothesis: "Soil development can be most profitably speeded up by planting alfalfa at a high density, cutting twice a year and resowing, after ploughing in, every three years".[27]

2. Topical RRA

A topical RRA aims to answer specific questions on a certain topic and as such has a narrower scope of investigation than an exploratory RRA. Examples of topics investigated by this type of RRA are listed in Table 3.

While focusing on one particular question, however, the topical RRA does *not* limit itself to only one facet of the issue. As in all RRAs a systems approach and a multidisciplinary team are used and all the techniques described above are likely to be included. As the topical RRA proceeds, the scope of investigation narrows further, while the depth of analysis increases. Thus the general, sometimes naive, inquiries of the literature search stage give way, through probing questions and analysis, to more considered, 'optimally informed' opinions in the final stage.

Figure 9 shows the relative duration of the techniques involved and how they are sequentially organized in a typical topical RRA. As can be seen, a possible end product of the process is an 'extended hypothesis'. Rather than a

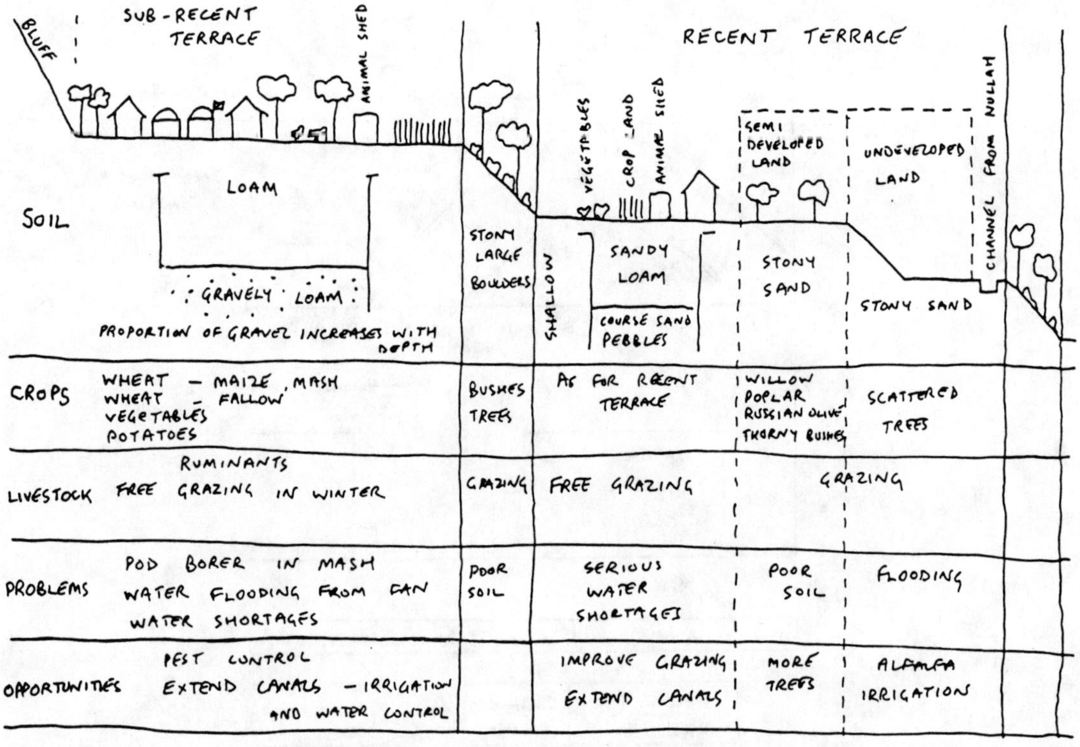

FIGURE 4. Transect of village in northern Pakistan.[33]

definitive answer to the question, this gives a concise description of the situation from the viewpoint of the researchers and summarises the primary causes they suggest are responsible for the problem being investigated. The hypothesis can then be used in one of two ways. First, as a 'working hypothesis' it can be assumed to be a true representation of the situation and actions can be taken, based on its findings and recommendations. Alternatively it can be considered a 'research hypothesis' and research set up to confirm or disprove it; this in turn may lead to recommendations for action.

The way in which ideas are 'filtered and focused' to arrive at the hypothesis (i.e., the thought processes behind the framework of Figure 9) will vary according to the topic of the RRA. Up to now no attempt has been made to produce a standardized format for this aspect of all topical RRAs. However, an RRA exercise carried out by McCracken,[35] attempts to bridge this gap and provide future practitioners with a practical structure on which to base their own individual initiatives. This RRA aimed to answer the key question 'Why are sugarcane yields in Fiji low?'. First, the secondary data survey, including analysis of international cane yield reports and farm data files of the Fiji Sugar Corporation, addressed the preliminary, broadbased questions of 'Is there a problem of low cane yields in Fiji?' and 'Are the yields uniformly low over Fiji?'. Having found the answers to these questions to be Yes and No, respectively, the study went on to try and identify the causal factors of the low yields and their spatial variation.

A range of the possible causal factors and processes were laid out as diagrammatic hypotheses (Figure 10) which were then refined and modified through SSIs with key informants and cane growers, direct observation in the field and the production of diagrammatic models. From the final version of the diagrammatic hypothesis, a short list of key causal factors was distilled and this then became the base for deeper analyses. All the information, ranging from the secondary data to the opinions expressed in the interviews, were brought together for each factor and a series of brief essays produced. The titles of these essays included: Cane Farming in Fiji, Patterns of Farming in Lovu (the survey area), Differences between Lovu Growers, Climate, Poor Management, and Poor Soil and Slope. Finally the filtering process culminated in a single extended hypothesis that brought together all the causal factors and put forward a concise description of the problem situation of Fiji cane yields.

Its basic thesis was:

> The cane yield of each farm depends on the balance between farm size and household size. Since the vast majority of farms are too small to support the average household by their cane production alone, at least some of the members must engage in off-farm work. The inevitable neglect of the cane means the majority of Fiji's cane producers obtain low yields. Medium-sized farms allow room for the production of vegetables, rice and goats in addition to the cane. This has two effects - the marketing of the produce lessens the household's cash flow problems and the need for much more regular farmwork keeps more family members on the farm and available for cane work when required. Such farms' cane yields are generally higher than

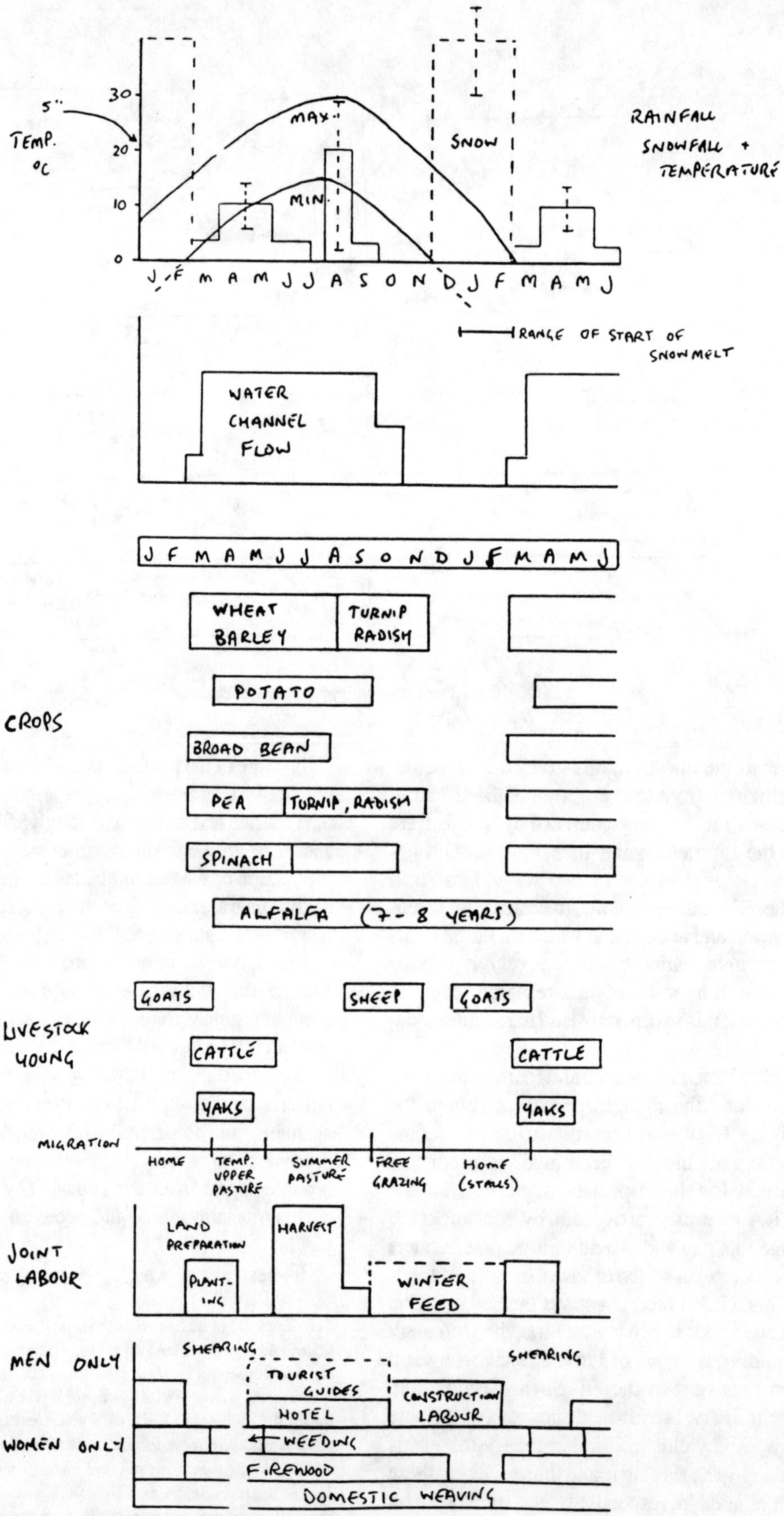

FIGURE 5. Seasonal calendar for village in northern Pakistan.[33]

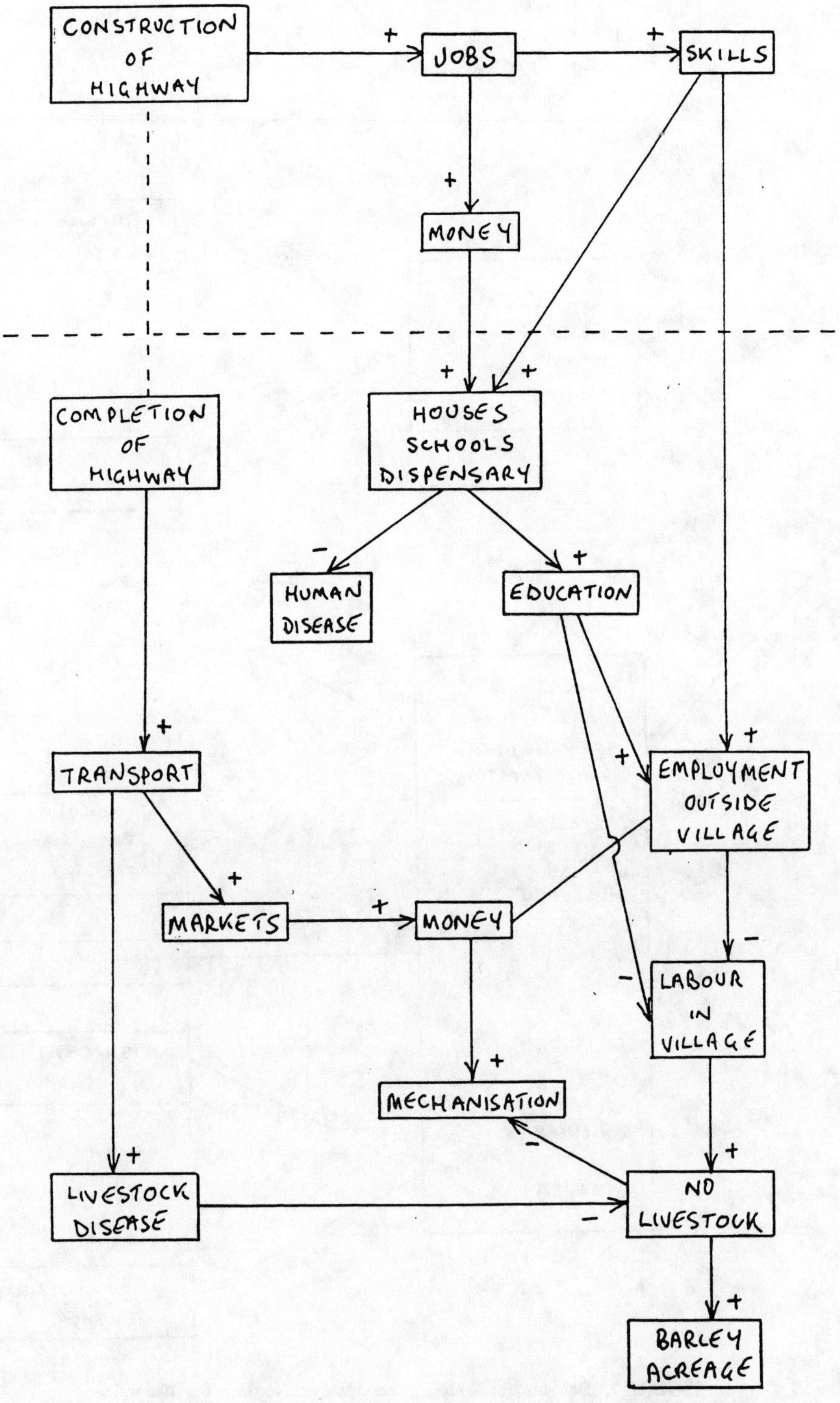

FIGURE 6. Impact flow diagram: effect on village of new highway.[33]

the national average. The largest farms, while benefitting from the opportunity for diversified production generally have cane areas too large to be managed by the household alone and thus require the expense of hired labor. Location of the farms is also an important factor. These largest farms are on the hilliest, remotest and least fertile land and so the growers incur extra costs of machinery hire for land levelling, transport hire for taking the cane to the mill, and large amounts of fertilizer. These large farms rarely produce high yields of cane[35]

3. Participatory RRA

All RRA exercises have at least some element of participation by the farmers and rural poor on whom they are targeted. At the very least they are partners in semi-structured interviewing. However, there will be many situations in which the primary goal is to involve the local people in crucial decision making. Then specifically designed participatory RRAs may be appropriate.

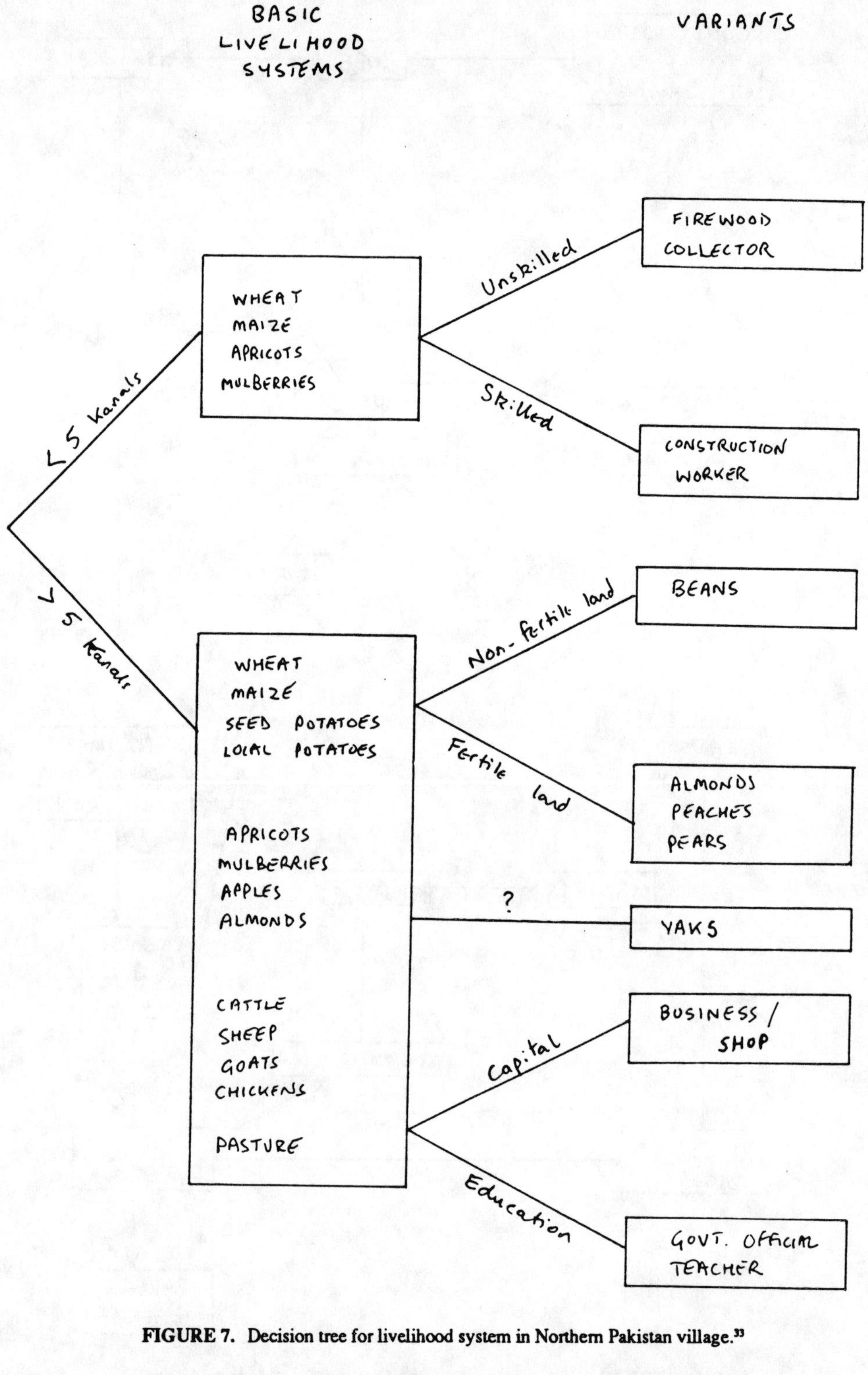

FIGURE 7. Decision tree for livelihood system in Northern Pakistan village.[33]

An example of a participatory RRA is the approach adopted by the Aga Khan Rural Support Programme (AKRSP).[36] They are concerned with developing several hundred villages in the Northern Areas of Pakistan in ways that capitalize on local knowledge and skills, and accord with local wishes. The first phase of the Programme, begun in 1983, focused on the development, in each village, of a single physical infrastructure project. The classical project cycle is accomplished through a series of interactive dialogues, termed the Diagnostic Survey. The First Dialogue is conducted by the project management team in the village with an assembly of villagers. They are asked to identify one infrastructure project which will increase village incomes, and which can be implemented and maintained by them. One condition is that the project must benefit at least 70 to 80% of the village population. Once a project is identified (per-

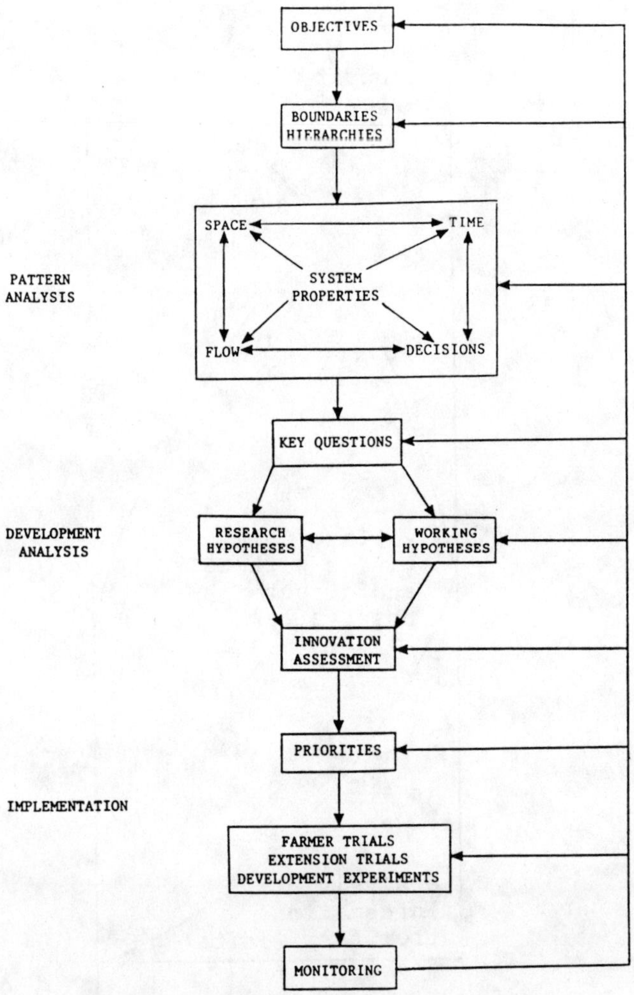

FIGURE 8. The procedure of Agroecosystem Analysis.

haps after much debate and many assemblies), the Second Dialogue occurs. This involves project engineers who survey the site, draw up plans and estimate costs, all with the participation of knowledgeable villagers nominated by the village at the First Dialogue. After the plans and estimates are checked by the senior engineer, project management returns to the village for a Third Dialogue. This consists of a full discussion of the rights and responsibilities of both AKRSP and the villagers. Terms of partnership are drawn up which explain the basis, amount and method of payment, and the common responsibilities of the village in managing the project and its finances, manpower and material. The project is initiated only if villagers can demonstrate their capacity for managing it on a permanent basis. A good part of the project cost is funded by a one-time grant by AKRSP to the village as a whole.

The Diagnostic Survey of the entire project area in Gilgit District (more than 300 villages) resulted in the initiation of 256 village-level infrastructure projects (generally one per village) at a cost of Rs 38.0 million (U.S. $2 million). So far 180 projects have been completed and an estimated 22,098 families will benefit. Only ten projects are considered to be moving 'slowly', and none has fallen into disuse through lack of repairs.

The Diagnostic Survey and the ensuing funding of the physical infrastructure projects provided an entry point for the AKRSP to assist with development of the villages. This has opened the way for the second phase of development, which focuses on realizing the agricultural potential of the villages through the judicious use of loans made against the villagers' savings.

4. Monitoring RRA

The aim of a monitoring RRA is to assess the impact of an intervention on a particular agroecosystem. Such intervention may have been the result of previous findings and recommendations from exploratory, topical or participatory RRA.

All the RRA techniques can be usefully applied in monitoring RRA. Secondary data review, particularly of the initial project documentation, is crucial to producing clear 'before and after' comparisons. The central feature of a monitoring RRA, however, is visits to the target area, which involve direct observation of the changes that have occurred and semi-structured interviews with the local inhabitants. Besides assessment of direct productivity indicators, the

TABLE 2
Innovation Assessment for a Village in Northern Pakistan

	Productivity	Stability	Sustainability	Equitability	Cost	Time for benefits	Feasibility	Priority
Innovation								
Development of 2nd terrace	+	+	++	++	H	M	+++	3
Development of 3rd terrace	+++	+	+(?o)	+	H	L	++	1
Artificial insemination	++	+	?	-	M	M	++	2
Catch crops	+	+	+	o	S	S	+	3
River bunds	o	+	+	+	H	S	++	4
Potato intensification	++++	+	-	+	S	S	++	2
Involvement of women	++	+	+	+++	S	M	+	2

TABLE 3
Examples of Topical RRAs Undertaken from Khon Kaen University[34]

Topic under investigation	No. of participants in RRA team
Causes and effects of trees in the paddy fields of northeast Thailand	4
Fuelwood situation in northeast Thailand — problems and processes of adjustment to its availability	7
Major factors explaining the various degrees of success in the operation of three dairy villages in northeast Thailand	10
Cropping patterns and the use of crop residues to supplement feed in dairy calf production in northeast Thailand	4
Extent of replacement of native black swine with non-native white in northeast Thailand, reasons why native swine production is lstill practised and whether such production will be sustained in the future	12
Socio-cultural and biophysical conditions allowing farmers in Surin to adopt peanuts after rice using residual soil moisture	6
How villagers in Srisaket have adjusted to annual flooding	5
Factors responsible for the varying degrees of usage of small-scale irrigation systems in northeast Thailand	4

interviews should provide information on changes in lifestyles, livelihoods, level of independence, contact with other areas, and also on the opinions of those affected.

Where the intervention being monitored has run into problems, and particularly where conflicts of interest have emerged, a workshop may be useful. This can bring together representatives of those affected in the different areas and at different levels, along with the planners and implementers of the intervention. Experiences can be compared and any knock-on effects traced. Social and economic, short and long term, local and widespread effects can be viewed together. Matrices can be drawn up to identify the relationships between these different components of the system and the positive and negative consequences for each. Following these discussions, recommendations can be agreed on how the intervention could be improved or any adverse effects alleviated.

An example of such a workshop was held in the Philippines, concerning the effects of a major development project — 'The Bicol Integrated Area Development III (Rinconada/Buhi-Lalo) Project'.[37] This project, as part of a larger scale programme of river basin development, had already been found to have adverse effects on the fishing and agricultural livelihoods of the lakeshore inhabitants, the transportation and domestic water supplies of a lakeside town, and the long-term productivity and stability of the lake itself. The workshop was a modification of the Agroecosystem Analysis

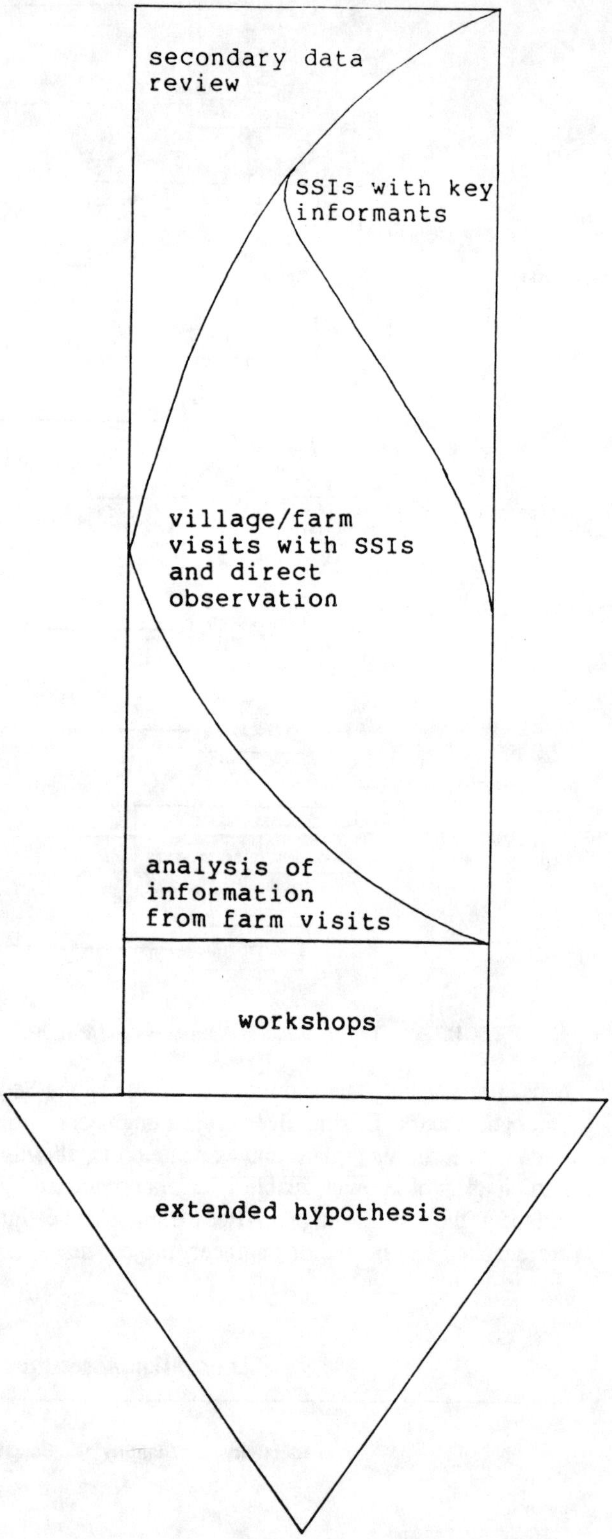

FIGURE 9. The process of a topical RRA.

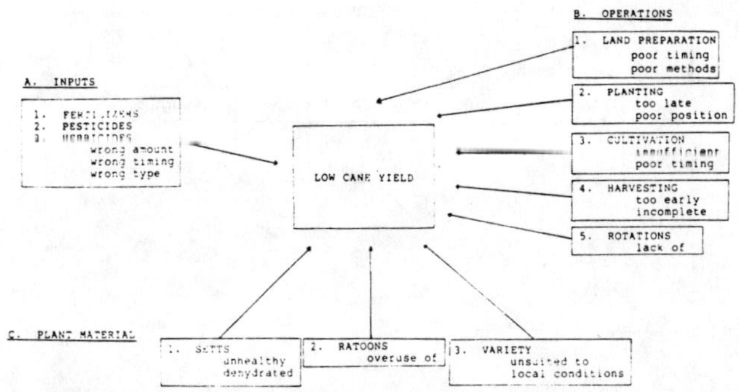

FIGURE 10. Diagrammatic working hypothesis for influence of farming practices on low cane yields in Fiji.

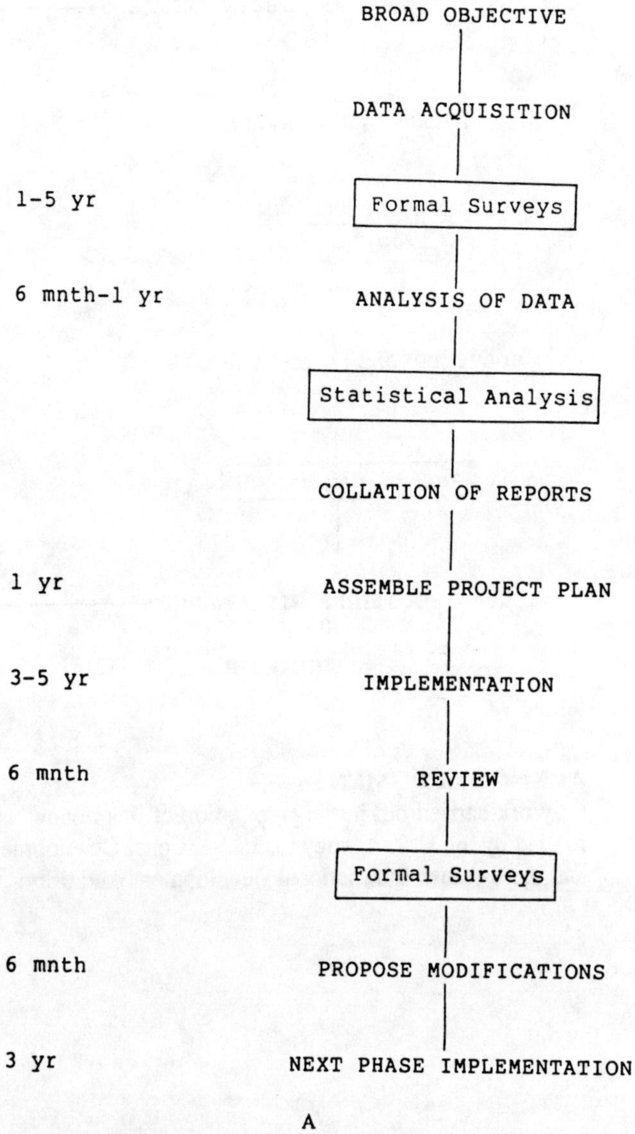

FIGURE 11. Models for project design and implementation. (A) Formal model. (B) Informal model.

approach. Over a period of 5 days, government officials, development agency personnel and representatives of the farmers and fishermen, came together to analyze a series of summary diagrams prepared during a brief field visit by a project team. The outcome was a set of key questions and hypotheses for research and development which were then assessed by the whole workshop to produce a plan of recommended changes and improvements.

A monitoring RRA, including such a workshop, need not take any longer than other forms of RRAs. Several such short surveys can be conducted over a period of time to monitor the progress of a project and to make corrections and improvements.

E. RRA in Project Design and Implementation

A traditional project cycle, particularly as it applies to large investment projects, goes through the sequential phases of data acquisition, analysis, planning, implementation, review and redesign (Figure 11A). It is ordered and methodical, but is often a costly and time consuming exercise. The logical progression is one which is designed to ensure that all factors and considerations are incorporated. But, as experience has shown, such an approach tends to become rigid and bureaucratized in practice. Critical questions are not asked and important insights are missed. At the other extreme, is a project cycle which simply moves rapidly from identifying problems to solving them and back again (Figure 11B). This approach is based on a very close relationship between development professionals and farmers and can only really be undertaken on a small and intimate scale.

This latter scheme relies entirely on RRAs. However RRA has its limitations. It will never, and indeed was never designed to, make redundant more traditional, formal and detailed surveys and analyses. RRAs and RRA techniques essentially complement more formal methods and while in some situations they may be substitutes, more often than not, they are preliminary exercises, leading up to more detailed analyses.

The advent of RRA has thus greatly enriched the availability of methods of analysis for rural development. Techniques can be chosen on the basis of the nature of the problem, the local situation and the resources to hand. In particular different techniques, both formal and informal, can be blended to produce a project cycle along the lines of Figure 12. This lies some way between the extremes of the schemes in Figure 11 and can be applied to a wide range of projects, both large and small in conception. In such a scheme the primary role of the RRA is to define and refine hypotheses which are then tested, either formally or informally, as part of the project cycle. Providing the cycle is iterative, flexible and open, it should be possible to combine speed with both rigor and sensitivity resulting in development that is not only productive but durable and equitable in its benefits.

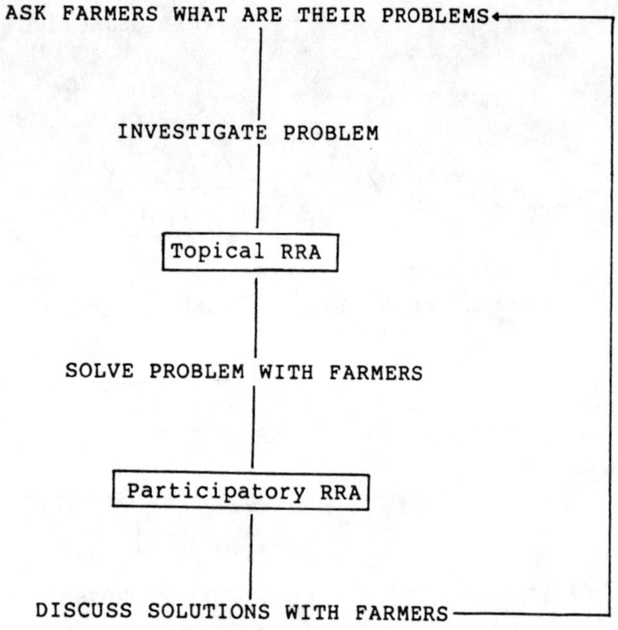

FIGURE 11B.

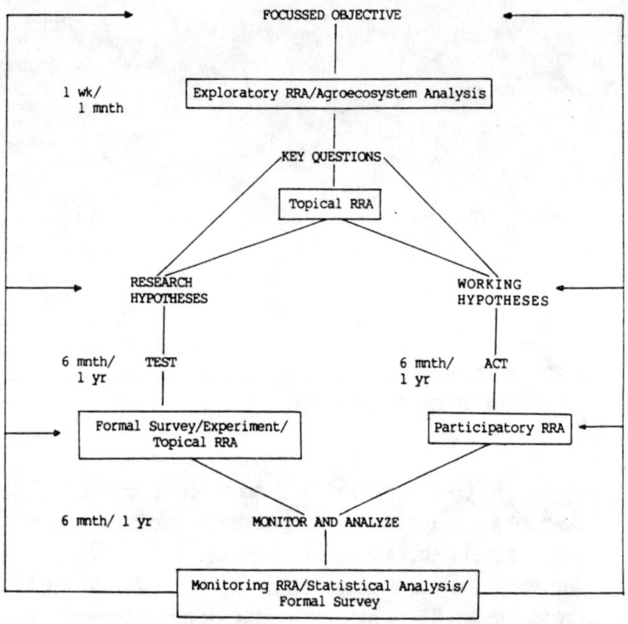

FIGURE 12. A model for project design and implementation which combines the use of Rapid Rural Appraisal and formal analysis and survey.

ACKNOWLEDGMENTS

Work carried out in the preparation of this review was funded by the U.S. Agency for International Development and the Swedish International Development Authority.

REFERENCES

1. **Conway, G. R.,** Agroecosystem analysis, *Agric. Admin.*, 20, 31, 1985.
2. **Conway, G. R.,** The properties of agroecosystems, *Agric. Syst.*, 24, 95, 1987.
3. **Chambers, R.,** Poverty in India: Concepts, Research and Reality; an Exploration, Institute for Development Studies, Sussex, UK, 1986.
4. **Ilchman, W. F.,** Decision rules and decision roles, *Afric. Rev.*, 2, 219, 1972.
5. *Agricultural Administration*, Special Issue on Rapid Rural Appraisal, 8 (6), 1981.
6. **Beebe, J.,** Rapid Appraisal: The Evolution of the Concept and the Definition of Issues, presented at Int. Conf. Rapid Rural Appraisal, Khon Kaen, Thailand, September 2 to 5, 1985.
7. **Conway, G. R.,** Rapid Rural Appraisal and Agroecosystem Analysis: A Case Study from Northern Pakistan, presented at Int. Conf. Rapid Rural Appraisal, Khon Kaen, Thailand, September 2 to 5, 1985.
8. **Gibbs, C. J. N.,** Rapid Rural Appraisal: An Overview of Concepts and Application, presented at Int. Conf. Rapid Rural Appraisal, Khon Kaen, Thailand, September 2 to 5, 1985.
9. **Grandstaff, T. B. and Grandstaff, S. W.,** A Conceptual Basis for Methodological Development, presented at Int. Conf. Rapid Rural Appraisal, Khon Kaen, Thailand, September 2 to 5, 1985.
10. **Limpinuntana, V.,** Conceptual Tools for Rapid Rural Appraisal in Agrarian Society, presented at Int. Conf. Rapid Rural Appraisal, Khon Kaen, Thailand, September 2 to 5, 1985.
11. **Samart, M.,** Some Dimensions of Variation in RRA Methodology at Khon Kaen University, presented at Int. Conf. Rapid Rural Appraisal, Khon Kaen, Thailand. September 2 to 5, 1985.
12. **Grandstaff, T. B. and Grandstaff, S. W.,** Semi-structured Interviewing, presented at Int. Conf. Rapid Rural Appraisal, Khon Kaen, Thailand, September 2 to 5, 1985.
13. **Rhoades, R. E.,** The Art of the Informal Agricultural Survey, presented at Int. Conf. Rapid Rural Appraisal, Khon Kaen, Thailand, September 2 to 5, 1985.

14. **Chambers, R.**, Shortcut Methods in Social Information Gathering for Rural Development Projects, presented at Int. Conf. Rapid Rural Appraisal, Khon Kaen, Thailand, September 2 to 5, 1985.
15. **Carson, B.**, Appraisal of Rural Resources Using Aerial Photography: An Example from a Remote Hill Region in Nepal, presented at Int. Conf. Rapid Rural Appraisal, Khon Kaen, Thailand, September 2 to 5, 1985.
16. **Subhadhira, S., Suphanchaimat, N., Smutkupt, S., Simaraks, S., Pakuthai, W., Supannapesat, K., and Petchsingha, P.**, Fuelwood Situation and Farmer's Adjustment in Northeastern Thai Villages, presented at Int. Conf. Rapid Rural Appraisal, Khon Kaen, Thailand, September 2 to 5, 1985.
17. **Thomas, D. E. and Suphanchaimat, N.**, The use of RRA in Cross-sectional and Longitudinal Studies, presented at Int. Conf. Rapid Rural Appraisal, Khon Kaen, Thailand, September 2 to 5, 1985.
18. **Galt, D.**, How Rapid Rural Appraisal and Other Socio-Economic Diagnostic Techniques Fit into the Cyclic FSR/E Process, presented at Int. Conf. Rapid Rural Appraisal, Khon Kaen, Thailand, September 2 to 5, 1985.
19. **de los Reyes, R.**, Sociotechnical Profile — A Tool for Rapid Rural Appraisal, presented at Int. Conf. Rapid Rural Appraisal, Khon Kaen, Thailand, September 2 to 5, 1985.
20. **Jamieson, N.**, The Paradigmatic Significance of Rapid Rural Appraisal, presented at Int. Conf. Rapid Rural Appraisal, Khon Kaen, Thailand, September 2 to 5, 1985.
21. **Hildebrand, P. E.**, Combining disciplines in Rapid Appraisal: the sondeo approach, *Agric. Admin.*, 8(6), 423, 1981.
22. **Rhoades, R. E.**, *Breaking New Ground: Agricultural Anthropology*, International Potato Center, Lima, 1984.
23. **Honadle, G.**, Rapid reconnaissance for development administration: mapping and moulding organisational landscapes, *World Dev.* 10, 633, 1982.
24. **Shaner, W. W., Philipp, P. F., and Schmehl, W. R.**, *Farming Systems Research and Development: Guidelines for Developing Countries*, Westview Press, Boulder, CO, 1982.
25. **Collinson, M. P.**, The exploratory survey: content, methods, and detailed guidelines for discussions with farmers, *CIMMYT Eastern Africa Economics Programme Farming Systems Newsletter*, April—June 1981.
26. **Alton, C. T. and Craig, I. A.**, The Rapid Assessment Technique (RAT): A Procedure for Identifying Farmer Problems and Development Opportunities, Northeast Rainfed Agricultural Development Project (NERAD), Khon Kaen, Thailand, 1987.
27. **Conway, G. R.**, *Agroecosystem Analysis for Research and Development*, Winrock International Institute for Agricultural Development, Bangkok, 1986.
28. **Gympantasiri, P., Wiboonpongse, A., Rerkasem, B., Craig, I., Rerkasem, K., Ganjapan, L., Titayawan, M., Seetisarn, M., Thani, P., Jaisaard, R., Ongpraset, S., Radanachaless, T., and Conway, G.R.**, *An Interdisciplinary Perspective of Cropping Systems in the Chiang Mai Valley: key questions for research*, Faculty of Agriculture, University of Chiang Mai, Thailand, 1980.
29. KEPAS, The Critical Uplands of Eastern Java: An agroecosystem analysis, Kelompok Penelitian Agro-Ekosistem, Agency for Agricultural Research and Development, Jakarta, Indonesia, 1985a.
30. KEPAS, *Swampland Agroecosystems of Southern Kalimantan*, Kelompok Penelitian Agro-Ekosistem, Agency for Agricultural Research and Development, Jakarta, Indonesia, 1985b.
31. KEPAS, *Agro-ekosistem Daerah kering di Nusa Jenggara Timur*, Kelompok Penelitian Agro-Ekosistem, Agency for Agricultural Research and Development, Jakarta, Indonesia, 1986.
32. **Conway, G. R., Alam, Z., Husain, T., and Mian, M. A.**, *Agroecosystem Analysis and Development for the Northern Areas of Pakistan*. Aga Khan Rural Support Programme, Gilgit, Pakistan, 1986.
33. **Conway, G. R., Husain, T., Alam, A. and Mian, M. A.**, Rapid Rural Appraisal for Sustainable Development: Experiences from the Northern Areas of Pakistan, presented at Conf. Sustainable Development, London, April 28—30, 1987.
34. Rapid Rural Appraisal Activities at Khon Kaen University, Institutional Report, Faculty of Agriculture, presented at Int. Conf. Rapid Rural Appraisal, Khon Kaen, Thailand, September 2 to 5, 1985.
35. **McCracken, J. A.**, Rapid Rural Appraisal: A Case Study of Sugarcane Yields in Fiji, M.S. thesis, University of London, 1986.
36. **Husain, T.**, Approaches to Understanding Mountain Environments, Research and Planning Experiences from the Gilgit District of Pakistan, Aga Khan Rural Support Programme, Gilgit, 1987.
37. **Conway, G. R. and Sajise, P. J., Eds.**, *The Agroecosystems of Buhi: Problems and Opportunities*, University of the Philippines, Los Baños, 1986.

22 FARMER-FIRST: A PRACTICAL PARADIGM FOR THE THIRD AGRICULTURE

Robert Chambers

EXPLANATIONS

This paper uses terms with the following meanings:

- CDR: complex, diverse and risk-prone. See also the third agriculture
- FF: farmer-first, referring to the new complementary paradigm of agricultural research and extension that reverses the learning and locations of TOT, with farm families playing a major part in technology development and choice
- Green revolution (GR) agriculture: the agriculture of fertile and well-watered areas in the South, notably the irrigated plains and deltas of Asia (Table 1)
- Industrial agriculture: the agriculture of the temperate and rich North, with high inputs and subsidies (Table 1)
- Normal professionalism: the thinking, values, methods and behavior dominant in professions and disciplines and reflecting "core" or "first" biases[1]
- The North: the richer, industrialized countries mainly in the temperate northern hemisphere
- Paradigm: a coherent and mutually supporting pattern of concepts, values, methods and action, amenable to wide application
- The South: the poorer, agricultural countries mainly in the tropics
- The third agriculture: the variously complex, diverse and risk-prone (CDR) agriculture of the South, mainly rainfed and on undulating land, found in hinterlands, mountains, hills, wetlands, and in the semi-arid, subhumid and humid tropics (Table 1).
- TOT: transfer-of-technology, referring to the normal basic paradigm of agricultural research and extension in which priorities are decided by scientists and funding bodies, and new technology is developed on research stations and in laboratories and then handed over to extension to transfer to farmers

I. THE GREAT CHALLENGE OF THE 1990

In agricultural development, production and poverty, the 1970s and 1980s have witnessed changes in reality and insight. By the mid-1980, agricultural production had risen sharply in the industrial agriculture of the rich North, and in the green revolution (GR) agriculture of the well-watered fertile plains of the South, but not much elsewhere, in the complex, diverse and risk-prone (CDR) 'third' agriculture of the South. Food surpluses had depressed world prices, with a glut on the market. Excepting Bangladesh, the most populous agricultural countries of Asia—Burma, China, India, Indonesia, Pakistan, Philippines and Thailand—had either achieved food and foodgrain self-sufficiency or had gotten close to it.[2] But for much of the third, non-GR agriculture of the South, there had been deepening crisis, with populations rising, land-holdings growing smaller, environments degrading and per capita food production remaining static or declining. According to one estimate, some 1.4 billion people were dependent on CDR agriculture, with roughly 100 million in Latin America, 300 million in Africa, and 1 billion in Asia;[3] and short of an AIDS or similar holocaust, theses were also the areas and countries where population growth rates would continue to be highest.

The problem now is not one of producing enough food in the world; it is a problem of who grows it, where it is grown, and who has access to it. With population growth and environmental fragility in CDR areas, the problem is also one of generating sustainable livelihoods for the much larger populations of the future, enabling them to live adequately and decently where they are.[4] The alternative is that they have to migrate, often in desperation, to GR and urban areas, where they depress wages and the incomes of other poor people, or to fragile mountain, forest or semi-arid environments where they may contribute to environmental degradation.

The great challenge for the 1990s in, then, to enable the third, CDR agriculture to transform itself into more sustain-

TABLE 1
Three Types of Agriculture Summarized

Type of agriculture	Industrial	Green revolution	Third 'CDR'
Place			
Main locations	Industrialized countries	Irrigated and high rainfall, high potential areas in the South	Rainfed tropics, hinterlands, most of sub-Saharan Africa, etc.
Climatic zone	Temperate	Mainly tropical	Mainly tropical
Status			
Condition	Overdeveloped	Developed	Underdeveloped
Current production as percentage of sustainable production	Far too high	Often near the limit	Low
Priority for production	Reduce production	Maintain production	Raise production
Character			
Topography usually	Flat or undulating	Flat	Undulating
Farming system, relatively	Simple	Simple	Complex
Environmental diversity relatively	Uniform	Uniform	Diverse
Relative stability	Low risk	Low risk	High risk
Use of external inputs	Very high	High	Low
R & D problems			
Similarity of farmers' and research station conditions	High	High	Low
Farmers consulted about research priorities	Richer farmers sometimes	Richer farmers sometimes	Rarely
Number of scientists/ extensionists per farming system	More	More	Fewer

Note: CDR = Complex, diverse, risk-prone.

able and productive systems, and to support many more people. To be sure, maintaining production and tackling poverty in GR areas is also vital. But the problems and solutions there are better known, although changing[5] and receive more attention. Moreover, the normal professionalism of agricultural science has served those areas better, but fits badly with the needs and priorities of the third agriculture (Table1).

II. NORMAL PROFESSIONALISM, TRANSFER-OF-TECHNOLOGY AND THE THIRD AGRICULTURE

Normal professionalism means the thinking, concepts, values and methods dominant in a profession. It is usually conservative, heavily defended, and reproduced through teaching, training, textbooks, professional rewards, and international professional meetings. Most professional mindsets change only slowly, sometimes long after the realities and priorities have changed. This is true in the social sciences as well as in the physical and biological sciences.

In agricultural research and extension, worldwide, the normal professional paradigm can be described as "transfer-of-technology" or TOT.[6] In this model, agricultural research priorities are determined by scientists and by funding agencies; scientists then experiment in-laboratory and on-station to generate new technology; and this is then handed over to extension to transfer to farmers. There have been many modifications and variants, but the TOT model is deeply embedded in normal professional thinking and prescription. It is reflected in teaching, in behavior in the field, and in the rhetoric of development.

The TOT mode has served industrial and GR agriculture rather well. Physical and economic conditions on research stations have been similar to those of resource-rich farms and farm families, which are typical of these two types of agriculture (Tables 2 and 3). The reductionism of normal agronomic research, in which only a few variables are manipulated, has led to simple packages suitable for unifrom controlled environments: E (the environment) has been made to fit high-yielding G (the genotype). Packages have served to standardize farming systems, and have fitted in with economies of scale associated with mechanization and subsidy. The outcome has been the well-known increases in productivity per unit of land in both industrial and GR agriculture.

However, the TOT model has not done well with the third agriculture. There have been limited successes, but no great production breakthroughs comparable with the green revolutions with wheat, maize and rice. The explanation lies partly in the contrasts between physical and economic conditions on research stations and those of the resource-poor farms and farm families which are typical of CDR conditions (Tables 2 and 3). It also lies in the disjuncture between the nature of CDR agriculture on the one hand, and the nature of normal professionalism on the other. This can be appreciated by examining CDR agriculture in more detail.

The complexity of any one CDR farming system has many aspects, and these also vary between farming systems.

TABLE 2
Typical Contrasts in Physical Conditions

	Research experiment station	Resource-rich farm (RRF)	Resource-poor farm (RPF)
Topography	Flat or sometimes terraced	Flat or sometimes terraced	Often undulating and sloping
Soils	Deep, fertile, few constraints	Deep, fertile, few constraints	Shallow, infertile, often severe constraints
Macro- and micronutrient deficiency	Rare, remediable	Occasional	Quite common
Plot size and nature	Large, square	Large	Small, irregular
Hazards	Nil or few	Few, usually controllable	More common — floods, droughts, animals grazing, crops, etc.
Irrigation	Usually available	Usually available	Often nonexistent
Size of management unit	Large, contiguous	Large or medium, contiguous contiguous	Small, often scattered and fragmented
Natural vegetation	Eliminated	Eliminated or highly controlled	Used or controlled at microlevel

Source: Chambers and Jiggins 1986, adapted from Chambers and Ghildyal, 1985:6.

TABLE 3
Typical Contrasts in Social and Economic Conditions

	Research experiment station	RRF family	RPF family
Access to seeds, fertilizers, pesticides and other purchased inputs	Unlimited, reliable	High, reliable	Low, unreliable
Source of seeds	Foundation stocks, and breeders' seed high quality	Purchased, high quality	Own seeds
Access to credit when needed	Unlimited	Good access	Poor access and seasonal shortages of cash when most needed
Irrigation, where facilities exist	Fully controlled by research station	Controlled by farmers or by others on whom s/he can rely	Controlled by others, less reliable
Labor	Unlimited, no constraint	Hired, few constraints	Family, constraining at seasonal peaks
Prices	Irrelevant	Lower than RPF for inputs; higher than RPF for outputs	Higher than RRF for inputs; lower than RPF for outputs
Priority for food production	Neutral	Low	High
Access to extension services	Good but one-sided	Good, almost all material designed for this category	Poor access; little relevant material

Source: Chambers and Jiggins 1986, adapted from Chambers and Ghildyal 1985:6.

Four deserve mention.First, physically, CDR farm holdings often comprise sloping lands with a variety of conditions of soil, slope, shade, aspect and water supply, and sometimes include lands in different ecological zones on the same holding, and with energy and nutrient linkages with common property resources. Second, in their internal linkages, CDR farming systems typically involve and rely on complex interactions between crops, livestock, grasses, trees, and sometimes fish and insects. Intercropping and agroforestry in their many forms are typical of this sort of complexity. Third, CDR farming systems are complex temporally, with many different processes and activities at different times of the year. Fourth, CDR farming systems entail several or many enterprises, often off-farm as well as on-farm; many species of useful plants and animals are husbanded, and often these are multipurpose and multi-product. Finally, compounding all these complexities, CDR farming systems are intimately interlinked with the farm household, its labor power, social structure and economy.

In addition, CDR agriculture often presents diversity of farming systems within short distances, corresponding with differences which are ecological, social and economic, for example in accessibility to markets. Finally, it is also often risky, being usually rainfed and subject to the vagaries of climate, without the stabilizing effects of reliable irrigation.

Normal agricultural science does not fit well with these characteristics. The complexity of CDR agriculture presents interactions difficult for scientists to manage and study. Some lie in the gaps between dominant disciplines (concerning agroforestry, tree fodders, crop residues, biological energy use, etc.): normal science homes in on its primary concern—crops for agronomists, livestock for animal scientists, trees for foresters—rather than their linkages. Some opportunities lie in complex simultaneous innovation, where several factors must be changed at the same time, as with developing rainfed rice-fish farming, or havesting soils, nutrients or water, or introducing a cover crop to inhibit weed growth, or much agroforestry where there are tree-crop, tree-livestock or tree-crop-livestock interactions. For scientists tied to respectable statistical methods, these complexities can be an unmanageable nightmare: for if they simplify them until they are measureable, they destroy the complexities which are their strength.

Precisely this bad fit of CDR agriculture with normal professionalism has served to conceal its potential. When the simple packages generated in the TOT mode are not adopted in CDR areas, the conclusion can easily be drawn that the areas themselves lack potential. So they are often referred to as "resource-poor" or "low-resource" areas. But a case can be made out that their sustainable potential as a multiple of present performance, is considerable[7,8] and far greater than that of GR agriculture, which is already near its limit.

This misfit has been compounded by diversity and by scientists' motivation. There is a problem of cost-effectiveness and rewards using normal R and D methods. Any innovation, such as a new variety or new practice, is likely to fit conditions and needs of far fewer farm families in CDR areas than in GR areas which are or can be made so much more uniform. This makes work harder to justify economically, and also reduces the prestige and incentives of the work for scientists looking for the big breakthroughs. This difficulty is compounded by the presence of far fewer scientists per farming system.[9] This reflects the past unpopularity of CDR agriculture, and its low status and low political priority. Irrigated green revolution agriculture has understandably been preferred by scientists and Ph.D students for reasons including accessibility, ease of control, and predictability of experiments, research papers, and Ph.Ds[10].

For the third, CDR agriculture, the TOT paradigm is, thus, in crisis. At the extreme, the research priorities and locations are wrong, the messages do not fit, the packages are rejected, and the bad experience is attributed either to farmers' ignorance (prescription—more and better extension), or to farm-level constraints (prescription—ease the farm-level constraints and simplify and control the farm to make it more like the research station).

The crisis is also one of direction. Often,CDR farmers reduce their risks by making their farming systems more complex. In terms of agroecology, this is analogous to the greater resilience in face of insults of complex compared with simple ecosystems. Normal TOT seeks to simplify, and thereby increases vulnerability, and emphasises purchased inputs which for CDR farmers often introduce problems of reliable access. For their part, CDR farm families tend to diversify (both to increase benefits from production and to provide buffering and redundancy to face contingencies) and to rely on factors of production that are under their control.

III. FARMER FIRST: THE COMPLEMENTARY PARADIGM

The crisis has led to questioning the very processes which generate agricultural technology, and to the exploration of new approaches. Increasingly during the 1980s, innovators in the agricultural and social sciences have been working with CDR farmers to find solutions to these problems. By concentrating on what they find to work, they have evolved a new paradigm for agricultural research and extension. The approaches of this paradigm have been given various labels: farmer-back-to-farmer;[11] farmer-first-and-last;[12] farmer participatory research;[13] and Approach Development.[14] The name does not much matter, but farmer participation is one key element. For inclusiveness and brevity, I shall try to capture the essence of these approaches with the title farmer-first (FF).

There are now many published sources of FF experience. They include *Experimental Agriculture*[15] with selected papers from the workshop on Farmers and Agricultural Research: Complementary Methods, held at the Institute of Development Studies, University of Sussex, in July 1987; and Paper of the Agricultural Administration (Research and Extension) Network of the Overseas Development Institute, London. Accessible examples include the work of Jacqueline Ashby and colleagues at CIAT in Colombia,[16] of Roland Bunch and World Neighbors,[17] of D.M. Maurya in India,[18] of David Norman and colleagues in Botswana,[19] of Robert Rhoades and others at CIP in Peru,[20] of Sumberg and Okali[21] on alley farming in Nigeria, and of Baker and others in Brazil;[22] In Asia, examples of FF and of movements in its direction include SUAN (the Southeast Universities' Agroecosystems Network), agroecosystems analysis (see Chapter 21) the pioneering rapid rural appraisal (RRA) work of the University of Khon Kaen,[23,24] the Northeast Rainfed Agricultural Development Project (NERAD), also in Thailand, and the Farming Systems Development Project, Eastern Visayas, in the Philippines.[25-27]

The essence of FF is reversals of parts of TOT that have tended to go unquestioned.[4] A reversal of explanation looks for reasons why farmers do not adopt new technology not in the ignorance of the farmer but in deficiencies in the technology and the process that generated it. A reversal of learning has researchers and extension workers learning from farmers. Location and roles are also reversed, with farms and farmers central instead of research stations, laboratories and scientists.

In this framework, much farming systems research can be seen as an extension of TOT: information has been obtained from farmers by outsiders, and analyzed by them to decide what would be good for the farmers, leading to the design of experiments for testing and adaptation. In contrast, FF reverses roles. Analysis, choice and experimentation are conducted by and with farmers themselves, with outsider professionals in a facilitating and support role.

In the latter 1980s, FF methods are evolving fast. Many forms and variants are being tried. Some of the contrasts with

TABLE 4
Transfer-of-Technology and Farmer-First Compared

	TOT	FF
Main objective	Transfer technology	Empower farmers
Analysis of needs and priorities by	Outsiders	Farmers assisted by outsiders
Transferred by outsiders to farmers	Precepts Messages Package of practices	Principles Methods Basket of choices
The "menu"	Fixed	A la carte
Farmers' behavior	Hear messages Act on precepts Adopt, adapt or reject package	Use methods Apply principles Choose from basket and experiment
Outsiders' desired outcomes emphasize	Widespread adoption of package	Wider choices for farmers Farmers' enhanced adaptability
Main mode of extension	Agent-to-farmer	Farmer-to-farmer
Roles of extension agent	Teacher Trainer	Facilitator Searcher for and provider of choice

TOT are presented in Table 4. While not all of these are found all the time, and some can be followed without others, they are mutually reinforcing and cohere as a paradigm contrasting with and complementary to TOT. While farmer participation is a widespread and crucial element, FF goes beyond that to influence decisions and methods which may not involve farmers directly and immediately, for example concerning on-station research.

One sequence which recurs in farmer participatory activities is an iterative process of farmers' analysis, choice, and experiment (FACE) followed by evaluation and extension. The main activities of farmers and roles of outsiders are

Farmers activities	New roles for outsiders
Analysis	Convenor, catalyst, adviser
Choice	Searcher and supplier
Experiment	Supporter and consultant

The actors and activities are presented diagrammatically in Table 5.

Let us consider the main activities in turn.

A. Analysis

This exploits farmers' comparative advantage in knowledge. Farmers are experts on their farming systems. Their analysis, if done well, can be expected automatically to screen out impractical irrelevances with a speed and accuracy to which no outsider could aspire, and should home in on their needs. In the process, farmers identify their priorities according to their own criteria. Outsiders can contribute by convening groups, asking key questions, drawing diagrams, and providing information and suggestions. They can take part, but they do not dominate: the main analysts are farmers themselves.

Farmers' analysis can be promoted and supported in many ways:

1. Sequences of farmers' group discussions and visits[19,22,25]
2. Inspection and discussion—visiting other farmers, research stations, or trial sites[16]
3. Innovator workshops, where farmer innovators meet and discuss and compare their new pracdtices[16, 28]
4. The use of key priming questions by outsiders, such as "What would an ideal variety look like to you?", "What would you like your landscape to look like in the future?", "What do you farmers talk about when you get together?", "Why do other farmers have dif-

TABLE 5
Activities in the Farmers' Analysis-Choice-Experiment (FACE) Approach

Farmers' activities	New roles for outsiders
Analysis	Convenor, catalyst, adviser
Choice	Searcher and supplier
Experiment	Supporter and consultant

The actors and activities are presented diagrammatically below

Actors	Actions					
CDR	Analyze	Identify needs, problems, opportunities, priorities	Request, demand	Experiment	Choose	Adapt
Farmers and farm families	Analyze					
Scientists, extensionists, Ngo field-workers, and other outsiders	Convene, catalyze, advise		Search and supply		Support, consult	

Note: The processes are iterative but for simplicity feedback loops have not been included.

ferent practices to you?", and the unhurried sequence "What was farming like when you were young, how has it changed, what problems have you faced, what have you tried to tackle them, and with what results?"

5. Visual aids to analysis such as seasonal diagramming,[29] aerial photographs and overlays, systems diagramming,[25] and charts representing farmers' information systematised by outsiders,[30] drawn on boards or on the ground

Methodological questions are many, and much remains to be learnt. Analysis can raise many different sorts of issues. In CDR areas, security of tenure is often a prerequisite for farmers taking a long view. Or, relations with Government Departments may turn out to be crucial. It may be necessary to tackle priorities such as these before those which are more directly agricultural. Or, analysis may lead straight to experimentation. Often, though, it will lead to search.

B. Search

Participatory analysis often generates demands for information and material. CDR farmers want and need wide choice and enhanced adaptability. The role of the outsider, whether researcher or extension agent, is to look for and supply a range of information about practices and potentials, and a range of genetic material. An example from the philippines is a research agenda geared towards meeting farmers' needs which included search for alternative live mulch, alternative leguminous trees, and alternative sources of leguminous cover crops.[31] The demand is not for the package of practices for normal research and extension, but for a basket of choices.

Methodological questions refer especially to the organisation of extension and research. Extension information systems have to be stood on their head, passing requests up first, before messages down. The difficulty of this reversal can be inferred from experience in the Philippines. Of the seven management information systems for agriculture and natural resources reported[32] six (for research management information, equipment infrastructure management, manpower management, financial management, publications mailing, and administrative support) appear designed to serve central management rather than farmers' needs for information. The seventh—a Research Information Storage and Retrieval System—with potential use to provide information and choices to farmers, was described only in the future tense, with the statement that financial support was needed to extend it to regions, suggesting that it was not yet in operation. As here, information systems normally serve the managers at the center before farmers at the periphery.

C. Choice

Presenting choice to farmers can take several forms:

1. Minikits (a well-known and well-established approach), containing several varieties of a crop, and several fertilizers, for farmers to test and choose from on their own
2. "Wait-and-see and pick-and-choose" (D. Rocheleau, personal communication). Planting a range of species, varieties or lines and giving farmers an opportunity to observe them and choose from them
3. Releasing small batches of advanced breeders' lines matched to the characteristics of farmers' landraces.[18]
4. Pre-screening of varietal materials by farmers, as with bush beans and cassava at CIAT in Colombia.[16]

There are methodological questions about how best to elicit and support farmers' criteria and choices. One example, of group discussion, can illustrate. In Colombia, difficulty was experienced with individual farmers making selections from 35 superficially similar varieties of snap beans; but as a group, farmers did better. They walked through rows of beans, examining bean plants and pods of each variety separately. Research staff asked them to indicate which varieties they considered should continue to be tested and which not. Farmers' discussion rapidly focused on quality characteristics related to market acceptability. In about an hour farmers identified two climbing varieties and two bush varieties which they considered outstanding by their criteria and six bush varieties they would test further.[16]

D. Experimentation

Finally, farmers themselves experiment, and adapt technology.[33-35] Here what is often most important is to transfer to them not packages and precepts, but principles and methods. A famous example of the transfer of a principle is the International Potato Center's experience with diffused light storage in potatoes. Farmers themselves discovered that sprouting in storage, a problem with new varieties, was inhibited by diffused light storage.

Scientists learned from the farmers, and transferred the principle internationally. But there was no standard store to be built; farm families did not adopt a design but applied a principle, in a myriad of locally adapted different ways. An example of the transfer of a method is provided by World Neighbors, who have a simple procedure for enabling farmers to conduct their own trials more systematically.[36]

Many methodological questions remain. One persistent problem is allowing and enabling farmers to "own" their experiments, and not to be dominated by outsiders. Enhancing farmers' capacity to experiment remains a major frontier on which much progress is needed and can be esxpected.

E. Evaluation and Extension

In the FF mode, evaluation is not by scientists' peers but by farmer adoption. For D. M. Maurya (personal communication) whether a line justifies the bulking of seed depends on whether the farmers who try it are asked for seed by other farmers. With farmers' inspections of one anothers' fields and trials, evaluation and extension merge. Extension is not top-down, as often in the T and V mode in practice, but lateral, from farmer to farmer, as with peanuts after rice in northeast Thailand,[37] with soil erosion control in the Philippines (personal communication, S. Fujisaka), and in the approach of World Neighbors.[17]

The FF paradigm is still evolving and will never have a final shape, since it is organic rather than a structure. All the same, there are recurring elements which hang together and support each other. One is the resonance between enhancing the adaptability of farmers through widening their choice and knowledge, and enhancing the adaptability of outsiders—scientists, extensionists and NGO staff—through widening theirs. For farmers the choices are of practices and plants; for outsiders, of approaches and methods. For farmers, the adaptability is to uncertain climatic and economic conditions; for outsiders it is to needs, opportunities and insights as they arise. For all, decentralization and reversals of authority to those "below" are entailed: to empower farmers to analyze choose, experiment and evaluate; and to empower outsiders, however junior, to use their initiative and choose their methods to fit local conditions.

FF thus has its own style, which is decentralized and democratic, in which there is mutual respect and service between outsiders and farmers. Personality is a key variable here, FACE may not be a bad acronuym, since the quality of the face-to-face interactions of farmers and outsiders are crucial. A personal impression is that those who have succdeeded in pioneering FF approaches have been sympathetic people who empathize with farmers and respect and like them. This cannot be expected of all outsiders, but the fascination and psychic rewards of working closely with farmers and learning from and with them are so high, that more and more outsiders may be attracted to this mode.

VI. REFLECTIONS FOR THE FUTURE

The argument for the FF paradigm to complement TOT has been developed here in terms of the third, CDR, agriculture, but its application is not necessarily so limited. It may increasingly fit the trends in GR agriculture towards complexity and diversity. Some of the new GR complexity comes from the range of inputs (seed, fertilizer, pesticide) and assocdiated practices that have become available and needed. Some also comes from the diversification of crops and sequences, for example with nonrice crops increasingly grown in a second season following rice in South and Southeast Asia, and with rotations such as rice-potato-wheat, cotton-wheat, and sugarcane-wheat. Further, the withdrawal or reduction of input subsidies in both GR and industrial agriculture may permit and encourage on-farm diversification towards complexity, as has happened in New Zealand. FF approaches and methods, devised and evolved to meet the special challenges of CDR agriculture, may in the 1990s be found to apply more and in GR and industrial agriculture, helping the 1990s to become a decade, worldwide, of diversification.

For the present, though, the higher priority appears to lie in CDR agriculture, evolving and testing methods, and striving for cost-effectiveness, spread and sustainability. This raises many questions, including these:

1. To what extent, and how, can the FF paradigm be parsimonious, that is, sparing in its demands on outsiders' time so that many more of the diverse farming systems can be served?
2. How can FF approaches and methods be assessed and evaluated, to identify what works, and how well it works and in what conditions?
3. How can FF pioneers in national and international agricultural research systems, and in national extension systems, be encouraged, supported and rewarded, in a sustained manner, with freedom to behave in new ways?
4. How can practitioners learn efficiently from their experience and pass it on to others?
5. How can new syllabi, textbooks and training courses be evolved to include FF experience and methods?

New paradigms are often rejected by those who have heavy personal investments in the old. In the case of FF, fortunately, conversion is not an all or nothing affair, in which scales fall from the eyes. The individual methods can be tried out piecemeal. Their strength, though, lies in their coherence and mutral reinforcement. It is too early to say what the ultimate potential of FF apporaches and methods will be. It is not too early to say that finding out that potential is a priority; for on it may depend the sustainable livelihoods of many millions of the poorest in the 1990s and the 21st century.

REFERENCES

1. **Chambers, R.,** Normal professionalism, new paradigms and development, *Discussion Paper* 227, Institute of Development Studies, University of Sussex, Brighton, UK, December 1986.
2. *FAO Trade Yearbook*, FAO,Rome, 1986.
3. **Wolf, E. C.,** *Beyond the Green Revolution: New Approaches for Third World Agriculture*, Worldwatch Paper 73, Worldwatch Institute, 1776 Massachusetts Ave NW, Washington, D.C., 1986.
4. *The Greening of Aid: Sustainable Livelihoods in Practice*, Conroy, Czech and Litvinoff, Eds., Earthscan Publications, London, 1988.
5. **Byerlee and Derek,** Maintaining the momentum in post-Green Revolution agriculture: a micro-level perspective from Asia, *MSU International Development Paper* No 10, Department of Agricultural Economics, Michigan State University, East Lansing, 1987.
6. **Chambers, R. and B. P. Ghildyal,** Agricultural research for resource-poor farmers: the farmer-first-and-last model, *Agricultural Administration*, 20, 1, 1985.
7. **Bunch, R.,** Case study of Guinope Integrated Development Programme, Guinope, Honduras, paper for Only One Earth: Conference on Sustainable Development, organised by the International Institute for Environment and Development, 3 Endsleigh Street, London, 28—30 April, 1987a.

8. **Chambers, R.,** Sustainable rural livelihoods: a strategy for people, environment and development, *Commissioned Study* 7, Institute of Development Studies, University of Sussex, Brighton, UK 1987.
9. **Chambers, R. and J. Jiggins,** Agricultural research for resource-poor farmers: a parsimonious paradigm, *IDS Discussion Paper* 220, Institute of Development Studies, University of Sussex, Brighton, UK, 1986.
10. **Gupta, A.,** Organising the poor client responsive research system: can tail wag the dog?, paper for the Conference on Farmers and Agricultural Research: Complementary Methods, Institute of Development Studies, University of Sussex, Brighton, UK, 26—31 July 1987.
11. **Rhoades, R. E. and R. H. Booth,** farmer-back-to-farmer: a model for generating acceptable agricultural technology, *Agricultural Administration,* 11, 127, 1982.
12. **Chambers, R. and B. P. Ghildyal,** Agricultural research for resource-poor farmers: the farmer-first-and-last model, *Agricultural Administration,* 20, 1, 1985.
13. **Farrington, J. and A. Martin,** Farmer participatory research: a review of concepts and practices, *Discussion Paper* 19, Agricultural Administration (Research and Extension) Network,Overseas Development Institute, Regent's College, Regent's Park, London NWI 4NS, 1987.
14. **Scheuermeier, U.,** *Approach Development: a contribution to participatory development of techniques based on a practical experience in Tinau Watershed Project, Nepal,* LBL, Landwirtschaftliche Beratungzentrale, CH-8315 Lindau, Switzerland.
15. **Farrington, J., Ed.,** *Experimental Agriculture,* 24, part 3, 1988.
16. **Ashby, J., C. Quiros and Y. Riviera,** Farmer participation in on-farm varietal trials, *Discussion Paper* 22, Agricultural Administration (Research and Extension) Network, Overseas Development Institute, Regent's College, Regent's Park, London NWI 4NS, December 1987.
17. **Bunch, R.,** *Two Ears of Corn: a guide to people-centered agricultural improvement,* World Neighbors, 5116 Portland Avenue, OK, City, 1985.
18. **Maurya, D. M., A. Bottrall, and J. Farrington,** Improved livelihoods, genetic diversity and farmer participation: a strategy for rice breeding in rainfed areas of India, *Experimental Agriculture,* 24, part 3, 311, 1988.
19. **Norman, D., D. Baker, G. Heinrich, and F. Worman,** Technology development and farmer groups: experience from Botswana, *Experimental Agriculture,* 24, part3, 321, 1988.
20. **Rhoades, R. E. and R. H. Booth,** Farmer-back-to-farmer: a model for generating acceptable agricultural technology, *Agricultural Administration,* 11, 127, 1982.
21. **Sumberg, J. and C. Okali,** Farmers, on-farm research and the development of new technology, *Experimental Agriculture,* 24, part 3, 333, 1988.
22. **Baker, G., H. Knipscheer, and J. de Souza Nieto,** The impact of regular research field hearings (RRFH) in on-farmer trials in northeast Brazil, *Experimental Agriculture,* 24, part 3, 281, 1988.
23. Khon Kaen University, *Proceedings of the 1985 International Conference on Rapid Rrual Appraisal,* Rural Systems Research and Farming Systems Research Project, Khon Kaen, Thailand, 1987.
24. **Lovelace, G., S. Subhadira, and S. Simaraks, eds.,** *Rapid Rural Appraisal in Northeast Thailand: Case Studies,* KKU-Ford Rural Systems Research Project, Khon Kaen University, Khon Kaen, Thailand, 1988.
25. **Lightfoot, C., O. de Guia, Jr., and F. Ocado,** A participatory method for systems-problem research: rehabilitating marginal uplands in the Philippines, *Experimental Agriculture,* 24, part 3, 301, 1988.
26. **Repulda, R., F. V. Quero, Jr., R. B. Ayaso, III, O. de Guia, Jr. and C. Lightfoot,** Doing research with resource-poor farmers: FSDP-EV Perspectives and Programs, paper for the Conference on Farmers and Agricultural Research: Complementary Methods, Institute of Development Studies, University of Sussex, Brighton, UK, 26—31 July 1987.
27. **Tung, L. and F.T. Balina,** *Proceedings: Philippines Upland Research and Extension Training Workshop,* ATI, NTC-Visayas, Visca, Baybay, Leyte, Philippines, 19—24 June 1988.
28. **Abedin, Z.and F. Hague,** Learning from farmer innovation and innovator workshops: experiences from Bangladesh. Paper for the Conference on Farmers and Agricultural Research: Complementary Methods, 26—31 July, Institute of Development Studies, University of Sussex, Brighton, UK.
29. **Conway, G.,** Diagrams for farmers, paper for the Conference on Farmers and Agricultural Research: Complementary Methods, 26—31 July, Institute of Development Studies, University of Sussex, Brighton, UK, 1987.
30. **Kabutha, C., and R. Ford,** Using RRA to formulate a village resources management plan, Mbusanyi, Kenya. In *RRA Notes* 2, International Institute for Environment and Development, London, 1988.
31. FARMIIS, *Farm and Resource Management Institute Information Service Newsletter,* Farm and Resource Management Institute, Visca, Baybay, Leyte, Philippines, December 1987.
32. **Valmayor, R. V., and C. R. Mamon,** Research Information Systems for Agriculture and Natural Resources in the Philippines. In International Service for National Agricultural Research Management, *International Workshop on Agricutural Research Management,* The Hague, Netherlands, October 1987.
33. **Johnson, A. W.,** Individuality and experimentation in traditional agriculture, *Human Ecology,* 1(2) 149, 1972.
34. **Richards, P.,** *Indigenous Agricultural Revolution: Ecology and Food Production in West Africa,* Hutchinson, London, 1985.
35. **Rhoades, P.,** Farmers and experimentation, *Discussion Paper* 21, Agricultural Administration (Research and Extension) Network, Overseas Development Institute, Regent's College, Regent's Park, London NWI 4NS, December 1987.
36. **Bunch, R.,** Small farmer research: the key element of permanent agricultural improvement, paper for the Conference on Farmers and Agricultural Research: Complementary Methods, 26—31 July, Institute of Development Studies, University of Sussex, Brighton, UK, 1987b.
37. **Jintrawet, A., S. Smutkupt, C. Wongsamun, R. Katawetin, and V. Kerdsuk,** Extension activities for peanuts after rice in Ban Sum Jan, Northeast Thailand: a case study in farmer-to-farmer extension methodology, Khon Kaen: Farming Systems Research Project, Khon Kaen University, Thailand, June 1985.

23 SMALL FARMERS' PARTICIPATION IN THE DESIGN OF TECHNOLOGIES

Jacqueline A. Ashby

I. INTRODUCTION

Small farmers in developing countries are often bypassed by the technology generation process. The lack of effective research for tackling the complex problems of adapting available technology to the highly diverse condition of small farmers means that many recommendations fail because farmers find them inappropriate to their needs and resources.

Substantial investments have been made in agricultural research that fails to reach small farmers. It is common to observe well-equipped experiment stations with modern agricultural techniques, literally surrounded by small farmers who grow the same crops using traditional technology. One reason for this is that small farmers in developing countries lack the organization to communicate their needs to technology designers, or to ensure that these needs are addressed when research agenda are planned and when resources for research are allocated.

To bridge this gap, efforts were initiated in the late seventies to bring researchers to the farmers' fields by conducting on-farm research or farming systems research. This effort has not, however, systematically involved small farmers as active participants in the planning, execution and evaluation of research. In particular, the concept of farmer participation in *planning* research has been neglected. Farmers usually participate in on-farm testing, either with a nominal presence as the grantor of land for experiments or, at best, as physical participants in implementing production operations for on-farm trials. When as a result, research conducted in farmer's fields is initiated and controlled completely by scientists, it merely constitutes a test of technological components in the physical environments of farms.

Knowing your clients and designing a program to meet their needs is an important management concept to which bureaucratic public service research organizations run by agricultural scientists have not been attuned in developing countries. Participatory management has been tried in other fields such as community health delivery systems, for example; but such techniques have yet to be applied to the management of agricultural research, where client involvement in the generation of technology is as critical as participation in the delivery system (i.e., extension system).

Conventional on-farm research or faming systems research methodologies have not come to grips with the need to institutionalize farmer-scientist collaboration in the planning, testing and evaluation of technologies. In developing countries the intellectual, cultural and social gap between the professional scientist based on an experiment station and the small farmer is great. It is difficult for agricultural scientists to conceive how semiliterate, barefoot farmers can participate in research, and to voluntarily involve farmers in the intellectual process of defining problems, setting priorities and identifying potential solutions.

Yet small farmers are active in the difficult task of adapting agricultural technology to the location-specific conditions of their farming environment. In every farming community, there exist innovative farmers who try out new farming techniques in an informal process of research and development called "farmer experimentation". In developed countries early efforts at agricultural technology generation were based on exploiting the knowledge of the best farmers and in promoting a process of horizontal transfer; that is, farmer-to-farmer innovation. In contrast, agricultural scientists in developing countries are trained to manage agricultural innovation as a process of vertical transfer; i.e., bringing in foreign technologies, adapting and transferring them to farmers. The initiative and indigenous technical knowledge embodied in farmer experimentation are valuable resources that remain untapped by technology development and transfer institutions. An important reason for the neglect of small farmer expertise is the lack of a formal methodology providing agricultural scientists with techniques for implementing farmer participation in ways that enable small farmers to utilize their expert knowledge of local conditions, their skills and their capacity for self-help.

This chapter discusses the issues of farmer participation by addressing three basic questions:

1. Why should small farmers participate in technology design? What are the benefits, both expected and observed, of farmer participation?
2. When should small farmers participate? What are the stages in technology design for agriculture when small farmers can make an input?
3. How can small farmers participate? What are the approaches, and practical techniques that are being developed and tested in field conditions to involve farmers in technology design?

Although farmer participation is increasingly recognized as an important principle for achieving adoptable technologies, practical implementation of farmer participation is at present an experimental and innovative field of endeavor. The chapter concludes with a brief discussion of the gaps in present knowledge, and suggestions for future work.

II. THE BENEFITS OF FARMER PARTICIPATION IN TECHNOLOGY DESIGN

In its present stage of development, farmer participation research focusses on a major problem faced by adaptive research for small farmers: how to cater for the enormous diversity of location-specific needs and constraints of this client group in technology design.

Six major areas of potential impact or benefit can be identified from the literature on farmer participation as reasons for adopting a participatory approach to technology design for agriculture:

1. Improving researchers' understanding of small farmers' needs toward better definition of research problems for adaptive (on-farm) research
2. Integrating indigenous technical knowledge and local innovations into adaptive research
3. Improving feedback about farmers' needs and objectives to orient applied (experiment-station) research
4. Accelerating transfer and adoption of improved technology among small farmers
5. Achieving cost-effectiveness and more efficient use of scarce human resources in on-farm research, through better linkages among farmers, researchers and extensionists
6. Developing organizational models, professional skills and values appropriate for demand-driven, problem-oriented technology design

Concrete evidence and experience as to the achievement of these benefits is for the most part restricted to the first three areas: improved problem definition, linkages between indigenous technical knowledge and formal scientific research, and feedback to orient applied research. To some extent, this reflects the innovative state-of-the-art in participatory research with small farmers: the majority of projects or programs reporting experiences with farmer participation in technology design are in the early stages of diagnosis and experimentation; few report results in terms of adoption; as yet almost no effort has been dedicated to documenting costs and time requirements. Moreover, the bulk of participatory research has occurred outside of national agricultural research programs in developing countries (in non-government supported projects, international agricultural research centers or universities) so that little is known about the costs or organizational requirements of farmer participation research in formal institutions on any appreciable scale.[1]

Technology design with farmer participation has been implemented in a number of different types of technology. Although most of the reported examples refer to the selection of genetic materials, there are instances of farmer participation in livestock and agroforestry research, in fertilizer research, in crop, soil and water management, in crop storage, and home gardens.[2-9] These experiences suggest that farmer participation has a high payoff for researchers working for client groups in small-farm systems which have evolved finely tuned adaptations to difficult local agroclimatic and edaphic conditions and resource constraints to meet specific needs. Local knowledge of these systems typically exceeds scientists' understanding of them.

An illustration is the traditional "bensani" system of weed management in India discussed by Maurya and Bottrall.[10] In the "bensani" system, 25- to 30-day-old rice is ploughed crosswise and planked to control a major weed (the *Echinochloa* species,) which is identical to the rice plant in its early stages of development. The weed plant develops internodes within 30 days, so that there are weak points just below the nodes that cause the weed shoots to break and detach during planking. The rice plant does not develop such nodes and internodes, and thus escapes breaking: during planking the rice bends and subsequently resumes its original erect position. Besides weed management, this practice ensures better water retention in the field and promotes tillering of the rice plant. The "bensani" technique is one of many farmer practices originally criticized by scientists, but which are being integrated into station and on-farm research as a result of efforts to involve farmers in the breeding and selection of varietal materials.[10] One of the major advantages of farmer participation is the integration of farmers' expertise in the management of their local system into the process of adapting new technology components into this system.

Much more challenging, however, is the situation where traditional management practices and knowledge about the local system are decreasingly viable because the system is breaking down or in disequilibrium. This situation is typical of areas where increased population pressure, deforestation, soil erosion, over-grazing and depleted soil fertility imply that small farmers will have to make major system-level changes in order to sustain a viable agriculture. An example from Zambia illustrates how farmers' own experimental initiatives to intensify a traditional shifting cultivation system provided an important foundation for the design of new technologies for testing on-farm and on-station.[6] The traditional chitemene system in northeast Zambia involves the felling and harvest of woody vegetation from a plot, followed by burning the collected wood on a sub-plot, which is cultivated in a six-year rotation. The plot is then left in fallow for several years. The trend is towards shorter fallows and sharp declines in productivity. One focus of agricultural research has been to replace the chitemene system with high-input maize-based systems of permanent cultivation which, however, were being adopted only by wealthier farmers. Study of other farmers' practices identified farmer experimentation with mounding as a way of incorporating plant biomass into the soil in the latter part of the traditional crop rotation, to prolong the useful life of the plot for bean or cassava production. Mounds were also being used by farmers as a conscious soil conservation practice for cultivating slopes. Mounding techniques developed by farmers were integrated with agroforestry research to incorporate trees providing high nutrient leafy biomass for mounding.

More effective adaptive research, oriented to farmers' problems which incorporates indigenous technical knowledge, builds linkages with farmer experimentation and generates rapid feedback to technology designers, is the result of farmer participation to date. Achieving these benefits depends critically on important innovations in the stages of technology design when small farmers participate.

III. FARMER PARTICIPATION IN DIFFERENT STAGES OF TECHNOLOGY DESIGN

In applied agricultural research, technology design usually proceeds through several different stages which can be broken out as follows:

- DIAGNOSIS: Identification of objectives, needs and problems
- PLANNING: Setting priorities among problems, definition of potential solutions and formulation of strategy to test solutions
- DESIGN of prototype technology
- EXPERIMENTATION to test and evaluate prototype technology, resulting in developed technology
- VALIDATION of developed technology, resulting in recommendations for use

These different stages may be carried out both on experiment stations and on farms. When farmer participation is implemented in on-farm or farming systems research, it normally occurs in the diagnostic stage, when problems are identified, and subsequently in the stage of validating results and developing recommendations.

In order to analyze farmer participation in this process, it is important to distinguish among different types of farmer participation in technology design.[2]

1. *Nominal farmer participation* occurs when the farmer's role is a passive one. The farmer is involved in research as a respondent to structured diagnostic questionnaires, designed, administered and analyzed by researchers; or as a collaborator contributing land and labor to on-farm trials designed and managed entirely by researchers, who also derive conclusions from trials without attempting to interact with farmers about their responses to the technology.
2. *Consultative farmer participation* is a type widely espoused by farming systems research. Diagnostic research involves informal interactions between researchers and farmers to identify problems to be addressed in technology design. Priorities among problems, further planning and design (outlined above) are decided by researchers (though it may include extensionists).[11] Experimentation usually involves researcher-managed trials of the type described above, often carried out to perform an initial screening of prototype technology on-farm. Subsequently, the technologies that emerge from this initial screening are validated in "farmer-managed" trials. Such trials can be more properly thought of as research-designed and farmer-implemented trials because researchers determine concepts in the design of the technology to be tested, and what changes in the farmers' existing practices or system are to be implemented.[2,12] Farmers' adoption of the technology is usually measured. Farmers opinions about the new technology may be sought during the validation stage of research.[13]
3. *Collegiate or decision-making farmer participation,* so termed because the farmer is an active participant placed squarely in the role of a colleague in the research process. The farmer acts not as a passive object who is studied and measured; but as a subject who studies, measures and intervenes in making decisions about the technology through participation in planning, design and experimentation, as well as in diagnosis and validation.[2,14,15]

Nominal farmer participation characterizes a technology design process, now widely criticized, which transfers the experiment station to farmers' fields. Consultative and collegiate types of farmer participation exist in practice as a continuum in which the latter is receiving increasing attention as a result of dissatisfaction with other approaches.[16,17]

Consultative farmer participation has two major shortcomings related to the stage at which hands-on involvement of farmers in technology design is activated in this approach. Although farmers participate early-on as informants and sources of ideas for problem diagnosis, the time-lag between identifying problems, designing potential tehnical solutions to farmers' problems (prototype technology) and obtaining farmers' reactions to developed technology during validation in farmer-implemented trials can be prolonged.[18] Most applied agricultural research programs involve evaluation of a large number of alternative solutions to farmers' problems, be they plant varieties, planting densities, pest and disease controls, farm implements, livestock management practices or other components. These are screened selectively on-station and on farm by researchers to identify the most promising options (often termed the researchers' "best-bet"). By the time farmers participate in on-farm trials, many options have been discarded and farmers' responses are obtained to those few which appear most promising from *the researchers' point of view*. The risk entailed by this approach is that researchers may already have excluded from on-farm trials technological options which may appear promising from *the farmers' point of view*. One reason for this is the widely-held assumption that farmer participation in testing technology is only appropriate for validating "developed" technologies in the experimental framework of farmer-implemented trials. Arguments in favor of this notion include shielding the farmer from risky experimentation; that multiple options are too difficult for farmers to evaluate so that trials should be kept simple and include only one or two new alternatives; and that farmers will lose confidence in research and extension services if many alternatives are seen to fail.[13,19]

A second shortcoming of consultative farmer participation is that as a technology becomes more developed (when for example, a set of cultural practices is evolved by on-station researchers and varieties are gradually selected for performance within this set of practices), it becomes more site-specific, and more specific to a sub-set of farmers' socioeconomic and cultural needs. The more developed, site and farmer-specific a technology, the more exacting and expensive it is to carry out adaptive research to fit the technology to farmer circumstances.[20]

Collegiate farmer participation originates from a different set of premises about the stages in technology design appropriate for involving farmers. The view that farmers are experimenters or researchers in their own right and that experiments should be jointly planned by scientists and farmers is central to this approach, and dates back to early advocates of on-farm and farming systems research.[21,22] Efforts to build a formal link with farmers' indigenous experimentation discussed earlier have involved farmers' in the early exploratory stages of screening technologies, when multiple options have been evaluated by farmers who made decisions about how to carry out the testing.[10,15,23,24]

An important objective of this work is to generate ideas for technology design which then guide applied on-station research. An example is the adaptation of potato storage research in Peru to the finding that unlike station researchers, farmers did not perceive storage losses in terms of wastage, but identified their priority problem as a tendency of seed potatoes to sprout and lose weight during storage. Research objectives and technology design were re-oriented to develop seed storage methods which inhibit sprouting.[9]

Such experiences have prompted researchers to question the appropriateness of formal trial methods compared with indigenous experimentation, particularly with reference to variety testing. For instance, can experiments be simplified to the extent of giving farmers planting material or training in a cultural practice, while researchers monitor farmers' opinions of varieties or adaptations of practices?[25,26] Limited experience to date indicates that information retrieval can be difficult with this approach, and that autonomous farmer experimentation may take a long time to arrive at reliable conclusions, in part due to slow rates of local seed multiplication in the case of varieties. Whether farmers make decisions about how to test technologies in a framework of formal trials or indigenous experimentation, the main thrust of this work has been to relax scientists' control over test parameters in order to generate better information about farmers' criteria for evaluating prototype technologies.

Farmer participation in planning, in setting priorities and in formulating strategy is viewed by some advocates as crucial to participatory research (otherwise considered a misleading title for technology generation that involves farmers, but is controlled by scientists). Participation in planning involves "explicitly enabling farmers themselves to identify what they want from scientists".[16]

An example of farmer participation in planning research for a farming system development project is reported by Lightfoot et al.[26] In this instance farmer meetings, farm visits for observation purposes and an informal survey were carried out to develop an analysis of problems with farmers and to enable farmers to arrive at a consensus about priorities for research. This process led researchers to study farmers' own experimental efforts to solve the priority problem. Experimental hypotheses were developed by farmers, elaborated by researchers in a debate with farmers, and resulted in tests carried out with farmers of several alternative (or prototype) solutions to the problem. In this participatory planning process, researchers are shown to act primarily as facilitators, stimulating farmers to exchange ideas about a problem that they perceive as important, and to identify important principles for solving the problem. Researchers assisted farmers to integrate and systematize their own efforts to find solutions and provided new inputs (in this example new legume species) critical to the successful application of principles for solving the problem identified by farmers.

Another example involving farmers in planning research is the Agricultural Technology Improvement Project (ATIP) in Botswana.[5] Farmer groups were formed initially as a method for trial management "and in order to create an opportunity for on-going dialogue about farmer problems and opportunities."[5] At village meetings, the full range of technology options available for testing were presented to farmers and discussed by them. Farmers selected which options they wanted to test. Monthly meetings were held after trial establishment to discuss progress, problems and farmers observations. In this "options-testing" process, the initial definition of trials was determined by farmers, resulting in a larger variety of trial types than had been defined by researchers on their own.[5]

An important issue raised by participatory planning and design with farmers is the degree of researcher initiative desirable in this process. The researcher has specialized technical knowledge of potential technical innovations about which farmers can have little or no knowledge, and can make suggestions which can broaden the scope of farmers' ideas, visions or hopes for the future. Farmers can identify and prioritize problems and potential solutions within the scope of their existing knowledge and experience, but how useful is this knowledge for planning research in situations where traditional knowledge is decreasingly viable?[1] One response to this challenge is to present farmers with multiple options in the exploratory stage of technology design. It has been argued that farming systems are a "moving target" and that technology design must build upon dynamic trends in existing systems.[18] One of the findings of the ATIP Project was that some trials identified by farmers who took part in selecting options, represented a substantial change from their traditional system.[5] Other experiences indicate the importance of exploring with farmers their vision of "futures possible" or alternative future uses of their land in participatory diagnosis.[6] In one example, important new breeding objectives were identified by asking farmers to describe and explain the desirable characteristics of an "ideal" new variety, which had not been identified when farmers were only asked to react to existing varietal options developed by breeders.[15] Participatory research has the potential to involve farmers in the development of new farming systems by enabling farmers and scientists to interact about the potential of innovative prototype systems.

Evidence to date of the results of involving farmers in identifying what they want from scientists is severely limited, at least in published sources, but experience suggests that farmer participation in planning and design of prototype technologies requires that the stages of technology design outlined above become greatly compressed in time. A relatively rapid interactive process is initiated once scientists become involved with farmers in dialogue about scenarios for prototype technologies. An important principle of farmer participation in technology design is therefore, to give the user the opportunity to evaluate options; this requires early hands-on testing and adaptation of prototype technology by all parties concerned. As a result, on-farm testing shifts from being a process for validating technology conceived of and prescreened by scientists on their own, to becoming an integral part of a rapid cycle of diagnosis-design-experimentation-feedback involving farmers and scientists in an exchange of ideas. This has important implications for methodology, which are discussed in the following section.

The earlier in the technology design process that farmer participation can be mobilized, the more likely it is that farmers' and researchers' ideas about desirable features of a technology will coincide. It is worthwhile, therefore, to consider at what stages in a research program information about farmers' reactions to proposed alternatives could make a difference to the design of technology within the following framework:

Early evaluations of multiple alternatives. — Farmer participation in pre-screening options at this stage can help researchers to sort out the "very good" and "very bad" options from the farmers' point of view. Even though researchers are likely to be concerned with broad adaptability at an early stage in the technology design process, while farmers are concerned with site-specific criteria, there is evidence that small farmers who share broadly comparable objectives and resources will identify desirable characteristics of technology in common. On the other hand, farmers with different resources may require different solutions to a similar problem. Farmer participation in pre-screening optional prototypes can help to fine-tune technology design to specific needs. Analysis of farmers' reasons for discriminating a good or bad technology from their point of view can identify important objectives which should be considered in the early stage of technology design.

Comparisons of most promising alternatives. — At a stage in research when a few alternatives to farmers' current technology have been identified, farmer participation can be implemented to determine not only what prototypes farmers perceive as most or least promising, but the specific reasons why one alternative is more, or less appealing than another to farmers.

Evaluation of developed technology. — Farmer participation in evaluating developed technology can be an important diagnostic tool and a means of initiating farmer-scientist dialogue. At this stage in the research process, the objective of farmer participation can be to test (not just validate) the best available option, as a vehicle for identifying its advantages and/or shortcomings, and to start a cycle of participatory diagnosis, planning and design.

IV. METHODS FOR FARMER PARTICIPATION IN TECHNOLOGY DESIGN

The disquiet among farming systems practitioners about the effectiveness of consultative farmer participation for communicating with small farmers about technology design means that many participatory research methods are being developed within the broad framework of farming systems research methodology. There is as a result an unfortunate tendency to classify as "participatory" any method which involves researchers in talking to farmers. A more precise definition requires participatory research methods to involve farmers in exercising some significant degree of control over what data is collected, and to what purposes or objectives it is analyzed (i.e., over the definition of relevant criteria for evaluating research results). Few published resources on farmer participation research are explicitly concerned with methods for enabling farmers to collect data and do their own research (see Bunch, 1987, and Villarreal, 1987 for exceptions).[3,27]

There are however, some emergent principles underlying the application of the many, different methods being generated by practitioners of participatory research in agriculture. These principles are appled to data collection at various stages in technology design, reflecting the tendency to collapse conventional stages of research into a dynamic, iterative process of diagnosis-design-and-adaptation of technology with farmers. In general, participatory research methods are

- **Interactive,** — emphasizing immediate two-way information flows both in the farmer-researcher/extensionist interface, and in farmer-to-farmer information exchange.
- **Flexible,** — tending to minimize researcher-control and maximize farmer intervention in research design.
- **Multiple,** — such that different techniques are applied simultaneously or overlapping in time (rather than sequentially).
- **Rapid,** — to maximize ability to respond to farmer initiatives or unanticipated areas of research.

Considerable emphasis has been placed on the development of procedures for improving the quality of farmer-scientist interaction in the diagnostic stages of farmer participation research. Techniques or methods such as studies of farmer experimentation, histories of local technological change, future scenarios of farmers, farmer taxonomies, teaching by farmers, mapping farmers' fields or systems diagramming with farmers focus on face-to-face communication and seek to improve scientists' abilities to learn from farmers and thereby build on indigenous experimentation.[28] Many of these techniques have been developed for use in tandem with formal and informal survey methods, as a way of providing researchers with access to farmers' subjective terms of reference for analyzing farming systems.

All of the above-mentioned approaches can be conducted with individual farmers, but the importance given to intensive interaction with farmers has led to increased emphasis on combining methods for individual participation with group work methods. Table 1 summarizes some of the advantages and disadvantages of group approaches to farmer participation. The major advantages relate to improving the quality of interaction with farmers, and increased efficiency in the use of staff time to sustain intensive communication with numbers of farmers. The disadvantages of working with farmer groups especially for diagnostic purposes relate to the potential for a group process to distort information, especially when researchers may be relatively new to the farming circumstances they are attempting to analyze with farmers.

Some of the problems encountered with group approaches are related to a methodological issue which has received relatively little attention to date in farmer participation research. That is: how to select which members of the farmer population should participate in technology design. This issue is equally important with respect to individual as well as group participation, but tends to be perceived as more problematic with respect to group formation, perhaps because the potential distortions or bias of information become more apparent through group dynamics.

TABLE 1
Advantages and Disadvantages of Group Participation

Advantages	Disadvantages
Group interactions stimulates discussion with and among farmers; may be productive of new ideas, of more reliable information	Groups can be dominated or inhibited, to produce false consensus and misleading information Farmers can become tired of time-consuming meetings
Group interaction helps to motivate farmers and sustain interest	Low-status or reticent individuals may be inhibited from expressing themselves in a group
Groups provide a quick, efficient forum for assessing opinions held by a majority of farmers	Special interests or problems may fail to be identified in a group setting
Discussions of differences of opinion in a heterogenous group can help to refine research objectives, targets	Heterogenous groups which include competing interests or needs may become stalled and unable to analyze problems constructively
Groups can provide an overview of variable results from on-farm trials to interpret these; are useful for feedback of results to farmers	Forming homogeneous groups that represent user populations meaningful for research purposes may be logistically difficult or time-consuming
Ratio of staff-time to farmer-contact can be more efficient	Group membership may fluctuate on a given occasion or between meetings; consistency of information may be problematic
Research-extension-farmer linkages can be maximized in groups	
Groups can be used to evaluate results with types of farmers under-represented in on-farm trials	

An example is a technique for group diagnosis of farming problems, termed participatory diagnosis, which is applied to prioritize crops, cropping systems, and specific farmer problems developed by the Colombian Institute for Agricultural Research. [29] Community members attend group meetings on the basis of open invitation, so that groups form on the basis of self-selection. The meeting dedicates several hours to brainstorming in small groups about agricultural problems. Problems are then ranked in order of importance within small groups, and scored numerically to arrive at a plenary consensus about the problems of most interest to the community. This approach is extremely effective in obtaining information about which there is a broad consensus in the community, and generates a direct input by farmers into the planning of extension activities. When however, a more detailed diagnosis of specific problems is required for planning research, the community self-selection method of group formation is found to mask information about the special needs or the priorities of different groups, such as women who may be underrepresented in a public meeting. Efforts are being initiated to identify specialized interest groups for research-oriented participatory diagnosis.[30]

Methods for working with groups may be particularly

important for implementing farmer participation in the planning and design stages of technology development, when researchers need rapid feedback about farmers' reactions to a number of alternative prototype technologies. For example, in Colombia farmer groups have been used to prescreen large numbers of varietal materials planted in regional trials, in order to identify farmers' criteria for selecting desirable materials. This screening used the technique of absolute evaluation, whereby farmers indicate on a like-dislike basis, their opinion of each variety, and the reasons for their selection. This prescreening of varieties ensures that materials obviously unacceptable to farmers do not enter on-farm trials; and also provides a forum for group discussion of how to design trials to test varieties on their farms.[15] A group forum called "Regular Research Field Hearings" (RRFH) has been implemented in Indonesia for on-farm livestock research. Experimental designs and components for on-farm testing are discussed with farmers, as well as specific problems in animal health, management and breeding, in regular monthly meetings which have had important functions not only for diagnosis, but for evaluating and monitoring progress.[31]

In these examples, group participation has proved a rapid technique for tapping the reactions of a number of farmers to a "menu" of technological options drawn up by researchers. In other instances, group participation has had an important function of organizing farmers to test ideas or prototype solutions identified by the farmers themselves, so that the group provides an important mechanism for strengthening indigenous experimentation and creating linkages with formal research.[23,32]

While methods for improving farmer-scientist interaction in the diagnostic stage have received considerable attention, much less effort has been devoted to systematizing techniques for involving farmers in planning, design and especially the analysis of experiments. In part, lack of attention to these aspects of farmer participation reflects the emphasis placed on farmers' manipulating, selecting and evaluating technological options in a trial framework free from the restrictions of formal experimentation.[8]

There is a need for statistical tests adapted to this type of "free" experimentation.[33] In participatory research with farmers, flexible experimental designs are used which dispense with controls for some factors. Others include large numbers of observations (sites) and imply high coefficients of variation in experimental variables, to accomodate farmers' adaptation of prototype treatments to specific circumstances. An important concept is the "adaptive trial" which has various applications. In one example, adaptive trials are agronomic trials performed on-station and on-farm, which are based on farmers' own experiments adapted for statistical analysis.[24] In other cases, prototypes are laid out on a few farm sites, and are evaluated by individual farmers or by groups, allowing farmers to observe and select components for implementation on their own plots. This implementation can then be monitored.[34,35] Another approach to experimental design that has proved useful for testing genetic materials with farmers is to superimpose a trial consisting of several different materials on a large number of farmers' fields; each trial (farmer) represents a replication. Farmers freely adapt existing practices to the new varieties; and select varieties according to their own objectives.[36]

TABLE 2
Farmers' Criteria for Selecting Desirable Cassava Varieties: Content Analysis of Three Group Evaluations of Nine Varieties, Pescador, Cauca, Colombia, 1987

Farmers' criteria	Frequency mentioned N	%
Starch content (quality for processing)	23	85
Fresh market quality[a]	22	81
Harvest date (earliness)[b]	17	63
Seed quality[c]	15	55
Plant architecture (foliage)[d]	12	44
Plant height	9	33
Easy to up-root	9	33
Distribution of roots[e]	7	26
Branching[f]	6	22
Resistance to pests	5	18

[a] Medium-sized root, dark skin, pink epidermis, white flesh, dry (not watery) flesh.
[b] Size of roots, presence of new leaves indicating maturity.
[c] Nodes close to each other; stake medulla white, not black; poor quality indicated by few nodes, widely separated, thickened stakes.
[d] Abundant foliage disliked.
[e] Roots with short pedunculo preferred; roots with no pedunculo related to storage losses; long pedunculo related to lower yield.
[f] Low branching types disliked (difficult to weed); high branching types difficult to harvest.

Source: Ashby and Pachico (1988)

Farmer evaluations can be carried out in many different types of trial design. These evaluations can be quantified to provide researchers' with data on the relative importance of different criteria used by farmers to evaluate alternatives, as illustrated in Table 2.[37]

It will be particularly important for future participation research to address the issue of how farmers can take part in formal experimentation for two reasons. First, although the acid test of technology developed and tested with farmer participation is whether or not farmers adopt it, the credibility of participatory methods within the scientific community will also require compatibility with accepted methods of scientific verification. Second, and more important, one of the objectives of participatory research which as yet is receiving little systematic attention, is to strengthen indigenous or "free" experimentation so that farmers can take on increased responsibility for location-specific adaptive research. This unquestionably will require the development of methods for experimental design and analysis which can be learnt and implemented by farmers. To date, participatory methods for technology design stress improving scientists' ability to work with farmers. Future efforts must address the need to improve farmers' abilities to work with scientists.

V. CONCLUSIONS

Farmer participation research requires intensive interaction between farmers and scientists. One of the chief obstacles to realizing the potential benefits of farmer participation research is likely to be institutional. This includes the skills and values scientists and extensionists require in order to carry out participatory technology design with farmers. Organizational models for participatory research in agriculture have yet to be systematically worked out. There are many indications that farmer participation in technology design requires a degree of decentralization and flexibility in plannig and execution of research which few formal agricultural research systems in developing countries have yet evolved. Improved feedback to technology designers, greater complementarities and interchange of functions between researchers and extensionists, and mobilization of farmer and community resources for research purposes have all been observed in farmer participation research. The payoff in terms of more focussed research-extension efforts and rapid adoption by farmers may be substantial. Future work must marshall the evidence.

REFERENCES

1. **Farrington, J. and A. Martin,** Farmer participation research: a review of concepts and practices. Discussion Paper No. 19, *Agricultural Administration (Research and Extension) Network,* Overseas Development Institute, London, 1987.
2. **Ashby, J.,** Methodology for the participation of small farmers in the design of on-farm trials. *Agricultural Administration,* 22, 1, 1986.
3. **Bunch, R.,** Small Farmer Research: The key element of permanent agricultural improvement, presented at the IDS Workshop on Farmers and Agricultural Research: Complementary Methods, University of Sussex, England, 1987.
4. **de Pedro, Jr. , C. Lightfoot, D. Apura, M. Acaba, and J. Cabiling,** Screening of sweet potato varieties by subsistence farmers in Basey, Samar, Philippines: a case of traditional experimentation in upland agriculture. *Annals of Tropical Research,* 8(4), 201, 1986.
5. **Norman, D.,** Farmer groups for technology development experiences from Botswana, presented at the IDS Workshop, Farmers and Agricultural Research: Complementary Methods, University of Sussex, England, 1987.
6. **Rocheleau, M. D.,** The user perspective and the agroforestry research and action agenda, presented at the IDS Workshop, Farmers and Agricultural Research: Complementary Methods, University of Sussex, England, 1987.
7. **Horton, D.,** Farming Systems Research: Twelve lessons from the Mantaro Valley Project. *Agricultural Administration,* 23, 93, 1986.
8. **Martin, A. and J. Farrington,** Abstracts of recent field experience with farmer participation research. *Agricultural Administration (Research and Extension) Network,* Network Paper No. 22, Overseas Development Institute, London, 1987.
9. **Rhoades, R. E. and R. H. Booth,** Farmer-back-to-farmer: A model for generating acceptable agricultural technology. *Agricultural Administration,* 11, 127, 1982.
10. **Maurya, D. and A. Bottrall,** Innovative approach of farmers for raising their farm productivity, presented at the IDS Workshop, Farmers and Agricultural Research: Complementary Methods, University of Sussex, England, 1987.
11. **Collinson, M.,** Farming systems research: diagnosing the problems, in *Research, Extension, Farmer. A Two Way Continuum for Agricultural Development,* Cernea, M., Coulter, J. and Russell, J., Eds., IBRD, Washington, 1985, chap. 9.
12. **Barker, R. and C. Lightfoot,** Farm experiments on trial. Farming Systems Research Paper Series, Paper No. 11, Kansas State University, Office of International Agriculture, 1985.
13. **Hildebrand, P. and F. Poey,** *On-farm Agronomic Trials in Farming Systems Research and Extension,* Lynne Reiner, Boulder, CO, 1985, chap. 7.
14. **Biggs, S.,** Interactions between resource-poor farmers and scientists in agricultural research. Presented at Second Study Workshop of the ISNAR Project on the Organisation and Management of On-farm Research, The Hague, 1987.
15. **Ashby, J., C. A. Quirós and Y. Rivera,** Farmer participation in on-farm varietal trials. *Agricultural Administration (Research and Extension) Network,* Discussion Paper No. 22, London, Overseas Development Institute, 1987.
16. **Chambers, R. and J. Jiggins,** Agricultural research for resource-poor farmers. I Transfer of technology and farming systems research, *Agricultural Administration and Extension,* 27, 35, 1987.
17. **Chambers, R. and J. Jiggins,** Agricultural research for resource poor farmers: a parsimonious paradigm, Institute of Development Studies, University of Sussex, Brighton, England, Discussion Paper No. 220, August, 1986.
18. **Maxwell, S.,** Farming systems research: Hitting a moving target. *World Development,* 14, 65, 1986.
19. CIMMYT., Planning technologies appropriate to farmers. Concepts and procedures, CIMMYT, Mexico, 1980.
20. **Menz, K. and H. C. Knipscheer,** The location specificity problem in farming systems research, *Agricultural Systems,* 7, 95, 1981.
21. **Gilbert, E. H., D. W. Norman, and F. E. Winch,** Farming systems research: a critical appraisal. MSU Rural Development Paper No. 6, Department of Agricultural Economics, Michigan State University, East Lansing, 1980.
22. **Harwood, R.,** Small farm development understanding and improving farming systems in the humid tropics. Westview Press, Boulder, CO, 1979, chap. 5.
23. **Lightfoot, C.,** Indigenous research and on-farm trials. *Agricultural Administration and Extension,* 24, 79, 1987.
24. **Box, L.,** Experimenting cultivators: a methodology for adaptive agricultural research, *Agricultural Administration (Research and Extension) Network,* Discussion Paper No. 23, London, Overseas Development Institute, December, 1987.
25. **Carey, E.,** El desarrollo de metodologías sencillas para la selección y diseminación de nuevas variedades, presented at the Taller Sobre Intercambio de Germoplasma, Cuarentena y Fitomejoramiento de Yuca y Batata, CIAT, CIP, UNDP, Palmira, 1987.
26. **Lightfoot, C., O. de Guia, Jr., A. Aliman and F. Ocado,** Letting farmers decide in on-farm research, presented at the IDS Workshop, Farmers and Agricultural Research: Complementary Methods, IDS, University of Sussex, England, 1987.
27. **Villarreal Farias, E. and F. Galván Castillo.** *Desarrollo de un método para optimar las tecnologías utilizadas por los pequeños productores de Secano, bajo el modelo productor experimentador,* INIFAP, Mexico, 1987.
28. **Chambers, R., A. Pacey, and L. A. Thrupp, Eds.,** *Farmer First. Farmer Innovation and Agricultural Research,* Intermediate Technology Publications, London, 1989, 218.

29. **Lopera, H., B. Peña A, J. Quirós D. and K. Verbaken,** Diagnóstico participativo, Experiencias con grupos de campesinos en el Norte de Antioquia. Instituto Colombiano Agropecuario, Subgerencia de Fomento y Servicios, División de Divulgación, Medellín, Colombia, 1985.
30. **Ashby, J. and V. Zapata,** Técnicas de interacción agrónomo-agricultor. Actitudes y habilidades para una interacción eficaz, IPRA Project, Working Paper No. 8, CIAT, Cali, Colombia, 1989.
31. **Knipscheer, H. C. and K. Suradisastra,** Farmer participation in Indonesian livestock farming systems by regular research field hearings (RRFH), *Agricultural Administration,* 22, 205, 1986.
32. **Fernandez, M. E.,** Participatory-action-research and the farming system approach with highland peasants. Technical Report Series, Small Ruminant Collaborative Program, Department of Rural Sociology, University of Missouri, 1986.
33. **Horton, D. and G. Prain,** CIP's experience with farmer participation in on-farm research, in *La Investigación de Frijol en Campos de Agricultores de América Latina,* Memorias de un Taller, CIAT, Cali, Colombia, 1987.
34. **Rocheleau, D.,** Criteria for re-appraisal and redesign: Examples from three agroforestry projects in Kenya. Working Paper No. 37, ICRAF, Nairobi, Kenya, 1985.
35. **Ashby, J.,** The effects of different types of farmer participation on the management of on-farm trials, *Agricultural Administration,* 25, 235, 1987.
36. **Pachico, D., J. Voss and J. Woolley,** *On-farm research in bean (Phaseolus vulgaris L.) production and improvement in the tropics,* A.V. Schoonhoven and O. Voysest, Eds., CIAT, Palmira, 1989, chap. 6.
37. **Ashby, J. and D. Pachico,** *Farmer Evaluations of Technology. A Handbook,* Workup Document No. 7, IPRA Project, CIAT, Cali, Colombia, October, 1989.

Index

INDEX

A

B

C